International and Uniform Plumbing Codes Handbook

International and Uniform Plumbing Codes Handbook

R. Dodge Woodson

McGraw-Hill

New York San Francisco Washington, D.C. Auckland Bogotá
Caracas Lisbon London Madrid Mexico City Milan
Montreal New Delhi San Juan Singapore
Sydney Tokyo Toronto

Library of Congress Cataloging-in-Publication Data

Woodson, R. Dodge (Roger Dodge), date.
   International and uniform plumbing codes handbook / R. Dodge Woodson.
      p.   cm.
   Includes index.
   ISBN 0-07-135899-4
   1. Plumbing—Handbooks, manuals, etc. 2. Plumbing—Standards—Handbooks,
manuals, etc. 3. Plumbing—Law and legislation—Handbooks, manuals, etc.
I. Title.
TH6125.W54   2000
696'.1'0218—dc21   00-041840

McGraw-Hill

*A Division of The **McGraw·Hill** Companies*

ISBN 0-07-135899-4

The sponsoring editor for this book was Larry S. Hager, and the production supervisor was Pamela A. Pelton. It was set in Times by Lone Wolf Enterprises, Ltd.

Printed and bound by R. R. Donnelley & Sons Company.

This book was printed on recycled, acid-free paper containing a minimum of 50% recycled de-inked fiber.

McGraw-Hill books are available at special quantity discounts to use as premiums and sales promotions, or for use in corporate training programs. For more information, please write to the Director of Special Sales, Professional Publishing, McGraw-Hill, Two Penn Plaza, New York, NY 10121. Or contact your local bookstore.

For Afton and Adam

Contents

Introduction

The plumbing codes have undergone considerable change since the 1997 editions were published. Some of the major codes have combined to create the International Plumbing Code. The Uniform Plumbing Code also has incorporated a number of changes in its 2000 edition. R. Dodge Woodson has used his decades of experience as a master plumber and plumbing contractor to distill these codes into a user-friendly guide for plumbers in the field. This is it, you are holding it in your hands.

There is no substitute for your local plumbing code book, but this invaluable tool will make your code book much easier to understand. Everything you need to know is in your code book, but can you understand it? Don't feel bad if you get confused by the code; many plumbers do. Woodson knows this. He has taught plumbing code classes at Central Maine Technical College, has been a master plumber and contractor for more than 20 years. He is the author of nearly every professional plumbing book McGraw-Hill has in print. Woodson even has been called as an expert witness in court cases involving plumbing issues. If the term *expert* fits any plumber, it's R. Dodge Woodson.

Code books tend to be cold, sterile, complex, and boring. This seems to be the nature of that type of book. Woodson's work, however, is conversational, easy to understand, warm, funny at times, and built around years and years of in-the-field experience. Reading the code requirements for an island vent is one thing; learning how to install such a vent in cramped quarters is a different matter. Pipe sizing is covered in the codes, but many plumbers don't know how to begin to size a system. With Woodson's conversational style of writing, readers learn quickly, easily, and without the stress of trying to decipher cryptic code books.

Look at the table of contents for this book. You will see that it is laid out in the same general order as those in your code books. Thumb through the pages and notice the tables, charts, and other illustrations. Read a paragraph here and there. Take note of Woodson's unique writing style. See how easy it is to comprehend. It won't take long to realize how valuable this code complement can be to you and your company.

Acknowledgments

I would like to acknowledge the contributions of the following organizations:

- International Code Council, Inc. and International Plumbing Code 2000;
- 2000 Uniform Plumbing Code (UPC) and International Association of Plumbing and Mechanical Officials (IAPMO).

CHAPTER 1
DEFINITIONS

Definitions are not exactly exciting reading, but they can play a vital role in the plumbing code. It would be easy to justify skipping this chapter, but I urge you not to. I've been plumbing since the mid-1970s, hold a master's license, and have owned my own plumbing company since 1979. In addition to field work and running my business, I've taught code and apprenticeship classes at Central Maine Technical Center. During all of these years and through all of this experience, I've seen countless plumbers who could not give the proper definition of a term. For example, do you know the difference between a stack vent and a vent stack? If you even *thought* of hesitating on this question, don't skip this chapter.

There are many industry terms that conflict with proper definitions. Local terms can get the job done, but they don't cut it on license testing, and they can make it difficult for you to communicate with suppliers. For example, most plumbers in my region have a pet name for a trap adapter. People in this area know what they are talking about, but if they move to another location, calling a trap adapter by another name could be a problem, especially when ordering parts.

If you are in charge of permit acquisitions, takeoffs and similar tasks, where using the right word or term can be crucial, you must be up to speed on definitions. We will use this chapter to learn and understand terms and definitions as they are set forth in the major plumbing codes. Don't feel that you have to memorize them, but become familiar enough with them to be comfortable when interpreting your local code book.

WORDS, TERMS, AND DEFINITIONS

ABS Acrylonitrile butadiene styrene

accepted engineering practice Any practice that conforms to accepted principles, tests or standards can be considered an accepted engineering practice. The accepted principles, tests or standards must be approved by technical or scientific authorities.

access Fixtures, appliances, and equipment that require access may be governed by one of two means of access. In essence, access refers to some means of making devices within reach. The means of access can be considered accessible (where the removal of a panel or plate is required for access) or ready access (where a device can be reached immediately and without the removal of a concealment device).

access cover An access cover is used to conceal plumbing that is required to be accessible. It is common for access covers to be secured with screws or bolts that can be removed easily.

accessible When a device is deemed accessible, it is within code requirements for the device to be concealed by a removable panel or plate. This is not to be confused with a device that is required to be readily accessible, in which case a concealment device is not allowed.

adapter fitting Any approved fitting used to connect pipes and fittings that would not otherwise fit together can be considered an adapter fitting.

administrative authority A variety of people and organizations can be considered an administrative authority. For example, your local plumbing inspector can be considered an administrative authority. In addition to individuals, code boards, code departments and code agencies can be considered administrative authorities. An authorized representative of an administrative authority also can carry the title of administrative authority.

air-admittance valve A one-way valve designed to allow air to enter a plumbing drainage system. Gravity closes the valve automatically sealing the vent terminal at zero differential pressure and under positive internal pressures. Many field plumbers call these devices mechanical vents. Air-admittance valves allow air to enter a drainage system without requiring a vent to the outside through a roof or wall. Air-admittance devices prevent sewer gas from entering a building. These valves commonly are installed under sinks and lavatories during remodeling work.

air break Don't confuse an air break with an air gap. Both can be found in drainage systems, but they are not the same. An air break is a piping arrangement where a drain from a fixture or device discharges indirectly into another fixture, such as a clothes washer discharging into a laundry sink.

air gap (drainage-type) There are air gaps in both drainage and water-distribution systems. In a drainage system, an air gap is the unobstructed vertical distance through the open air, between the outlet of a waste pipe and the flood level rim of the receptacle receiving the discharge. [Example: A condensate pipe terminates above a floor drain. The distance from the discharge pipe to the floor drain would be considered the air gap.]

air gap (water-distribution-type) The unobstructed vertical distance through open air between the lowest opening from pipe or faucet supplying water to a tank, plumbing fixture, or other device and the flood level rim of a receptacle.

alternative engineered design It is possible to create plumbing systems that perform in accordance with the intent of the plumbing code, even though the system may not be piped according to the code. When this is done, the system is known as an alternative engineered design. As long as the system protects public health, safety and welfare, it can be approved by a local administrative authority.

anchors See *supports*.

antisiphon Devices designed to prevent siphonage can be called antisiphon devices.

approved Anything meeting the required standards of an administrative authority can be considered approved.

approved testing agency Groups or organizations primarily established to test plumbing to standards approved by the administrative authority.

area drain Devices installed to collect storm water from an open area, such as an areaway, are called area drains.

aspirator Devices which, when supplied with water or other fluid under positive pressure that passes through an integral orifice or construction, cause a vacuum. Also called suction devices. Similar to ejectors.

backflow Water, other liquids, mixtures, or substances entering a potable water system from a source not intended to mix with the potable system is known as backflow. [Example: A water hose that is connected to a hose bibb with an attached fertilizer watering device could create a potentially deadly backflow if the contents of the watering device were sucked into the potable water system. The simple installation of a backflow preventer can avert such disasters.]

backflow connection Any type of plumbing connection that is not protected from backflow can be considered a backflow connection.

backflow preventer Device designed to prevent backflow into a potable water system.

backpressure The imbalance of water pressure within a water-distribution system that causes backflow is called backpressure.

backpressure, low head Pressure that is less than, or equal to 4.33 pounds per square inch (psi) or the amount of pressure exerted by a 10-foot column of water.

backsiphonage Contaminated backflow into a potable water system is called backsiphonage. This can occur when the pressure in a potable water system falls below atmospheric pressure of the plumbing fixtures or devices.

backwater valve Some sewers and building drains are subject to backflow. Backwater valves are installed to prevent drainage and waste from backing up into the building drain or sewer. You can think of a backwater valve as a type of check valve. Drainage and waste can flow out of the pipe in the proper direction, but it cannot back up into the pipe beyond the backwater valve.

ballcock A water-supply valve that is operated by means of a float and used to fill a tank with water. Ballcocks most often are found in toilet tanks. Modern versions of ballcocks are equipped with antisiphon devices that prevent water in a toilet tank from being sucked back into the potable water supply system.

base flood elevation Reference points established to determine the peak elevation of a potential flood based on the likelihood of a flood within a 100-year period are known as base flood elevations. The base point considers the wave height of any flooding that might occur. All base flood elevation points are established within the guidelines of local building code requirements.

bathroom Any room equipped with a bathing unit, such as a bathtub or shower, is considered a bathroom.

bathroom group A bathroom group consists of any group of plumbing fixtures that includes a water closet, a lavatory, and a bathing unit, such as a bathtub or shower that might or might not include a bidet or emergency floor drain. All fixtures in a bathroom group are located together on the same floor level.

battery of fixtures Wherever two, or more, similar fixtures, are installed side by side, and discharge into a common horizontal waste or soil branch, a battery of fixtures is created. [Example: A number of urinals on the wall of a public restroom.]

bedpan steamer A fixture used to scald bedpans or urinals by the direct application of steam or boiling water. Also known as a *bedpan boiler*.

bedpan washer and sterilizer Plumbing fixtures installed for the purpose of washing bedpans where the contents of the fixture are flushed into the sanitary drainage system. Also included in this classification are fixtures that drain into the sanitary drainage system but that also provide for disinfecting utensils by scalding them with steam or hot water. (See *bedpan steamer*.)

bedpan washer hose Devices installed adjacent to a water closet or clinical sink, and are supplied with hot and cold water for the purpose of cleaning bedpans.

boiler blowoff The emptying of discharge or sediment from a boiler is done through a boiler blowoff.

branch Any part of a piping system that is not a riser, a main, or a stack.

branch, fixture See *fixture branch*.

branch, horizontal See *horizontal branch*.

branch interval The distance along a soil or waste stack that corresponds to the story height of a building, but is not less than 8 feet within which the horizontal branches from one floor or story of a building are connected to a stack.

branch vent Any vent that connects one or more individual vents with a vent stack or a stack vent.

brazed joint A connection made by joining metal parts with alloys that melt at temperatures higher than 840°F, but lower than the melting temperature of the parts being joined.

building A structure that is occupied or intended for occupancy by people.

building drain The lowest drainage piping that receives the discharge from soil, waste, and other drainage pipes inside a building. Building drains extend up to 30 inches beyond the walls of a structure and convey their contents into what becomes the building sewer, once the pipe is extended beyond the 30-inch limit.

building drain, combined Some jurisdictions allow a building drain to convey both sewage and storm water into a single pipe. When this is the case, the pipe is known as a combined building drain.

building drain, sanitary A building drain that conveys only sewage.

building drain, storm A building drain that conveys storm water or other drainage, but never conveys sewage.

building sewer A pipe that begins about 30 inches outside of a building and that conveys sewage from the building to a public sewer or a private sewage disposal system.

building sewer, combined One that conveys both sewage and storm water.

building sewer, sanitary One that conveys only sewage.

building sewer, storm One that conveys storm water or other drainage, but never conveys sewage.

building subdrain Any portion of a drainage system that does not drain by gravity into a building sewer.

building supply A pipe that supplies water to a building from a water meter or other water source. A building supply often is referred to as water service.

building trap A device installed in the building drain or building sewer to prevent the circulation of air between the drainage system of the building and the building sewer. These devices are no longer common.

certified backflow assembly tester Someone who is approved to test and maintain backflow assemblies to the satisfaction of the administrative authority.

cesspool A lined excavation in the earth that collects discharge from drainage systems and retains organic matter while allowing liquids to seep through the bottom and sides of the lining and be absorbed in the ground.

chemical waste See *special wastes.*

circuit vent Any vent connecting to a horizontal drainage branch that vents two to eight traps, or trapped fixtures, connected in a battery of fixtures.

cistern A storage tank that normally is used to collect and store storm water for uses not associated with potable water.

cleanout An access opening in a drainage system that facilitates the removal of obstructions in the piping. The most common type of cleanout is a removable plug or cap, but a removable fixture such as a water closet, or a removable trap on a plumbing fixture such as a sink, can also be considered a cleanout.

clinic sink Sinks having a flush rim, an integral trap with a visible trap seal, and the basic flushing and cleansing characteristics of a water closet that are intended to receive the discharge from bedpans.

code Regulations set forth and adopted by local jurisdictions to dictate proper plumbing procedures as enforced by the administrative authority.

code official An individual authorized to enforce local code requirements.

combination fixture A sink or laundry tray that has two or three compartments in a single unit, or a fixture, that combines a sink with a laundry tray.

combination thermostatic/pressure-balancing valve Mixing valves that can sense water temperature from the outlet location and maintain it by regulating the temperature of incoming hot and cold water.

combination waste and vent systems Systems designed to accept the drainage of sinks and floor drains without the standard use of vertical vents. As an alternative to routine vertical venting, these systems utilize horizontal wet venting in oversized pipes.

combustible construction Any structure containing building materials that will ignite and burn at a temperature of 1392°F, or less.

common A word used to describe any part of a plumbing system that is meant to serve more than one fixture, building, system, or appliance.

common vent A vent that serves more than one fixture and that is connected at the junction of the fixture drains or to a fixture branch that serves the fixtures.

concealed fouling surface Any surface of a plumbing fixture that is not readily visible and that is not scoured or cleansed with each operation of the fixture.

conductor Storm water piping found inside a building that conveys storm water from a roof to a storm drain, a combined building sewer or some other approved location.

confined space An area that has a volume less than 50 cubic feet per 1000 Btu/h of the aggregate input rating of a number 2 fuel-burning appliance installed in the area.

construction documents Graphics, blueprints, specifications, descriptions, and other materials required to obtain a building permit. It is expected that the documents will be drawn to scale when relevant.

contamination Any impairment of water quality in a potable water system that may cause public health problems such as poisoning or disease.

continuous vent Any vertical vent that is a continuation of the drain for which it serves as the vent.

continuous waste A piping arrangement that connects drains from multiple fixtures to a common trap, such as a double-bowl kitchen sink connected to a single trap.

CPVC Chlorinated polyvinyl chloride

critical level The minimum height at which a backflow preventer or vacuum breaker can be installed above the flood-level rim of a fixture or receptor. Any space below the critical level is assumed to present a risk of backflow. When obvious markings are not evident to establish a critical level, the bottom of the device is considered to be the critical level.

cross connection Any connection or arrangement that allows the possibility for the contamination of a potable water system.

dead end Any branch of a soil, waste or vent pipe, or building drain or sewer that extends for a length of two feet or more and ends with a plug, cap, or other closed fitting.

department having jurisdiction Any agency or organization including, but not limited to the administrative authority that has the authority to interpret and enforce the plumbing code.

depth of water seal A measurement of liquid, usually water, that would have to be removed from a trap to allow air to pass through the trap.

developed length The full length of a section of piping, including fittings, when measured along the centerline of the pipe and fittings.

diameter Except where otherwise stated, diameter is considered to be the nominal diameter as designated commercially.

discharge pipe Any pipe that conveys the discharge from plumbing fixtures and/or appliances.

domestic sewage Liquid and waterborne wastes that come from the ordinary living processes that do not contain industrial wastes and can be disposed of satisfactorily. There is no need for special treatment to prepare domestic sewage for disposal. Domestic sewage can be disposed of in a sanitary public sewer or private sewage disposal treatment without any preliminary treatment.

downspout A piping arrangement, extending down the exterior of a building, that carries storm water from a roof to a building storm drain, combined building sewer, or other means of satisfactory disposal.

drain Pipes that carry wastewater or waterborne wastes in a building drainage system are drains.

drainage fittings Special fittings used in a drainage system that are recessed and tapped to eliminate ridges on the inside of installed pipe. They differ from standard cast-iron fittings in that cast-iron fittings have a bell and spigot design that does not offer the same smooth service created with drainage fittings.

drainage system Any piping located within private or public buildings that conveys sewage, rainwater, or other liquid waste to a point of disposal. Excluded from the drainage system are any main or public sewer system, sewage treatment or disposal plant.

drainage system, gravity A drainage system that drains entirely by gravity to a building sewer.

drainage system, sanitary A drainage system that carries sewage but not storm, surface, or ground water.

drainage system, storm A drainage system that conveys storm water, surface water, condensate waste, and other similar liquids.

durham system A special type of soil or waste drainage system that is created with threaded pipe or tubing that utilizes recessed drainage fittings.

effective opening This term has multiple meanings. Generally, it refers to the minimum cross-sectional area at a point where water supply discharge is measured or expressed in terms of the diameter of a circle. In cases where the opening is not circular, the diameter measurement of a circle of equivalent cross-sectional area is used. In addition to these conditions, effective opening also can apply to an air gap.

emergency floor drain A floor drain that does not receive the discharge of any drain or indirect-waste pipe and that protects against damage from accidental spills, fixture overflows, and leakage.

essentially nontoxic transfer fluids Many types of fluids are considered essentially nontoxic when they have a Gosselin rating of 1. The most common fluids of this type include propylene glycol, polydimethylsiloxane, mineral oil, hydrochlorofluorocarbon, chlorofluorocarbon, and hydrofluorocarbon refrigerants, as well as FDA-approved boiler-water additives for steam boilers.

essentially toxic transfer fluids Any soil, waste, gray water, or fluids that have a Gosselin rating of 2 or more are considered to be essentially toxic transfer fluids. These fluids can include ethylene glycol, hydrocarbon oils, ammonia refrigerants, and hydrazine.

existing installations Plumbing work and systems installed prior to the effective date of the current plumbing code, for which a permit was issued.

existing work See *existing installations.*

faucet A device attached to the end of a water-supply pipe that makes it possible to draw water held in the pipe.

fixture See *plumbing fixture.*

fixture branch This term has two meanings. A fixture branch is a drain that serves two or more fixtures and discharges into another drain or stack. When related to a water-supply system, a fixture branch is a water-supply pipe that runs from a water-distribution pipe to a fixture.

fixture drain A section of drain pipe that runs from the trap of a fixture to a junction with another drain pipe.

fixture fitting A fitting that controls the volume and/or directional flow of water to a fixture. Fixture fittings generally are attached to a fixture but can simply be accessible from the fixture.

fixture supply A pipe or tube that connects a fixture to a branch water supply or to a main water-supply pipe.

flammable vapor (or fumes) The concentration of flammable constituents in air that exceeds 25 percent of its lower flammability limit.

flooded A condition that occurs when a fixture is filled with liquid that rises to the flood-level rim.

flood-level rim The upper edge of a fixture where water will overflow if its height is greater than the edge of the fixture.

flood zone, flood-hazard zone (A zone) An area said to be prone to flooding but not subject to high-velocity waters or wave action.

flood zone, flood-hazard zone (V zone) Areas that are prone to flooding where wave action can exceed a height of three feet or where high-velocity wave runup or wave-induced erosion is likely to occur. Typically, these areas are associated with tidal waters.

flow pressure A measurement of water pressure in a pipe that is near a faucet or water outlet. The flow pressure is established when the faucet or water outlet is in a full-open position.

flushometer tank A device installed within an air accumulator vessel for the purpose of discharging a predetermined quantity of water to fixtures for flushing purposes.

flushometer valve A valve that provides a predetermined quantity of water to fixtures for flushing purposes and is actuated by direct water pressure. The valve gradually closes to reseal fixture traps and to avoid water hammer.

flush tank A tank that usually is controlled by a ballcock and equipped with a flush valve that holds water that is released on demand, to flush the contents of a bowl or other portion of a fixture, such as a water closet or urinal.

flush valve A valve located at the base of a flush tank that provides for the flushing of water closets and similar fixtures.

gang or group shower Two or more showers in a common area.

grade An amount of slope or fall of a pipe in reference to a horizontal plane. Grade is also called pitch. While grade or pitch can be a factor in various types of piping, it is most commonly considered when working with drainage systems.

grease interceptor A device, with a rated flow exceeding 50 gallons per minute (gpm), located outside of a building. Also, an interceptor of at least 750-gallon capacity that is capable of serving one or more fixtures that are remotely located.

grease-laden waste An effluent discharge that is a by-product of food processing, food preparation or other source where grease, fats, and oils enter an automatic dishwasher pre-rinse station, sink, or other appurtenance.

grease trap A device with a rated flow of 50 gpm or less, is located within a building, and serves one to four fixtures.

hangers See *supports*.

high hazard See *contamination*.

horizontal branch Any pipe that extends laterally from a soil or waste stack or building drain, with or without vertical sections or branches, that receives the discharge from one of more fixture drains and conducts its contents to a soil or waste stack or to a building drain.

horizontal pipe A pipe or fitting that makes an angle of less than 45° with the horizontal.

hot water Water with a temperature equal to or greater than 120°F.

house drain See *building drain.*

house sewer See *building sewer.*

house trap See *building trap.*

indirect-waste pipe Pipes that discharge into a drainage system through an air break or air gap without attaching directly to the drainage piping.

individual sewage disposal system Any approved system that uses a septic tank, cesspool or mechanical treatment to dispose of domestic sewage in a way that does not rely on a public sewer or public treatment facility.

individual vent A single vent that vents a fixture trap and that either connects to a vent system above the fixture being served or terminates into open air.

individual water supply Any water supply, other than an approved public water supply, that serves one or more families.

industrial waste All liquid or waterborne waste from industrial or commercial processes, except domestic sewage.

insanitary A condition that creates a risk to public health and sanitary principles.

interceptor A device that separates and retains deleterious, hazardous or undesirable matter from normal waste. Interceptors can be operated automatically or manually and must allow normal waste and sewage to pass through.

invert The lowest portion of the inside of a horizontal pipe.

joint, brazed A joint made by connecting metal parts with alloys that melt at temperatures higher than 840°F but lower than the melting temperature of the parts being connected.

joint, expansion A piping arrangement that allows for the expansion and contraction of the piping system. Loops, return bends, and return offsets are used to create expansion joints. The primary need for this type of arrangement is in locations where there may be a rapid change of temperature, such as in power plants and steam rooms.

joint, flexible A type of joint between two pipes that will allow one pipe to be moved without moving the other pipe.

joint, mechanical See *mechanical joint.*

joint, slip A joint that is made by means of a washer or a special type of packing compound in which one pipe is slipped into the end of an adjacent pipe.

joint, soldered A joint obtained by connecting metal parts using metallic mixtures or alloys that melt at a temperature up to and including 840°F.

labeled Materials, fixtures, equipment, and devices bearing the label of an approved agency.

lavatories in sets Two or three lavatories that are served by a single trap.

leader Exterior drainage pipe that conveys storm water from a roof or gutter drain to an approved means of disposal.

lead-free pipe and fittings Pipes and fittings containing no more than 8.0 percent lead.

lead-free solder and flux Solder and flux containing no more than 0.2 percent lead.

liquid waste Discharge from any fixture, appliance, or appurtenance, in connection with a plumbing system that does not receive fecal matter.

listed See *labeled.*

listing agency Any agency approved by the administrative authority that is responsible for the listing and/or labeling of materials and the ongoing inspection, testing and reporting of the materials.

local vent stack A type of vent used in connection with bedpan washers. The vent is a vertical pipe to which connections are made from the fixture side of traps and through which vapor or foul air is removed from the fixture being vented.

lot A single or individual parcel or area of land that is legally recorded or validated by other means acceptable to the administrative authority on which is situated a building, or is the site of any work regulated by the code. This includes yards, courts, and unoccupied spaces legally required for the building or works, which is owned by, or is in the lawful possession of the owner of the building or works.

low hazard See *pollution.*

macerating toilet system An assembly that consists of a water closet and sump with a macerating pump that is designed to collect, grind, and pump wastes from the water closet and up to two other fixtures connected to the sump.

main A principal pipe artery to which branches are connected.

main sewer See *public sewer.*

main vent A principal pipe artery of a vent system to which the vent branches can be connected.

manifold See *plumbing appurtenance.*

may A permissive term.

mechanical joint Typically, a joint that is made by applying compression along the centerline of the pieces being joined. The joint may be part of a coupling, fitting or adapter. Mechanical joints are not screwed, caulked, threaded, soldered, solvent-cemented, brazed, or welded.

mechanical vent See *air-admittance valve*.

medical gas system A complete system used to deliver medical gases for direct patient application from a central supply system through piping networks with pressure and operating controls, alarm-warning systems and related components, extending to station outlet valves at patient-use points.

medical vacuum systems A system consisting of central-vacuum-producing equipment with pressure and operating controls, shutoff valves, alarm-warning systems, gauges, and a network of piping extending to, and terminating with suitable station inlets at locations where patient suction might be required.

mobile home park sewer Part of a horizontal piping drainage system that begins two feet downstream from the last mobile home site and conveys sewage to a public sewer, private sewer, individual sewage disposal system, or some other point of disposal.

nonpotable water Water that is not safe for drinking purposes, personal or culinary use.

nuisance A nuisance can be any inadequate or unsafe water supply or sewage disposal system. If work regulated by the code is done in a way that is dangerous to human life or detrimental to health and property, the act is a nuisance. Any public nuisance known in common law, or in equity jurisprudence also is a nuisance.

occupancy The purpose for which a building, or portion thereof, is utilized or occupied.

offset Any combination of elbows or bends in a line of piping that brings one section of the pipe out of line but into a line parallel with the other section.

oil interceptor See *interceptor*.

open air Fresh air outside a structure.

PB Polybutylene.

PE Polyethylene.

person A legal term meaning a person, heir, executor, administrators, or assigns and also shall include a firm, corporation, municipal or quasi-municipal corporation, or governmental agency. Singular includes plural, and male includes female.

pipe Any cylindrical conduit or conductor conforming to the particular dimensions commonly known as pipe size.

plumbing Any business, trade, or work that has to do with the installation, removal, alteration, or repair of plumbing and drainage systems, or parts thereof.

plumbing appliance A special class of plumbing fixtures meant to perform special functions. Fixtures that depend upon motors, controls, heating elements, pressure- or temperature-sensing elements all can be appliances.

plumbing appurtenance A device that is an adjunct to a basic piping system and plumbing fixtures. Appurtenances do not demand any additional water supply and do not add any discharge load to a fixture or drainage system.

plumbing fixture A receptacle or device that is either permanently or temporarily connected to a water-distribution system of the premises and demands a supply of water therefrom, discharges waste water, liquidborne waste materials or sewage, either directly or indirectly, to the drainage system of the premises, or requires both a water supply connection and a discharge to the drainage system of the premises.

plumbing official See *administrative authority.*

plumbing system Includes a water supply and its distribution pipes, plumbing fixtures, traps, water-treating equipment, water-using equipment, soil pipes, waste pipes, vent pipes, sanitary sewers, storm sewers, and building drains. The system also can include connections, devices, and appurtenances within a structure or premises.

pollution Impairment of the quality of potable water in an amount sufficient to cause disease or harmful physiological effects, and conforming in bacteriological and chemical quality, to the requirements of the Public Health Service Drinking Water Standards or the regulations of the public health authority having jurisdiction.

potable water Water that is safe and suitable for drinking, culinary purposes, and domestic purposes.

PP Polypropylene

pressure Typical amount of force exerted by a homogeneous liquid or gas, per unit of area, on the wall of a container or conduit.

pressure-balancing valve A mixing valve that receives both hot and cold water and keeps the pressure stable by compensating for fluctuations in either hot or cold water.

pressure, residual The usable amount of water pressure available at a fixture or water outlet after allowances have been made for pressure drops caused by friction loss, head, and other reasons for decreased pressure.

pressure, static The amount of pressure present when there is no flow.

private When the word "private" is used in the plumbing code, it refers to plumbing fixtures that are not intended for use by the general public. Fixtures installed in residences, rooms of hotels and motels, and other facilities intended for use by either a family or an individual, are considered private.

private sewage disposal system Most often consists of a septic tank that allows effluent to discharge into a subsurface septic field. However, any sewage disposal system that meets code criteria, and that does not discharge into a public sewer, can be considered a private system.

private sewer Any pipe that receives drainage from more than one building drain and then conveys the drainage to a public sewer or private sewage disposal system.

public use Any use that is not defined as private.

public sewer A common sewer that is controlled by any public authority.

public water main A primary water-supply pipe that is controlled by any public authority.

PVC Polyvinyl chloride.

quick-closing valve Any valve or faucet that closes automatically when released manually or that is controlled by a mechanical means for closing quickly.

ready access Means of direct access to a fixture or device. To qualify as ready access, access must be possible without the need for removal of a panel, opening of a door, or any other obstruction. Additionally, access must be possible without the need of a ladder or other similar device.

receptor An approved fixture or device used to accept the discharge from indirect waste pipes that is readily cleaned.

reduced-pressure principal backflow preventer Any backflow preventer that contains two independently acting check valves. These check valves are: (1) internally force-loaded to a normally closed position and separated by an intermediate chamber in which there is an automatic relief means of venting to the atmosphere; and (2) internally loaded to a normally open position between two tightly closing shutoff valves. This type of backflow prevention has a means for testing the tightness of the checks and opening of relief means.

registered design professional An architect or engineer who is registered or licensed to practice their profession within the guidelines of the governing agency.

regulating equipment Any valve or control used in a plumbing system that is required to be either accessible or readily accessible.

relief valve, pressure A valve that is pressure-actuated and held closed by a spring, or other means, that serves to relieve pressure automatically when a set pressure is reached.

relief valve, temperature A relief valve that opens when a set temperature is reached.

relief valve, temperature and pressure A relief valve that opens when a set pressure or a set temperature is reached.

relief valve, temperature A relief valve that opens when a set temperature is reached.

relief vent Any vent that provides air circulation between drainage and vent systems.

remote outlet When used for sizing water piping, a remote outlet is the farthest outlet dimension, measuring from the meter, of either the developed length of the cold-water piping or through the water heater to the farthest outlet on the hot-water piping.

rim The unobstructed open edge of a fixture.

riser Any water-supply pipe that extends vertically for one full story, or more, to convey water to branches or fixtures.

roof drain Any drain that is installed to receive water from a roof and then convey the water to a suitable discharge location.

rough-in Any part of a plumbing system that is installed prior to the installation of plumbing fixtures.

sand interceptor See *interceptor*.

SDR Standard dimensional ratio.

seepage pit Any lined excavation in the ground that accepts discharge from a septic tank and then allows the effluent to seep into the earth from the bottom and sides of the seepage pit.

self-closing faucet Any faucet that closes automatically, once deactivation of the opening means occurs.

separator See *interceptor*.

septic tank Any approved container, usually made of concrete, that is buried in the ground and accepts the discharge from a drainage system or pipe. Septic tanks must be watertight and designed to retain solids, digest organic matter via a period of detention, and allow effluent to flow into a septic field or other approved discharge destination.

sewage Liquid waste that contains animal or vegetable matter in suspension or solution and may include liquids containing chemicals in solution.

sewage ejector A device that lifts or pumps sewage by entraining the sewage in a high-velocity jet of steam, air, or water.

sewage pump A mechanical device, other than an ejector, that is permanently installed to remove sewage or liquid waste from a sump.

sewer, building See *building sewer*.

sewer, public See *public sewer*.

sewer, sanitary A sewer that conveys sewage without combining it with storm, surface, or ground water.

sewer, storm A sewer that conveys storm water, surface water, condensate, cooling water, and similar liquid wastes.

shall A mandatory term.

shielded coupling Any approved elastomeric sealing gasket that is equipped with an approved outer shield and a mechanism for tightening.

shock arrestor See *water-hammer arrestor*.

single-family dwelling A structure that is constructed for the purpose of housing the property owner. The structure must be the only dwelling located on a parcel of ground with typical accessory buildings but no other homes.

size and type of tubing See *diameter*.

slip joint An adjustable tubing connection that consists of a compression nut, a friction ring, and a compression washer, all designed to fit a threaded adapter fitting or standard taper pipe thread.

slope See *grade*.

soil pipe Any pipe that conveys sewage containing fecal matter to a building drain or building sewer.

soldered joint A connection created by joining metal parts with metallic mixtures or alloys that melt at a temperature below 800°F and above 300°F.

special wastes Any waste that requires special handling or treatment.

spill-proof vacuum breaker An assembly of one check valve, that is force-loaded closed, and an air-inlet vent valve that is forced-loaded open to atmosphere that is downstream of the check valve, and located between, and including two tightly closing shutoff valves and a test cock.

stack A vertical pipe that is part of a soil, waste, or vent system that rises to a height of at least one story.

stack vent A continuation of a soil or waste stack above the highest horizontal drain connected to the stack.

stack venting Using a stack vent to vent a soil or waste stack.

sterilizer, boiling-type A nonpressure-type fixture utilized for boiling devices for disinfection.

sterilizer, instrument A device used to sterilize instruments.

sterilizer, pressure A device used for sterilization that is a pressure vessel using steam under pressure for the sterilization process.

sterilizer, pressure instrument washer A pressure vessel fixture that washes and sterilizes instruments.

sterilizer, utensil A sterilizer used to sterilize utensils.

sterilizer, water A device used to sterilize water and then store it.

sterilizer vent A pipe or stack that is connected indirectly to a drainage system at a lower terminal to receive vapors from nonpressure sterilizers or exhaust vapors from pressure sterilizers. The vent then transports the vapors directly to open air. Other names for a sterilizer vent include vapor, steam, atmospheric, or exhaust vent.

storm drain See *building drain, storm.*

storm sewer See *sewer, storm.*

structure Anything that is built or constructed, or any portion thereof.

subsoil drain A drain used to collect subsurface water or seepage water and convey it to an approved disposal location.

sump A container, or pit, located below the normal grade of a gravity system, used to accept sewage or liquid waste that will be pumped out of the holding device.

sump pump An electric pump that works automatically to remove the contents of a sump that does not contain raw sewage.

sump vent A vent from a pneumatic sewage ejector, or similar device, that extends to open air.

supports Devices used to support and/or secure pipes, fixtures, and equipment.

swimming pool Any structure, container, or device that contains an artificial body of water for the purposes of swimming, diving, or recreational bathing, that has a depth of two feet or more at any point.

tailpiece A pipe or tubing that connects the outlet of a plumbing fixture to a trap.

tempered water Water with a temperature range from 85°F to 120°F.

thermostatic valve Also known as a temperature control valve, this valve is designed to mix hot and cold water, while compensating for temperature fluctuations, to maintain an even water temperature at the point of delivery.

trap A fitting or device that holds water, to prevent the emission of sewer gas, without materially affecting the flow of sewage or waste water through a pipe.

trap arm A section of pipe that extends from a fixture drain trap to a drain.

trap primer A device and system used to maintain a suitable amount of water in a trap that seldom sees use and potentially would dry up without the aid of the primer.

trap seal The vertical distance between the weir and the top of the dip of a trap.

type-A dwelling unit Any dwelling unit designed and built for accessibility in accordance with the provisions of CABO/ANSI A117.1.

type-B dwelling unit Any dwelling unit designed and built in accordance with the provisions of CABO/ANSI A117.1.

unconfined space Any room, space, or area that has a volume equal to at least 50 cubic feet per 1000 Btu/h of the aggregate input rating of all fuel-burning appliances installed in the room, space, or area.

unstable ground Earth that does not provide a uniform bearing for the barrel of a sewer pipe between the joints at the bottom of the pipe trench.

vacuum Pressure that is less than that exerted by the atmosphere.

vacuum breaker See *backflow preventer*.

vacuum-relief vent A device that doesn't allow excessive pressure to develop in a pressure vessel.

vent A pipe used to ventilate a plumbing system to prevent trap siphonage and back-pressure, or to equalize the air pressure within the drainage system.

vented appliance categories Category I is an appliance that operates with a non-positive vent static pressure and with a vent gas temperature that avoids excessive condensate production in the vent. Category II is an appliance that operates with a non-positive vent static pressure and with a vent gas temperature that might cause excessive condensate production in the vent. Category III is an appliance that operates with a positive vent static pressure and with a vent gas temperature that avoids excessive condensate production in the vent. Category IV is an appliance that operates with a positive vent static pressure and with a vent gas temperature that might cause excessive condensate production in the vent.

vent pipe See *vent*.

vent stack A vertical pipe installed to provide air circulation to a drainage system.

vent system A pipe, or system of piping used to ventilate a plumbing system to prevent trap siphonage and backpressure, or to equalize the air pressure within the drainage system.

vertical pipe A pipe or fitting installed in a vertical position or at an angle of less than 45° with the vertical.

wall-hung water closet A water closet mounted on a wall and does not touch the floor.

waste Discharge from any fixture, appliance, area, or appurtenance that does not contain fecal matter.

waste fitting A combination of components that convey the sanitary waste from the outlet of a fixture to the sanitary drainage system connection.

waste pipe A pipe that conveys only waste.

water-conditioning or -treating device A device that conditions or treats a water supply to change its chemical content or to remove suspended solids through filtration.

water-distributing pipe A pipe in a building that conveys potable water from the building supply pipe to plumbing fixtures and water outlets.

water-hammer arrestor In a water supply system, a device used to absorb the pressure surge that occurs when water flow is stopped suddenly.

water main A water-supply pipe that provides water for public or community use.

water outlet Any discharge opening through which water is supplied to a fixture into the atmosphere, except into an open tank that is part of a water-supply system, to a boiler or heating system, or to any devices or equipment requiring water to operate but that are not part of a plumbing system.

water pipe, riser Any water supply pipe that rises at least one full story to convey water to branches or a group of fixtures.

water pipe, water-distribution pipe A pipe within a building that conveys water from a water-service pipe, or from a water meter, when the meter is located in the building, to a point of utilization.

water pipe, water service A pipe from a water main, or other potable water source, or from a meter, when the meter is at the public right of way, to the water-distribution system in the building being served.

water supply system Components used to create a water supply system can include a water service pipe, water-distribution pipes and all necessary connecting pipes, fittings, and control valves, as well as all appurtenances in, or adjacent to, a structure.

welded joint or seam A joint or seam created by joining metal parts in a plastic molten state.

welder, pipe A person who specializes in the welding of pipes and who holds a valid certificate of competency from a recognized testing laboratory.

well, bored Any well created by boring a hole in the earth with an auger and fitting with a casing.

well, drilled Any well made with a drilling machine and fitted with a casing or screen.

well, driven Any well created when a pipe is driven into the earth.

well, dug A well created by digging a large-diameter hole in the ground and installing a casing.

wet vent Any vent that serves as both a vent and drain.

whirlpool bathtub A bathtub fitted with a circulating piping system designed to accept, circulate, and discharge bath water.

yoke vent A pipe that connects upward from a soil or waste stack to a vent stack, for the purpose of preventing pressure changes in the stacks.

CHAPTER 2
ADMINISTRATIVE POLICIES
AND PROCEDURES

Administrative polices and procedures make the plumbing code effective. Without the proper procedures and administration, the plumbing code would be little more than an organized outline for good plumbing procedures. To be effective, the code must be enforced. To be fair, the rules for the administration of the code must be clear to all who work with it. Administrative policies dictate the procedure for code enforcement, interpretation, and implementation. There is a fine line between administration and enforcement. This chapter deals with administration. Chapter 3 addresses the issue of code enforcement. Although the two matters are broken into two separate chapters, there will be some overlap of material. Administration and enforcement are so closely related, it is necessary to commingle the two from time to time.

The rules, regulations, and laws used to make the plumbing code are structured around facts. These facts are the result of research into ways to protect the health of our nation. As a plumber, you are responsible for the health and sanitation of the public. A mistake or a code violation could result in widespread illness or even death.

In some jurisdictions, the plumbing code is composed of rules. In other areas, the plumbing code is a compilation of laws. There is a big difference between a rule and a law. When you are working with a rule-based code, you are subject to various means of punishment for violating the rules. The punishment may mean the suspension or revocation of your license. Cash fines may be required for violations, but there is no jail time. In jurisdictions using a law-based code you could find yourself behind bars for violating the plumbing code.

The procedure required to obtain a plumber's license is not easy. Some people believe there are too many restraints in licensing requirements for plumbers, but these people are not aware of the heavy responsibility plumbers must bear. The public often perceives a plumber as someone who works in sewers, has a poor education, and is slightly more than a common laborer. This is a false perception.

Professional plumbers are much more than sewer rats. Today's plumbers generally are well-educated and have the ability to perform highly technical work. The mathematical demands on a plumber could perplex many well-educated people. The mechanical and physical abilities of plumbers are often outstanding.

While drain cleaning is a part of the trade, so is the installation of $2,500 gold faucets. The plumbing trade is not all sewage and water. Plumbers are known for their fabled high incomes, and it is true that good plumbers make more money than many professionals. The trade can offer a prosperous living, but it must be worked for.

Whether you are a plumber or an apprentice, you have much to be proud of. The plumbing trade is more than a job. As a professional plumber, you will have the satisfaction of knowing you are helping to maintain the health of the nation and the integrity of our natural resources.

Not long ago, there were no plumbing codes or code enforcement. People could pollute our lakes and streams with their ineffective cesspools and outhouses. The plumbing code is designed to stop pollution and health hazards. The code is part of any plumber's life and career. Learning the code can be a laborious task, but the self-satisfaction obtained when you master the code is well worth the effort.

When you earn your master's license, you will have the opportunity to run your own shop. Having your own plumbing business can be quite profitable. To realize these profits, you first must learn the code and the trade. Then, you must pass the tests for your journeyman and master's license. During your learning stages, you are earning a good wage and providing a vital service to the community.

Few professions allow you to earn a good living while you are gaining the skills necessary to master your craft. When you attend college, you must pay for your education. With plumbing, you get paid to learn. After your training, you are unlimited in the wealth you can build from your own business.

The first step toward financial independence as a plumber is a clear understanding of the plumbing code. Unlicensed plumbers are not allowed to work in many jurisdictions. Your license is your ticket to respectable paychecks and a solid future. Now, let's see how the administrative policies and procedures for the plumbing code will affect you.

WHAT DOES THE PLUMBING CODE INCLUDE?

The plumbing code includes all major aspects of plumbing installations and alterations. Design methods and installation procedures are a cornerstone to the plumbing code. Sanitary piping for the disposal of waste, water, and sewage is controlled by the code. Potable water supplies fall under the jurisdiction of the plumbing code. Storm water, gas piping, chilled-water piping, hot-water piping, and fire sprinklers all are dealt with in the plumbing, mechanical, and building codes. Any installation of fuel gas distribution piping and equipment, fuel gas-fired water heaters, and water heater venting systems are to be regulated by the International Fuel Gas Code.

The plumbing code is meant to ensure the proper design and installation of plumbing systems and to ensure public health and safety. The plumbing code is intended to be interpreted by the local code enforcement officer. The interpretation of the code officer might not be the same as yours, but it is the code officer's option to determine the meaning of the code, under questionable circumstances.

HOW THE CODE PERTAINS TO EXISTING PLUMBING

The plumbing code requires any alterations or repairs to an existing plumbing system to conform to the general regulations of the code as they would apply to new installations. No alteration or repair shall cause an existing plumbing system to become unsafe. Further, the alterations or repairs shall not be allowed to have a detrimental effect on the operation of the existing system.

For example, if a plumber is altering an existing system to add new plumbing, the plumber must make all alterations in compliance with code requirements. It would be a violation of the code to add new plumbing to a system that was not sized to handle the additional load of the increased plumbing.

There are provisions in the codes to allow existing conditions that are in violation of the current code to be used legally. If an existing condition was of an approved type prior to the current code requirements, that existing condition may be allowed to continue in operation, as long as it is not creating a safety or health hazard.

If the use or occupancy of a structure is being changed, the change must be approved by the proper authorities. It is a violation of the plumbing code to change the use or occupancy of a structure without the proper approvals. For example, it would be a breach of the code to convert a residential dwelling to a professional building without the approval of the code enforcement officer.

SMALL REPAIRS

Small repairs and minor replacements of existing plumbing may be made without bringing the entire system into compliance with the current plumbing code standards. These changes must be made in a safe and sanitary method and must be approved.

For example, it would be permissible to repair a leak in a half-inch pipe without changing the half-inch pipe to a three-quarter-inch pipe, even if the current code required a three-quarter-inch pipe under the present use. It also would be allowed to replace a defective S trap with a new S trap, even though S traps are not in compliance with the current code requirements. In general, if you are only doing minor repair or maintenance work, you are not required to update the existing plumbing conditions to current code requirements.

It is incumbent upon the owner of a property to keep the plumbing system in good and safe repair. The owner may designate an agent to assume responsibility for the condition of the plumbing, but it is mandatory that the plumbing be kept safe and sanitary at all times.

RELOCATION AND DEMOLITION OF EXISTING STRUCTURES

If a building is moved to a new location, the building's plumbing must conform to the current code requirements of the jurisdiction where the structure is located. In the event a structure is to be demolished, it is the owner's, or the owner's designated agent's, responsibility to notify all companies, persons, and entities having utilities connected to the structure. These utilities can include, but are not limited to, water, sewer, gas, and electrical connections.

Before the building can be demolished or moved, the utilities connected to the property must disconnect and seal their connections in an approved manner. This applies to water meters and sewer connections, as well as other utilities.

MATERIALS

All materials used in a plumbing system must be approved for use by the code enforcement officer. These materials shall be installed in accordance with the requirements of the local code authority. The local code officer has the authority to alter the provisions of the plumbing code, as long as the health, safety, and welfare of the public is not endangered.

A property owner, or that owner's agent, may request a variance from the standard code requirements when conditions warrant a hardship. It is the code officer's decision whether to grant the variance. The application for a variance and the final decision of the code officer shall be in writing and filed with the code enforcement office.

PRODUCTS AND MATERIALS REQUIRING THIRD-PARTY TESTING AND THIRD-PARTY CERTIFICATION

PRODUCT OR MATERIAL	THIRD-PARTY CERTIFIED	THIRD-PARTY TESTED
Potable water supply system components and potable water fixture fittings	Required	—
Sanitary drainage and vent system components	Plastic pipe, fittings and pipe-related components	All others
Waste fixture fittings	Plastic pipe, fittings and pipe-related components	All others
Storm drainage system components	Plastic pipe, fittings and pipe-related components	All others
Plumbing fixtures	—	Required
Plumbing appliances	Required	—
Backflow prevention devices	Required	—
Water distribution system safety devices	Required	—
Special waste system components	—	Required
Subsoil drainage system components	—	Required

FIGURE 2.1 These materials and products must be certified by a third party. (*Courtesy of International Code Council, Inc. and International Plumbing Code 2000.*)

The use of previously used materials shall be open to the discretion of the local code officer. If the used materials have been reconditioned, tested, and are in working condition, the code officer may allow their use in a new plumbing system.

Alternative materials and methods not specifically identified in the plumbing code may be allowed under certain circumstances. If the alternatives are equal to the standards set forth in the code for quality, effectiveness, strength, durability, safety, and fire resistance, the code officer may approve the use of the alternative materials or methods. Where the requirements of reference standards or manufacturer's installations instructions do not conform to minimum provisions of this code, the provision of this code shall apply.

Before alternative materials or methods are allowed for use, the code officer can require adequate proof of the properties of the materials or methods. Any costs involved in testing or providing technical data to substantiate the use of alternative materials or methods shall be the responsibility of the permit applicant.

CODE OFFICERS

Code officers are responsible for the administration and enforcement of the plumbing code. They are appointed by the executive authority for the community. Code officers may not be held liable on a personal basis when working for a jurisdiction. Legal suits brought forth against code officers, arising from on-the-job disputes, will be defended by the legal representative for the jurisdiction.

The primary function of code officers is to enforce the code. Code officers also are responsible for answering questions pertinent to the materials and installation procedures used in plumbing. When application is made for a plumbing permit, the code officer is the individual who receives the application. After reviewing a permit application, the code officer will issue or deny a permit.

After the code officer issues a permit, it is the code officer's duty to inspect all work to ensure it is in compliance with the plumbing code. When code officers inspect a job, they are looking for more than just plumbing. These inspectors will be checking for illegal or unsafe conditions on the job site. If unsafe conditions exist on the site, or the plumbing is found to be in violation of the code, the code officer will issue a notice to the responsible party.

Code officers normally perform routine inspections personally. Inspections, however, may be performed by authoritative and recognized services or individuals, other than code officers. The results of all inspections shall be documented in writing and certified by an approved individual.

If there is ever any doubt as to the identity of a code officer, you may request to see the inspector's identification. Code officers are required to carry official credentials while discharging their duties.

Another aspect of the code officer's job is the maintenance of proper records. Code officers must maintain a file of all applications, permits, certificates, inspection reports, notices, orders, and fees. These records are required to be maintained for as long as the structure they apply to is still standing, unless otherwise stated in other rules and regulations.

PLUMBING PERMITS

Most plumbing work, other than minor repairs and maintenance, requires a permit. These permits must be obtained before any plumbing work begins. The code

Plumbing Permit Fees

Permit Issuance

1. For issuing each permit ....*_____
2. For issuing each supplemental permit ....*_____

Unit Fee Schedule (in addition to items 1 and 2 above)

1. For each plumbing fixture on one trap or a set of fixtures on one trap (including water, drainage piping and backflow protection therefor)...*_____
2. For each building sewer and each trailer park sewer ....*_____
3. Rainwater systems – per drain (inside building)..*_____
4. For each cesspool (where permitted)..*_____
5. For each private sewage disposal system...*_____
6. For each water heater and/or vent ....*_____
7. For each gas-piping system of one to five outlets ....*_____
8. For each additional gas piping system outlet, per outlet ....*_____
9. For each industrial waste pretreatment interceptor including its trap and vent, except kitchen-type grease interceptors functioning as fixture traps................................*_____
10. For each installation, alteration or repair of water piping and/or water treating equipment, each ...*_____
11. For each repair or alteration of drainage or vent piping, each fixture ....*_____
12. For each lawn sprinkler system on any one meter including backflow protection devices therefor .....*_____
13. For atmospheric-type vacuum breakers not included in item 12:
    1 to 5...*_____
    over 5, each..*_____
14. For each backflow protective device other than atmospheric type vacuum breakers:
    2 inch (51 mm) diameter and smaller ....*_____
    over 2 inch (51 mm) diameter...*_____
15. For each graywater system ....*_____
16. For initial installation and testing for a reclaimed water system ....*_____
17. For each annual cross-connection testing of a reclaimed water system (excluding initial test) ......*_____
18. For each medical gas piping system serving one to five inlet(s)/outlet(s) for a specific gas ....*_____
19. For each additional medical gas inlet(s)/outlet(s) ....*_____

Other Inspections and Fees

1. Inspections outside of normal business hours..*_____
2. Reinspection fee...*_____
3. Inspections for which no fee is specifically indicated..*_____
4. Additional plan review required by changes, additions or revisions to approved plans (minimum charge – one-half hour)....................................*_____

* Jurisdiction will indicate their fees here.

FIGURE 2.2 Most jurisdictions have a schedule of fees for required permits. *(Reprinted from the* 2000 Uniform Plumbing Code (UPC) *with the permission of the International Association of Plumbing and Mechanical Officials (IAPMO).)*

enforcement office provides forms to individuals wishing to apply for plumbing permits. The application forms must be completed properly and submitted to the code enforcement officer.

Permits are to be obtained by the person or agent who will install all or part of any plumbing system. The applicant must meet all qualifications required of a permit applicant. It is also required that the full name and address of the applicant be stated in the permit application.

The permit application shall give a full description of the plumbing to be done. This description must include the number and type of plumbing fixtures to be installed, the location where the work will be done, and the use of the structure housing the plumbing.

The code officer may require a detailed set of plans and specifications for the work to be completed. Duplicate sets of the plans and specs may be required, so that copies can be placed on file in the code enforcement office. If the description of the work deviates from the plans and specifications submitted with the permit application, it may be necessary to apply for a supplementary permit.

Permit Issuance

1. For issuing each permit . . . . $ _____
2. For issuing each supplemental permit . . . . _____

Unit Fee Schedule (in addition to Item 1 or 2 above)

1. For each plumbing fixture or trap or set of fixtures on one trap
   (including water, drainage piping and backflow protection thereof) . . . . _____
2. For each building sewer and each trailer park sewer . . . . _____
3. Rainwater systems—per drain (inside building) . . . . _____
4. For each cesspool (where permitted) . . . . _____
5. For each private sewage disposal system . . . . _____
6. For each water heater and/or vent . . . . _____
7. For each industrial waste pretreatment interceptor including its trap and vent,
   excepting kitchen-type grease interceptors functioning as fixture traps . . . . _____
8. For installation, alteration or repair of water-piping and/or water-treating
   equipment, each . . . . _____
9. For repair or alteration of drainage or vent piping, each fixture . . . . _____
10. For each lawn sprinkler system on any one meter including backflow
    protection devices therefor . . . . _____
11. For atmospheric-type vacuum breakers not included in Item 2:
    1 to 5 . . . . _____
    over 5, each . . . . _____
12. For each backflow protective device other than atmospheric-type vacuum breakers:
    2 inches (51 mm) and smaller . . . . _____
    Over 2 inches (51 mm) . . . . _____

Other Inspections and Fees

1. Inspections outside of normal business hours . . . . _____ per hour
   (minimum charge two hours)
2. Reinspection fee assessed under provisions of Section 107.3.3 . . . . _____ each
3. Inspections for which no fee is specifically indicated . . . . _____ per hour
   (minimum charge—one-half hour)
4. Additional plan review required by changes, additions or revisions to
   approved plans (minimum charge—one-half hour) . . . . _____ per hour

---

FIGURE 2.3 Most jurisdictions have a schedule of fees for required permits. *(Reprinted from the* 2000 *Uniform Plumbing Code (UPC)* with the permission of the International Association of Plumbing and Mechanical Officials (IAPMO).) (Continued)*

The supplementary permit will be issued after a revised set of plans and specs have been given to the code officer and approved. The revised plans and specifications must show all changes in the plumbing that are not in keeping with the original plans and specs.

Plans and specifications may not be required for the issuance of a plumbing permit. If plans and specs are required, however, they may require a riser diagram and a general blueprint of the structure. The riser diagram must be very detailed, indicating pipe size, direction of flow, elevations, fixture-unit ratings for drainage piping, horizontal pipe grading, and fixture-unit ratings for the water distribution system.

If the plumbing to be installed is an engineered system, the code officer may require details about computations, plumbing procedures, and other technical data. Any application for a plumbing permit to install new plumbing might require a site plan, which must identify the locations of the water service and sewer connections. The location of all vent stacks and their proximity to windows or other ventilation openings must be shown.

If new plumbing is being installed in a structure served by a private sewage disposal system, there are even more details to be included in the site plan. When a private sewage system is used the plan must show the location of the system and all technical information pertaining to the proper operation of the system.

When a plumbing permit is applied for, the code officer will process the application in a timely manner. If the application is not approved, the code officer will notify the applicant, in writing, of the reasons for denial. If the applicant fails to follow through with the issuance of a permit within six months from the date of application, the permit request can be considered void. Once a permit is issued, it may not be assigned to another person or entity. Permits will not be issued until the appropriate fees are paid. Fees for plumbing permits are established by the local jurisdiction.

Plumbing permits bear the signature of the code officer or an authorized representative. The plans submitted with a permit application will be labeled as approved plans by the code officer. One set of the plans will be retained by the code enforcement office and one set of approved plans must be kept on the job site. The approved plans kept on the job must be available to the code officer or an authorized representative for inspection at all reasonable times.

It is possible to obtain permission to begin work on part of a plumbing system before the entire system has been approved. For example, you might be given permission to install the underground plumbing for a building before the entire plumbing system is approved. These partial permits are issued by the code officer, with no guarantee the remainder of the work will be approved. If you proceed to install the partial plumbing, you do so at your own risk.

Time limits are involved after a plumbing permit is issued. If work is not started within six months of the date a permit is issued, the permit may become void. If work is started, but then stalled or abandoned for a period of six months, the permit may be rendered useless. A permit may be revoked if the code officer finds that the permit was issued using false information. Misrepresentation in the application for a permit or in the plans submitted for review is reason for the revocation of a plumbing permit.

All work performed must be done according to the plans and specifications submitted to the code officer in the permit application process. All work must be in compliance with the plumbing code. Code officers are required to conduct inspections of the plumbing being installed during the installation and upon completion of the installation. The details of these inspections are described in chapter 3.

MULTIPLE PLUMBING CODES

There are many plumbing codes in use. Each local jurisdiction generally takes an existing code and amends it to local needs. The three primary plumbing codes are similar in many ways and very different in other ways. In addition to the three primary codes, other codes exist. To be sure of your local code requirements, you must check with the local code enforcement office. This book is written to explain good plumbing procedures. Various jurisdictions, however, have different opinions about what constitutes good plumbing procedures. Some states, counties, or towns adapt an existing code without much revision. Other areas make significant changes in the established code that is used as a model. It would not be unheard of to find a jurisdiction working with regulations from multiple plumbing codes. In light of these facts, always check with your local authorities before performing plumbing work.

CHAPTER 3
REGULATIONS, PERMITS, AND CODE ENFORCEMENT

When working with the plumbing code, you must be aware of the requirements and procedures involved with regulations, permits, and enforcement. These three elements work together, ensuring the proper use of the plumbing code. This chapter is going to teach you about all three key elements, but it also will do much more. You are going to learn about pipe protection, health, safety, pipe connections, temporary toilet facilities, and more. There even will be tips on how to work with plumbing inspectors, instead of against them.

REGULATIONS

What are regulations? Regulations are the rules or laws used to regulate activity—in this case, the performance of plumbing. The plumbing code is governed by rules in some states and laws in others. If you violate the regulations in a rule-base state, you may face disciplinary action, but not jail. In states where the plumbing regulations are laws, you risk going to jail if you violate the regulations.

It might be difficult to imagine going to jail for violating a plumbing regulation, but it could happen. Certain violations could result in personal injury or death. To protect yourself and others, it is important to understand and abide by the regulations governing the plumbing trade.

EXISTING CONDITIONS

The regulations we are going to examine first have to do with existing conditions. This is an area where many people have difficulty determining their responsibility to the plumbing code. Generally, any existing condition that is not a hazard to health and safety is allowed to remain. When existing plumbing is altered, however, it might have to be brought up to current code requirements.

Although the code normally is based on new installations, it can also apply to existing plumbing that is being altered. These alterations can include repairs, renovations, maintenance, replacement, and additions. The question of when work on existing plumbing must meet code requirements is one that plagues many plumbers. Let's address this question and clear it up.

The code is concerned only with changes being made to existing plumbing. As long as the existing plumbing is not creating a safety or health hazard, and is not being altered, it does not fall under the scrutiny of most plumbing codes. If you are altering an existing system, the alterations must comply with the code requirements, but there may be exceptions to this rule. For example, if you were replacing a kitchen sink and there was no vent on the sink's drainage, code would require that you vent the fixture. Where undue hardship exists in bringing an existing system into compliance, the code officer may grant a variance.

In the case of the kitchen sink replacement, such a variance may be in the form of permission to use a mechanical vent. Whenever you encounter a severe hardship in making old plumbing come up to code, talk with your local code officer. The code officer should be able to offer some assistance, either in the form of a variance or in advice about how to accomplish your goal.

Since the code does come into play with repairs, maintenance, replacement, alterations, and additions, let's see how it affects each of these areas. If you are repairing a plumbing system, you must be aware of code requirements. If no health or safety hazards exist, nonconforming plumbing may be repaired to keep it in service.

Do you need to apply for a plumbing permit to replace a faucet? No, faucet replacement does not require a permit, but it does require the replacement to be made with approved materials and in an approved manner. You do need a permit to replace a water heater. Even if the replacement heater is going in the same location and connecting to the existing connections, you must apply for a permit and have your work inspected. An improperly installed water heater can be a serious hazard, capable of causing death and destruction. Failure to comply with the code in these circumstances could ruin your life and the lives of others.

Routine maintenance of a plumbing system must be done according to the code, but it does not require a permit. Alterations to an existing system may require the issuance of a permit, depending upon the nature of the alteration. In any case, alterations must be done with approved materials and in an approved manner. When adding to a plumbing system, you normally will need to apply for a permit. Adding new plumbing will come under the authority of the plumbing code and generally will require an inspection.

You can get yourself into deep water when adding on to an existing system. When you add new plumbing to an old system, you must be concerned with the size and ability of the old plumbing. Increasing the number of fixture units entering an old pipe might force you to increase the size of the old pipe. This can be very expensive, especially when the old pipe happens to be the building drain or sewer. Before you install any new plumbing on an old system, verify the size and ratings of the old system. If your new work will rely on the old plumbing to work, the old plumbing must meet current code requirements.

CHANGE-IN-USE REGULATIONS

Beware of change-in-use regulations. If you will be performing commercial plumbing, this regulation can have a particularly serious effect on your plumbing costs and methods. If the use of a building is changed, the plumbing also may have to be changed. Change-in-use regulations come into play most often on commercial properties, but they could affect a residential building.

Assume for a moment that you receive a request to install a three-bay sink in a convenience store. You discover that the store's owner is having the sink installed so he can perform food preparation for a new deli in the store. This store has never been equipped for food preparation and service. What complications could arise from this situation? First, zoning may not allow the store to have a deli. Secondly, if the use is allowed, the plumbing requirements for the store may soar. There could be a need for grease traps, indirect wastes, and a number of other possibilities. When you are asked to perform plumbing that involves a change of use, investigate your requirements before committing to the job.

The remaining general regulations of the plumbing code are easily understood. By reading your local code book, you should have no trouble in understanding them. Now, let's move on to permits.

PERMITS

Permits generally are required for various plumbing jobs. When a permit is required, it must be obtained before any work is started. Minor repairs and drain cleaning do not require the issuance of permits. In most cases, plumbing permits can be obtained only by master plumbers or their agents. In some cases, however, homeowners may be allowed to receive plumbing permits for work to be done by themselves, on their own homes. Permits are obtained from the local code enforcement office, which provides the necessary forms for permit applications.

The information required to obtain a permit will vary from jurisdiction to jurisdiction. You may be required to submit plans, riser diagrams and specifications for the work to be performed. At a minimum, you likely will be required to adequately describe the scope of work to be performed, the location of the work, the use of the property, and the number and type of fixtures being installed.

The degree of information required to obtain a permit is determined by the local code officer. It is not unusual for the code officer to require two sets of plans and specifications for the work to be performed. The detail of the plans and specs is also left up to the judgment of the code officer. Requirements may include details of pipe sizing, grade, fixture units, and any other information the code officer may deem pertinent.

If your project will involve working with a sewer or water service, expect to be asked for a site plan, which should show the locations of the water service and sewer. If you will be working with a private sewage disposal system, its location should be indicated on the site plan. After your plans are approved, any future changes to the plans must be submitted to and approved by the code officer.

Plumbing permits must be signed by the code officer or an authorized representative. If you submitted plans with your permit application, the plans will be labeled with appropriate wording to prove they have been reviewed and approved. If it is found later that the approved plans contain a code violation, the plumbing must be installed according to code requirements, even if the approved plans contain a nonconforming use. Most jurisdictions require a set of approved plans to be kept on the job site and available to the code officer at all reasonable times.

If plans are required, they must be approved before a permit is issued. All fees associated with the permit must be paid prior to the issuance of the permit. These fees are established by local jurisdictions. After a permit is issued, all work must be done in the manner presented during the permit application process.

It is possible to get a partial permit, which approves a portion of work proposed for completion. When time is of the essence, it may be possible to obtain these partial permits, but there is risk. Assume you obtained permission and a permit to install your underground plumbing but had not yet been issued a permit for the remainder of it. With freezing temperatures coming on, you decide to install your groundworks so the concrete floor can be poured over the plumbing before freezing conditions arrive. This is a good example of how and why partial approvals are good, but look at what could happen.

You have installed your underground plumbing and the concrete is poured. After a while, the code officer notifies you that the proposed above-grade plumbing is not in acceptable form and will require major changes. These changes will affect the location and size of your underground plumbing. What do you do now? Well, you probably are going to spend some time with a jackhammer or concrete saw. The underground plumbing must be changed, or the above-ground plumbing must be redesigned to work with the groundworks. In either case, you have trouble and expense that would have been avoided if you had not acted on a partial approval. Partial approvals have their place, but use them cautiously.

How can a plumbing permit become void? If you do not begin work within a specified time, usually six months, your permit will be considered abandoned. When this happens, you must repeat the entire process to obtain a new permit. Plumbing permits can be revoked by the code official. If it is found that facts given during the permit application were false, the permit may be revoked. If work stops for an extended period of time, normally six months, a permit may be suspended.

CODE ENFORCEMENT

Code enforcement generally is performed on the local level by code enforcement officers. These individuals frequently are referred to as inspectors. It is their job to interpret and enforce the regulations of the plumbing code. Since code enforcement officers interpret the code, there may be times when a decision appears to contradict the code book. The code book is a guide, not the last word—the last word comes from the code enforcement officer. This is an important fact to remember. Regardless of how *you* interpret the code, the code officer's decision is final.

INSPECTIONS

Every job that requires a permit also requires inspection. Many jobs require more than one inspection. In the plumbing of a new home, there might be as many as four inspections. One inspection would be for the sewer and water service installation. Another inspection might be for underground plumbing. Then you would have a rough-in inspection of the pipes that are to be concealed in walls and ceilings. When the job is done, there will be a final inspection.

These inspections must be done while the plumbing work is visible. A test of the system generally is required, either with pressure from air or water. The inspection usually is done by the local code officer, but not always. The code officer may accept the findings of an independent inspection service. Any test results submitted to a code office from an independent testing agency must come from an agency that is approved by the code jurisdiction. Before independent inspection results will be accepted, the inspection service must be approved by the code officer. This is also the case when independent inspection services are used to inspect prefabricated construction.

Plumbing inspectors generally are allowed the freedom to inspect plumbing at any time during normal business hours. These inspectors cannot enter a property without permission, unless they obtain a search warrant or other proper legal authority. Permission for entry frequently is granted by the permit applicant when the permit is signed.

Backflow preventers are to be inspected annually. The backflow assemblies and air gaps both must be inspected to see that they will operate properly. Backflow preventers and double-check-valve assemblies must be inspected when they are installed, immediately after any repairs or relocation, and on a routine, annual basis. All testing and inspection must be done in accordance with the requirements of local codes.

WHAT INSPECTORS LOOK FOR

When plumbing inspectors look at a job, they look at many aspects of the plumbing. They will inspect to see that the work is installed in compliance with the code and that the plumbing will likely last for its normal lifetime. Inspectors will check to see that all piping is tested properly and that all plumbing is in good working order.

WHAT POWERS DO PLUMBING INSPECTORS HAVE?

Plumbing inspectors can be considered the plumbing police. These inspectors have extreme authority over any plumbing-related issue. If plumbing is found to be in violation of the code, plumbing inspectors may take several forms of action to rectify the situation.

Normally, inspectors will advise the permit holder of the code violations and allow a reasonable time for correction of the violations. This advice will come in the form of written documents and will be recorded in an official file. If the violations are not corrected, the code officer will take further steps. Legal counsel may be consulted. After a legal determination is made, action may be taken against the permit holder in violation of the plumbing code. This could involve cash fines, license suspension, license revocation, and in extreme cases, jail.

Code officers have the power to issue a stop-work order. This order requires all work to stop until code violations are corrected. These orders are not used casually; they are used when an immediate danger is present or possible. Code officers do have a protocol to follow in the issuance of stop-work orders. If you ever encounter a stop-work order, stop working. These orders are serious, and violation of a stop-work order can deliver more trouble than you ever imagined.

When code officers inspect a plumbing system and find it to be satisfactory, they will issue an approval on the system. This allows the pipes to be concealed and the system to be placed into operation. In certain circumstances, code officers may issue temporary approvals. These temporary approvals are issued for portions of a plumbing system when conditions warrant them.

When a severe hazard exists, plumbing inspectors have the power to condemn property and force occupants to vacate the property. This power would be used only under extreme conditions, where a health or safety hazard was present.

Code officers are empowered to authorize the disconnection of utility services to a building in the case of an emergency or when there is a need to eliminate an immediate danger to life or property. Code officers are to attempt to notify residents and owners of the intent to disconnect utility service when time allows. If residents and owners cannot be notified prior to disconnection, they shall be notified in writing as soon as practical thereafter.

WHAT CAN YOU DO TO CHANGE A CODE OFFICER'S DECISION?

If you believe you have received an unfair ruling from a code officer, you may make a formal request to have the decision changed. In doing this, you must make your request to an appeal board. Your reasons for an appeal must be valid and pertain to specific code requirements. Your appeal could be based on what you believe is an incorrect interpretation of the code. If you believe the code does not apply to your case, you have reason for an appeal. There are other reasons for appeal, but you must specify why the appeal is necessary and how the decision you are appealing is incorrect.

TIPS ON HEALTH AND SAFETY

Health and safety are two key issues in the plumbing code. These two issues are, by and large, the reasons for the plumbing code's existence. The plumbing code is designed to ensure health and safety to the public. Public health can be endangered by faulty or improperly installed plumbing. Code officers have the power and duty to condemn a property where severe health or safety risks exist. It is up to the owner of each property to maintain the plumbing in a safe and sanitary manner.

When it comes to safety, there are many more considerations than just plumbing pipes. Most safety concerns arise in conjunction with plumbing but not from the plumbing itself. It is far more likely that a safety hazard will result from the activities of a plumber on other components of a building. An example could be cutting so much of a bearing timber that the structure becomes unsafe. Perhaps a plumber removes a wire from an electric water heater and leaves it exposed and unattended; this could result in a fatal shock to someone. The list of potential safety risks could go on for pages, but you get the idea. It is your responsibility to maintain safe and sanitary conditions at all times.

One aspect of maintaining sanitary conditions includes the use of temporary toilet facilities on job sites. It is not unusual for the plumbing code to require toilet facilities to be available to workers during the construction of buildings. These facilities can be temporary, but they must be sanitary and available.

PIPE PROTECTION

It is the plumber's responsibility to protect plumbing pipes. This protection can take many forms. Here we are going to look at the basics of pipe protection. You will gain insight into pipe protection needs that you might never have considered before.

Backfilling

When backfilling over a pipe, you must take measures to prevent damage to the pipe. The damage can come in two forms, immediate damage and long-term damage. If you are backfilling with material that contains large rocks or other foreign objects, the weight or shape of the rocks and objects might puncture or break the pipe. The long-term effect of having large rocks next to a pipe could result in stress breaks. It is important to use only clean backfill material when backfilling pipe trenches.

Even the weight of a large load of backfill material could damage the pipe or its joints. Backfill material should be added gradually. Layers of backfill should not be more than six inches deep before they are compacted. Each layer of this backfill should be compacted before the next load is dumped. Backfill material both under and beside a pipe shall be compacted for pipe support.

Flood Protection

If a plumbing installation is made in an area subject to flooding, special precautions must be taken. High-water levels can float pipes and erode the earth around them. If your installation is in a flood area, consult your local code officer for the proper procedures to protect your pipes. Floodproofing must be in accordance with the requirements of the *International Building Code.*

Penetrating An Exterior Wall

When a pipe penetrates an exterior wall, it must pass through a sleeve. The sleeve should be at least two pipe-sizes larger than the pipe passing through it. Once the pipe is installed, the open space between the pipe and the sleeve should be sealed with a flexible sealant. By caulking around the pipe, you eliminate the invasion of water and rodents. If the penetration is through a fire-resistive material, the space around the sleeve must be sealed with an approved, fire-resistive material.

Freezing

Pipes must be protected against freezing conditions. Outside, this means placing the pipe deep enough in the ground to avoid freezing. The depth will vary with geographic locations, but your local code officer can provide you with minimum depths. Normally, exterior water supply piping must be installed at least six inches below the frost line and not less than 12 inches below finished grade level. Above-ground pipes, in unheated areas, must be protected with insulation or other means of protection from freezing.

Corrosion

Pipes that tend to be affected by corrosion must be protected. This protection can take the form of a sleeve or a special coating applied to the pipe. For example, copper pipe can have a bad chemical reaction when placed in contact with concrete. If you have a copper pipe extending through concrete, protect it with a sleeve. The sleeve can be a foam insulation or some other type of noncorrosive material. Some soils are capable of corroding pipes. If corrosive soil is suspected, you may have to protect entire sections of underground piping.

Seismic Zones

In seismic zones 3 and 4, hubless cast-iron pipes in sizes five inches or larger, suspended in exposed locations over public or high-traffic areas must be supported on both sides of the coupling if the length of the pipe exceeds four feet.

Firestop Protection

Any DWV or stormwater piping penetrations of fire-resistance materials and enclosures must be protected in accordance with all code requirements. Plans and specifications must detail clearly how penetrations of fire-resistive materials and spaces will be firestopped for adequate protection before a permit will be issued. All firestopping materials must be code-approved. There are ratings that pertain to firestopping. For example, an F rating is the time period that the penetration firestop system limits the spread of fire through the penetration when tested in accordance with ASTM E 814. If you run across a T rating, you will be dealing with the time period that the penetration firestop system, including the penetrating item, limits the maximum temperature rise of 325°F above its initial temperature through the penetration on the nonfire side, when tested in accordance with ASTM E 814.

Combustible Installations. All ABS and PVC DWV piping installations must be protected in accordance with the appropriate fire-resistant rating requirements in the building code that list the acceptable area, height, and type of construction for use in specific occupancies to assure compliance and integrity of the fire-resistance rating prescribed.

All penetrations must be protected by an approved penetration firestop installation. The systems must have a F rating of at least one hour, but not less than the required fire-resistance rating of the assembly being penetrated. Systems that protect floor penetrations must have a T rating of at least one hour, but not less than the required fire-resistance rating of the floor being penetrated. Floor penetrations contained within the cavity of a wall at the location of the floor penetration do not require a T rating. No T rating is required for floor penetrations by piping that is not in direct contact with combustible material.

When piping penetrates a rated assembly, noncombustible piping must not be connected to combustible piping, unless it can be shown the transition complies with all codes. Before any piping is concealed, the installation must be inspected and approved.

Noncombustible Installations.   The basic rules we just covered for plastic pipe apply to metallic pipe. There are some differences. For example, concrete, mortar, or grout may be used to fill the annular spaces around cast iron, copper, or steel piping. The nominal diameter of the penetrating item should not exceed six inches, and the opening size should not exceed 144 square inches. Thickness of the firestop should be the same as the assembly being penetrated. Unshielded couplings are not to be used to connect noncombustible piping unless it can be demonstrated that the fire-resistive integrity of the penetration is maintained.

The Inspection Process

The inspection process for firestopping is handled by the administrative authority. An external examination covers the assembly type, insulation type and thickness, type and size of any sleeve, type and size of penetrant, size of opening, orientation of penetrant, annular space and rating. The elements shall be inspected for compliance with the approved drawing submitted.

An internal examination usually is made with a contractor present and prepared to make repairs. The contractor is asked to cut into the firestop enough to reveal the type and amount of backing materials. Then the contractor repairs the cut and the code officer inspects the repair. Assuming that all goes well, the inspection then is approved.

PIPE CONNECTIONS

Pipe connections can require a variety of adapters when combining pipes of different types. It is important to use the proper methods when making any connections, especially when mating different types of pipes together. Many universal connectors are available to plumbers today. These special couplings allow the connection of a wide variety of materials. Threaded joints on iron-pipe-size (IPS) pipe and fittings must have standard taper pipe threads. Threads on tubing must be of an approved type. When flared joints are made for soft copper tubing they must be made with approved fittings. The tubing must be reamed to the full inside diameter, resized to round, and expanded with a proper flaring tool.

Male and female adapters have long been an acceptable method of joining opposing materials, but today the options are more numerous. You can use compression fittings and rubber couplings to match many types of materials to each other. Special insert adapters allow the use of plastic pipe with bell-and-spigot cast iron.

CONDENSATE DISPOSAL

Liquid combustion by-products of condensing appliances must be collected and discharged to an approved plumbing fixture or disposal area in accordance with the manufacturer's installation instructions. All piping used to handle the condensate must be made of an approved, corrosion-resistant material and must not be smaller than the drain connection on the appliance. Condensate piping must maintain a minimum horizontal slope in the direction of discharge that is at least $\frac{1}{8}$-unit vertical in 12 units horizontal.

Equipment containing evaporators of cooling coils must be provided with a condensate drainage system. The system must be designed and installed in accordance with code requirements. Condensate drainage from all cooling coils and evaporators must be conveyed from the drain pan outlet to an approved place of disposal. It must not be allowed to dump into a street, alley, or other areas so as to cause a nuisance.

The components used to create a condensate drainage system may be made of any of the following materials:

- ABS
- Copper
- Galvanized Steel
- Polybutylene
- Polyethylene
- PVC
- CPVC

All components used to build a condensate drainage system must be rated for the pressure and temperature requirements of the system. Waste lines must not be less than $\frac{3}{4}$-inch in internal diameter. The size is not allowed to decrease from the drainage pan connection to the place of disposal. If multiple drain pipes are connected to a single disposal pipe, the piping must be sized to meet total flow requirements. Horizontal sections of piping must be installed uniformly and with proper pitch.

If there is risk of damage to any building component, a secondary drainage system is required. There are different ways to comply with this requirement. One way is to install an auxiliary pan with a separate drain under the coils on which condensation will occur. This pan must discharge to a conspicuous location to alert people that the primary drain is not functioning. This pan must have a minimum depth of 1.5 inches and it must be not less than 3 inches larger than the unit or the coil dimensions in width and length. The pan must be made of corrosion-resistant material. When metal pans are used, they must have a minimum thickness of 0.0276 inches.

Another option is to use a separate overflow drain pipe that is connected to the main drain pan. This drain, too, must discharge into a conspicuous location. The secondary drain must connect to the drain pan at a location higher than the primary discharge pipe.

A third option is to use an auxiliary drain pan without a separate drain. This pan must be equipped with a water-level detection device. The device will be designed and installed to cut off the equipment being served if the water reaches an unacceptably high level.

All condensate drains must be trapped using the appliance manufacturer's recommendations. Equipment efficiencies must conform with the *International Energy Conservation Code.*

WORKING WITH THE SYSTEM INSTEAD OF AGAINST IT

Code officers are expected to enforce the regulations set forth by the code. Plumbers are expected to work within the parameters of the code. Naturally, plumbers and code officers will come into contact with each other on a regular basis. This contact can lead to some disruptive actions.

The plumbing code is in place to help people, not hurt them. It is not meant to ruin your business or to place you under undue hardship in earning a living. The code really is no different from our traffic laws. Traffic laws are there to protect all of us, but some people resent them. Some plumbers resent the plumbing code. They view it as a vehicle for the local jurisdiction to make more money while they, the plumbers, are forced into positions to possibly make less money.

When you learn and understand the plumbing code and its purpose, you will learn to respect it. You should respect it; it shows the importance of your position as a plumber to the health of our entire nation. Whether or not you agree with the code, you must work within the constraints of it. This means working with the inspectors.

When inspectors choose to play hardball, they hold most of the cards. If you develop an attitude problem, you can pay for it for years to come. Even if you know you are right about an issue, give the inspector a place to escape; none of us enjoys being ridiculed in our profession.

The plumbing code is largely a matter of interpretation. If you have questions, ask your code officer for help. Code officers are generally quite willing to give advice. It is only when you walk into their offices with a chip on your shoulder that you are likely to hit the bureaucratic wall. Like it or not, you must learn to comply with the plumbing code and to work with code officers. The sooner you learn to work with them on amicable terms, the better off you will be.

Little things can mean a lot. Apply for your permit early. This prevents the need to hound the inspector to approve your plans and issue your permit. Many jurisdictions require at least a 24-hour advance notice for an inspection request, but even if your jurisdiction doesn't have this rule, be considerate and plan your inspections in advance. By making life easier for the inspector, you will be helping yourself.

CHAPTER 4
FIXTURES

There is much more to fixtures than meets the eye. Fixtures are a part of the final phase of plumbing. When you are planning a plumbing system, you must know which fixtures are required, what type of fixtures are needed, and how they must be installed. This chapter will guide you through the myriad of fixtures available and how they can be used.

WHAT FIXTURES ARE REQUIRED?

The number and type of fixtures required will depend on local regulations and the use of the building in which they are being installed. Your code book will provide you with information about what is required and how many of each type of fixture is needed. These requirements are based on the use of the building housing the fixtures and the number of people who might use the building. Let me give you some examples.

SINGLE-FAMILY RESIDENCE

When you are planning fixtures for a single-family residence, you must include certain fixtures. If you choose to install more than the minimum, that's fine, but you must install the minimum number of required fixtures. The minimum number and type of fixtures for a single-family dwelling are as follows:

- One toilet
- One bathing unit
- One lavatory
- One kitchen sink
- One clothes washer hookup

MULTIFAMILY BUILDINGS

The minimum requirements for a multifamily building are the same as those for a single-family dwelling, but the requirements are that each dwelling in the multifamily building must be equipped with the minimum fixtures. There is one exception—the laundry

MINIMUM NUMBER OF PLUMBING FACILITIES*

	OCCUPANCY	WATER CLOSETS (Urinals, see Section 419.2) Male	Female	LAVATORIES	BATHTUBS/ SHOWERS	DRINKING FOUNTAINS (see Section 410.1)	OTHERS
A S S E M B L Y	Nightclubs	1 per 40	1 per 40	1 per 75	—	1 per 500	1 service sink
	Restaurants	1 per 75	1 per 75	1 per 200	—	1 per 500	1 service sink
	Theaters, halls, museums, etc.	1 per 125	1 per 65	1 per 200	—	1 per 500	1 service sink
	Coliseums, arenas (less than 3,000 seats)	1 per 75	1 per 40	1 per 150	—	1 per 1,000	1 service sink
	Coliseums, arenas (3,000 seats or greater)	1 per 120	1 per 60	Male 1 per 200 Female 1 per 150	—	1 per 1,000	1 service sink
	Churches[b]	1 per 150	1 per 75	1 per 200	—	1 per 1,000	1 service sink
	Stadiums (less than 3,000 seats), pools, etc.	1 per 100	1 per 50	1 per 150	—	1 per 1,000	1 service sink
	Stadiums (3,000 seats or greater)	1 per 150	1 per 75	Male 1 per 200 Female 1 per 150	—	1 per 1,000	1 service sink
I N S T I T U T I O N A L	Business (see Sections 403.2, 403.4 and 403.5)	1 per 50		1 per 80	—	1 per 100	1 service sink
	Educational	1 per 50		1 per 50	—	1 per 100	1 service sink
	Factory and industrial	1 per 100		1 per 100	(see Section 411)	1 per 400	1 service sink
	Passenger terminals and transportation facilities	1 per 500		1 per 750	—	1 per 1,000	1 service sink
	Residential care	1 per 10		1 per 10	1 per 8	1 per 100	1 service sink
	Hospitals, ambulatory nursing home patients[c]	1 per room[d]		1 per room[d]	1 per 15	1 per 100	1 service sink per floor
	Day nurseries, sanitariums, nonambulatory nursing home patients, etc.[c]	1 per 15		1 per 15	1 per 15[e]	1 per 100	1 service sink
	Employees, other than residential care[c]	1 per 25		1 per 35	—	1 per 100	—
	Visitors, other than residential care	1 per 75		1 per 100	—	1 per 500	—
	Prisons[c]	1 per cell		1 per cell	1 per 15	1 per 100	1 service sink
	Asylums, reformatories, etc.[c]	1 per 15		1 per 15	1 per 15	1 per 100	1 service sink
	Mercantile (see Sections 403.2, 403.4 and 403.5)	1 per 500		1 per 750	—	1 per 1,000	1 service sink
R E S I D E N T I A L	Hotels, motels	1 per guestroom		1 per guestroom	1 per guestroom	—	1 service sink
	Lodges	1 per 10		1 per 10	1 per 8	1 per 100	1 service sink
	Multiple family	1 per dwelling unit		1 per dwelling unit	1 per dwelling unit	—	1 kitchen sink per dwelling unit; 1 automatic clothes washer connection per 20 dwelling units
	Dormitories	1 per 10		1 per 10	1 per 8	1 per 100	1 service sink
	One- and two-family dwellings	1 per dwelling unit		1 per dwelling unit	1 per dwelling unit	—	1 kitchen sink per dwelling unit; 1 automatic clothes washer connection per dwelling unit[f]
	Storage (see Sections 403.2 and 403.4)	1 per 100		1 per 100	(see Section 411)	1 per 1,000	1 service sink

a. The fixtures shown are based on one fixture being the minimum required for the number of persons indicated or any fraction of the number of persons indicated. The number of occupants shall be determined by the *International Building Code*.

b. Fixtures located in adjacent buildings under the ownership or control of the church shall be made available during periods the church is occupied.

c. Toilet facilities for employees shall be separate from facilities for inmates or patients.

d. A single-occupant toilet room with one water closet and one lavatory serving not more than two adjacent patient rooms shall be permitted where such room is provided with direct access from each patient room and with provisions for privacy.

e. For day nurseries, a maximum of one bathtub shall be required.

f. For attached one- and two-family dwellings, one automatic clothes washer connection shall be required per 20 dwelling units.

FIGURE 4.1 These charts list the minimum facilities necessary that, in most cases, a plumber will encounter during a career. *(Courtesy of International Code Council, Inc. and International Plumbing Code 2000.)*

hookup. With a multifamily building, laundry hookups are not required in each dwelling unit. It is sometimes required that a laundry hookup be installed for common use when the number of dwelling units is 20. For each interval of 20 units, you must install a laundry hookup when this code is in effect.

For example, in a building with 40 apartments, you would have to provide two laundry hook-ups. If the building had 60 units, you would need three hookups. This ratio is not always the same. Sometimes a dwelling-unit interval is 10 rental units and the local requirements could require one hookup for every 12 rental units, but no less than two hookups for buildings with at least 15 units.

NIGHTCLUBS

When you get into businesses and places of public assembly, such as nightclubs, the ratings are based on the number of people likely to use the facilities. n a nightclub, the minimum requirements are usually:

- Toilets—one toilet for every 40 people
  (*Note: Fixtures located in a unisex toilet or bathing room can be counted in determining the minimum required number of fixtures for assembly and mercantile occupancies only.*)
- Lavatories—one lavatory for every 75 people
- Service sinks—one service sink is required
- Drinking fountains—one drinking fountain for every 500 people
- Bathing units—none

DAY-CARE FACILITIES

The minimum number of fixtures for a day-care facility are usually:

- Toilets—one toilet for every 15 people
- Lavatories—one lavatory for every 15 people
- Bathing units—one bathing unit for every 15 people
- Service sinks—one service sink is required
- Drinking fountains—one drinking fountain for every 100 people

In contrast, some local codes require the installation of toilets and lavatories only in day-care facilities. The ratings for these two fixtures are the same as above, but the other fixtures required by standard code requirements may not be required in all regions. This type of rating system will be found in your local code and will cover all the normal types of building uses.

In many cases, facilities must be provided in separate bathrooms, to accommodate each sex. When installing separate bathroom facilities, the number of required fixtures will be divided equally between the two sexes, unless there is cause and approval for a different appropriation. Separate facilities are not required for private facilities.

Some types of buildings do not require separate facilities. For example, some jurisdictions do not require the following buildings to have separate facilities: Residential properties, small businesses, where fewer than fifteen employees work, or where fewer than fifteen people are allowed in the building at the same time.

Minimum Plumbing Facilities[1]

Each building shall be provided with sanitary facilities, including provisions for the physically handicapped as prescribed by the Department having jurisdiction. For requirements for the handicapped, ANSI A117.1-1992, Accessible and Usable Buildings and Facilities, may be used.

The total occupant load shall be determined by minimum exiting requirements. The minimum number of fixtures shall be calculated at fifty (50) percent male and fifty (50) percent female based on the total occupant load.

Type of Building or Occupancy[2]	Water Closets[14] (Fixtures per Person)		Urinals[5, 10] (Fixtures per Person)	Lavatories (Fixtures per Person)		Bathtubs or Showers (Fixtures per Person)	Drinking Fountains[3, 13] (Fixtures per Person)
Assembly Places – Theatres, Auditoriums, Convention Halls, etc. – for permanent employee use	Male 1: 1-15 2: 16-35 3: 36-55 Over 55, add 1 fixture for each additional 40 persons.	Female 1: 1-15 3: 16-35 4: 36-55	Male 0: 1-9 1: 10-50 Add one fixture for each additional 50 males.	Male 1 per 40	Female 1 per 40		
Assembly Places – Theatres, Auditoriums, Convention Halls, etc. – for public use	Male 1: 1-100 2: 101-200 3: 201-400 Over 400, add one fixture for each additional 500 males and 1 for each additional 125 females.	Female 3: 1-50 4: 51-100 8: 101-200 11: 201-400	Male 1: 1-100 2: 101-200 3: 201-400 4: 401-600 Over 600 add 1 fixture for each additional 300 males.	Male 1: 1-200 2: 201-400 3: 401-750 Over 750, add one fixture for each additional 500 persons.	Female 1: 1-200 2: 201-400 3: 401-750		1: 1-150 2: 151-400 3: 401-750 Over 750, add one fixture for each additional 500 persons.
Dormitories[9] School or Labor	Male 1 per 10 Add 1 fixture for each additional 25 males (over 10) and 1 for each additional 20 females (over 8).	Female 1 per 8	Male 1 per 25 Over 150, add 1 fixture for each additional 50 males.	Male 1 per 12 Over 12 add one fixture for each additional 20 males and 1 for each 15 additional females.	Female 1 per 12	1 per 8 For females, add 1 bathtub per 30. Over 150, add 1 per 20.	1 per 150[12]
Dormitories – for staff use	Male 1: 1-15 2: 16-35 3: 36-55 Over 55, add 1 fixture for each additional 40 persons.	Female 1: 1-15 3: 16-35 4: 36-55	Male 1 per 50	Male 1 per 40	Female 1 per 40	1 per 8	
Dwellings[4] Single Dwelling Multiple Dwelling or Apartment House	1 per dwelling 1 per dwelling or apartment unit			1 per dwelling 1 per dwelling or apartment unit		1 per dwelling 1 per dwelling or apartment unit	
Hospital Waiting rooms	1 per room			1 per room			1 per 150[12]
Hospitals – for employee use	Male 1: 1-15 2: 16-35 3: 36-55 Over 55, add 1 fixture for each additional 40 persons.	Female 1: 1-15 3: 16-35 4: 36-55	Male 0: 1-9 1: 10-50 Add one fixture for each additional 50 males.	Male 1 per 40	Female 1 per 40		
Hospitals Individual Room Ward Room	1 per room 1 per 8 patients			1 per room 1 per 10 patients		1 per room 1per 20 patients	1 per 150[12]
Industrial[6] Warehouses Workshops, Foundries and similar establishments – for employee use	Male 1: 1-10 2: 11-25 3: 26-50 4: 51-75 5: 76-100 Over 100, add 1 fixture for each additional 30 persons	Female 1: 1-10 2: 11-25 3: 26-50 4: 51-75 5: 76-100		Up to 100, 1 per 10 persons Over 100, 1 per 15 persons[7, 8]		1 shower for each 15 persons exposed to excessive heat or to skin contamination with poisonous, infectious, or irritating material	1 per 150[12]

FIGURE 4.2 Above are specific requirements and ratios of plumbing fixtures for most installations. *(Reprinted from the 2000 Uniform Plumbing Code (UPC) with the permission of the International Association of Plumbing and Mechanical Officials (IAPMO).) (Continued)*

Local codes may not require separate facilities in the following buildings: offices with less than 1,200 square feet, retail stores with less than 1,500 square feet, restaurants with less than 500 square feet, self-serve laundries with less than 1,400 square feet, and hair salons with less than 900 square feet.

EMPLOYEE AND CUSTOMER FACILITIES

There are some special regulations pertaining to employee and customer facilities. For employees, toilet facilities must be available to employees within a reasonable distance and with relative ease of access. The general code requires these facilities to be in the immediate work area; the distance an employee is required to walk to the facilities may not exceed 500 feet. The facilities must be located in a manner so that employees do not

Type of Building or Occupancy[2]	Water Closets[14] (Fixtures per Person)		Urinals[5,10] (Fixtures per Person)	Lavatories (Fixtures per Person)		Bathtubs or Showers (Fixtures per Person)	Drinking Fountains[3,13] (Fixtures per Person)
Institutional – Other than Hospitals or Penal Institutions (on each occupied floor)	Male 1 per 25	Female 1 per 20	Male 0: 1-9 1: 10-50 Add one fixture for each additional 50 males.	Male 1 per 10	Female 1 per 10	1 per 8	1 per 150[12]
Institutional – Other than Hospitals or Penal Institutions (on each occupied floor) – for employee use	Male 1: 1-15 2: 16-35 3: 36-55 Over 55, add 1 fixture for each additional 40 persons.	Female 1: 1-15 3: 16-35 4: 36-55	Male 0: 1-9 1: 10-50 Add one fixture for each additional 50 males.	Male 1 per 40	Female 1 per 40	1 per 8	1 per 150[12]
Office or Public Buildings	Male 1: 1-100 2: 101-200 3: 201-400 Over 400, add one fixture for each additional 500 males and 1 for each additional 150 females.	Female 3: 1-50 4: 51-100 8: 101-200 11: 201-400	Male 1: 1-100 2: 101-200 3: 201-400 4: 401-600 Over 600 add 1 fixture for each additional 300 males.	Male 1: 1-200 2: 201-400 3: 401-750 Over 750, add one fixture for each additional 500 persons	Female 1: 1-200 2: 201-400 3: 401-750		1 per 150[12]
Office or Public Buildings – for employee use	Male 1: 1-15 2: 16-35 3: 36-55 Over 55, add 1 fixture for each additional 40 persons.	Female 1: 1-15 3: 16-35 4: 36-55	Male 0: 1-9 1: 10-50 Add one fixture for each additional 50 males.	Male 1 per 40	Female 1 per 40		
Penal Institutions – for employee use	Male 1: 1-15 2: 16-35 3: 36-55 Over 55, add 1 fixture for each additional 40 persons.	Female 1: 1-15 3: 16-35 4: 36-55	Male 0: 1-9 1: 10-50 Add one fixture for each additional 50 males.	Male 1 per 40	Female 1 per 40		1 per 150[12]
Penal Institutions – for prison use Cell Exercise Room	1 per cell 1 per exercise room		Male 1 per exercise room	1 per cell 1 per exercise room			1 per cell block floor 1 per exercise room
Restaurants, Pubs and Lounges[11]	Male 1: 1-50 2: 51-150 3: 151-300 Over 300, add 1 fixture for each additional 200 persons	Female 1: 1-50 2: 51-150 4: 151-300	Male 1: 1-150 Over 150, add 1 fixture for each additional 150 males	Male 1: 1-150 2: 151-200 3: 201-400 Over 400, add 1 fixture for each additional 400 persons	Female 1: 1-150 2: 151-200 3: 201-400		
Schools – for staff use All schools	Male 1: 1-15 2: 16-35 3: 36-55 Over 55, add 1 fixture for each additional 40 persons	Female 1: 1-15 2: 16-35 3: 36-55	Male 1 per 50	Male 1 per 40	Female 1 per 40		
Schools – for student use Nursery	Male 1: 1-20 2: 21-50 Over 50, add 1 fixture for each additional 50 persons	Female 1: 1-20 2: 21-50		Male 1: 1-25 2: 26-50 Over 50, add 1 fixture for each additional 50 persons	Female 1: 1-25 2: 26-50		1 per 150[12]
Elementary	Male 1 per 30	Female 1 per 25	Male 1 per 75	Male 1 per 35	Female 1 per 35		1 per 150[12]
Secondary	Male 1 per 40	Female 1 per 30	Male 1 per 35	Male 1 per 40	Female 1 per 40		1 per 150[12]
Others (Colleges, Universities, Adult Centers, etc.)	Male 1 per 40	Female 1 per 30	Male 1 per 35	Male 1 per 40	Female 1 per 40		1 per 150[12]
Worship Places Educational and Activities Unit	Male 1 per 150	Female 1 per 75	Male 1 per 150	1 per 2 water closets			1 per 150[12]
Worship Places Principal Assembly Place	Male 1 per 150	Female 1 per 75	Male 1 per 150	1 per 2 water closets			1 per 150[12]

FIGURE 4.2 Above are specific requirements and ratios of plumbing fixtures for most installations. (*Reprinted from the* 2000 Uniform Plumbing Code (UPC) *with the permission of the International Association of Plumbing and Mechanical Officials (IAPMO).*) (*Continued*)

have to negotiate more than one set of stairs for access to the facilities. There are some exceptions to these regulations, but in general, these are the rules.

It is expected that customers of restaurants, stores, and places of public assembly will have toilet facilities. This is usually based on buildings capable of holding 150 or more people. Buildings with an occupancy rating of fewer than 150 people are not usually required to provide toilet facilities, unless the building serves food or beverages. When facilities are required, they can be placed in individual buildings, or in a shopping mall situation, in a common area, not more than 500 feet from any store or tenant space. These

1. The figures shown are based upon one (1) fixture being the minimum required for the number of persons indicated or any fraction thereof.

2. Building categories not shown on this table shall be considered separately by the Administrative Authority.

3. Drinking fountains shall not be installed in toilet rooms.

4. Laundry trays. One (1) laundry tray or one (1) automatic washer standpipe for each dwelling unit or one (1) laundry tray or one (1) automatic washer standpipe, or combination thereof, for each twelve (12) apartments. Kitchen sinks, one (1) for each dwelling or apartment unit.

5. For each urinal added in excess of the minimum required, one water closet may be deducted. The number of water closets shall not be reduced to less than two-thirds (2/3) of the minimum requirement.

6. As required by ANSI Z4.1-1968, Sanitation in Places of Employment.

7. Where there is exposure to skin contamination with poisonous, infectious, or irritating materials, provide one (1) lavatory for each five (5) persons.

8. Twenty-four (24) lineal inches (610 mm) of wash sink or eighteen (18) inches (457 mm) of a circular basin, when provided with water outlets for such space, shall be considered equivalent to one (1) lavatory.

9. Laundry trays, one (1) for each fifty (50) persons. Service sinks, one (1) for each hundred (100) persons.

10. General. In applying this schedule of facilities, consideration shall be given to the accessibility of the fixtures. Conformity purely on a numerical basis may not result in an installation suited to the need of the individual establishment. For example, schools should be provided with toilet facilities on each floor having classrooms.

    a. Surrounding materials, wall and floor space to a point two (2) feet (610 mm) in front of urinal lip and four (4) feet (1219 mm) above the floor, and at least two (2) feet (610 mm) to each side of the urinal shall be lined with non-absorbent materials.

    b. Trough urinals shall be prohibited.

11. A restaurant is defined as a business which sells food to be consumed on the premises.

    a. The number of occupants for a drive-in restaurant shall be considered as equal to the number of parking stalls.

    b. Employee toilet facilities shall not be included in the above restaurant requirements. Hand washing facilities shall be available in the kitchen for employees.

12. Where food is consumed indoors, water stations may be substituted for drinking fountains. Offices, or public buildings for use by more than six (6) persons shall have one (1) drinking fountain for the first one hundred fifty (150) persons and one (1) additional fountain for each three hundred (300) persons thereafter.

13. There shall be a minimum of one (1) drinking fountain per occupied floor in schools, theatres, auditoriums, dormitories, offices or public building.

14. The total number of water closets for females shall be at least equal to the total number of water closets and urinals required for males.

FIGURE 4.2 Above are specific requirements and ratios of plumbing fixtures for most installations. *(Reprinted from the* 2000 Uniform Plumbing Code (UPC) *with the permission of the International Association of Plumbing and Mechanical Officials (IAPMO).) (Continued)*

central toilets must be placed so that customers will not have to use more than one set of stairs to reach them.

Toilet facilities in buildings other than assembly or mercantile uses cannot be installed more than one story above or below the employee's regular working area and the path of travel to the toilet facilities cannot exceed 500 feet. There is a potential exception for maximum travel distance when the building is used for factory or industrial uses.

When toilet facilities for employees are located in covered malls, the travel distance must not exceed 300 feet. There are exceptions to this rule, so check your local requirements. Facilities in covered malls are based on total square footage. Toilet facilities must be installed in each individual store or in a central toilet area located no more than 300 feet from the source of travel for individuals using the facility. Travel distance is measured from the main entrance of any store or tenant space. Some jurisdictions use a square-footage method to determine minimum requirements in public places. For example, retail stores are rated as having an occupancy load of one person for every 200 square feet of floor space. This type of facility is required to have separate facilities when the store's square footage exceeds 1,500 square feet. A minimum of one toilet is required for each facility when the occupancy load is up to 35 people. One lavatory is required in each facility, for up to 15 people. A drinking fountain is required for occupancy loads up to 100 people. Drinking fountains may not be installed in public restrooms.

HANDICAP FIXTURES

Handicap fixtures are not cheap; you cannot afford to overlook them when bidding a job. The plumbing code usually will require specific minimum requirements for handicap-accessible fixtures in certain circumstances. It is your responsibility to know when handicap facilities are required. There are also special regulations pertaining to how handicap fixtures shall be installed. We are about to embark on a journey into handicap fixtures and their requirements.

When you are dealing with handicap plumbing, you must mix the local plumbing code with the local building code. These two codes work together in establishing the minimum requirements for handicap plumbing facilities. When you step into the field of handicap plumbing, you must play by a different set of rules. Handicap plumbing is like a different code, all unto its own.

WHERE ARE HANDICAP FIXTURES REQUIRED?

Most buildings frequented by the public are required to have handicap-accessible plumbing fixtures. The following handicap examples are based on general code requirements.

Single-family homes and most residential multi-family dwellings are exempt from handicap requirements. A common standard for most public buildings is the inclusion of one toilet and one lavatory for handicap use.

Hotels, motels, inns, and the like are required to provide a toilet, lavatory, bathing unit, and kitchen sink, where applicable, for handicap use. Drinking fountains also may be required Drinking fountains must not be installed in public restrooms. This provision will depend on both the local plumbing and building code. If plumbing a gang-shower arrangement, such as in a school gym, at least one of the shower units must be handicap-accessible. Door sizes and other building code requirements must be observed when dealing with handicap facilities. There are local exceptions to these rules; check with your local code officers for current, local regulations.

INSTALLATION CONSIDERATIONS

When it comes to installing handicap plumbing facilities, you must pay attention to the plumbing code and the building code. In most cases, approved blueprints will indicate the requirements of your job, but in rural areas, you might not enjoy the benefit of highly detailed plans and specifications. When it comes time for a final inspection, the plumbing must pass muster along with the open space around the fixtures. If the inspection is failed, your pay is held up and you are likely to incur unexpected costs. This section will apprise you of what you may need to know. It is not all plumbing, but it is all needed information when working with handicap facilities.

HANDICAP TOILET FACILITIES

When you think of installing a handicap toilet, you probably think of a toilet that sits high off the floor. But do you think of the grab bars and partition dimensions required around the toilet? Some plumbers don't, but they should. The door to a privacy stall for a handicap toilet must provide a minimum of 32 inches of clear space.

The distance between the front of the toilet and the closed door must be at least 48 inches. It is mandatory that the door open outward, away from the toilet. Think about it, how could a person in a wheelchair close the door if the door opened into the toilet? These facts might not seem like your problem, but if your inspection doesn't pass, you don't get paid.

The width of a water closet compartment for handicap toilets must be a minimum of five feet. The length of the privacy stall shall be at least 56 inches for wall-mounted toilets and 59 inches for floor-mounted models. Unlike regular toilets, which require a rough-in of 15 inches to the center of the drain from a side wall, handicap toilets require the rough-in to be at least 18 inches off the side wall.

Then, there are the required grab bars. Sure, you might know that grab bars are required, but do you know the mounting requirements for the bars? Two bars are required for each handicap toilet. One bar should be mounted on the back wall and the other will be installed on the side wall. The bar mounted on the back wall must be at least three feet long. The first mounting bracket of the bar must be mounted no more than six inches from the side wall. Then, the bar must extend at least 24 inches past the center of the toilet's drain.

The bar mounted on the side wall must be at least 42 inches long. The bar should be mounted level and with the first mounting bracket located no more than one foot from the back wall. The bar must be mounted on the side wall that is closest to the toilet. This bar must extend to a point at least 54 inches from the back wall. If you do your math, you will see that a 42-inch bar is pushing the limits on both ends. A longer bar will allow more assurance of meeting the minimum requirements.

When a lavatory will be installed in the same toilet compartment, the lavatory must be installed on the back wall. The lavatory must be installed in a way that its closest point to the toilet is no less than 18 inches from the center of the toilet's drain. When a privacy stall of this size and design is not suitable, there is an option.

Another way to size the compartment to house a handicap toilet and lavatory is available. There may be times when space restraints will not allow a stall with a width of five feet. In these cases, you may position the fixture differently and use a stall with a width of only 3 feet. In these situations, the width of the privacy stall may not exceed 4 feet.

The depth of the compartment must be at least 66 inches, when wall-mounted toilets are used. The depth extends to a minimum of 69 inches with the use of a floor-mounted water closet. The toilet requires a minimum distance from side walls of 18 inches to the center of the toilet drain. If the compartment is more than three feet wide, grab bars are required, with the same installation methods as described before.

If the stall is made at the minimum width of three feet, grab bars, with a minimum length of 42 inches, are required on each side of the toilet. These bars must be mounted no more than one foot from the back wall, and they must extend a minimum of 54 inches from the back wall. If a privacy stall is not used, the side wall clearances and the grab bar requirements are the same as listed in these two examples. To determine which set of rules to use, you must assess the shape of the room when no stall is present.

If the room is laid out in a fashion such as the first example, use the guidelines for grab bars as listed there. If, on the other hand, the room tends to meet the description of the last example, use the specifications in that example. In both cases, the door to the room may not swing into toilet area.

HANDICAP FIXTURES

Handicap fixtures are designed especially for people with less physical ability than the general public. The differences in handicap fixtures might appear subtle, but they are important. Let's look at the requirements a fixture must meet to be considered a handicap fixture.

In assembly and mercantile occupancies, unisex toilet and bathing rooms must be provided in accordance with the local code. An accessible unisex toilet room is required when an aggregate of six or more male or female water closets are required. In buildings with mixed occupancy, only those water closets required for the assembly or mercantile occupancy shall be used to determine the unisex toilet room requirement. A unisex bathing room is required in recreational facilities where separate-sex bathing rooms are provided. There is an exception. When a separate-sex bathing room has only one shower or bathtub, a unisex bathing room is not required.

Unisex toilet facilities must comply with all code requirements. The facilities must consist of only one water closet and only one lavatory. Unisex bathing rooms are to be considered a unisex toilet room. As usual, there are exceptions. A separate-sex toilet room that contains no more than two water closets, no urinals, or one water closet and one urinal will be considered a unisex toilet room. When unisex toilet and bathing rooms are installed, they must be located on an accessible route. The rooms must not be located more than one story above or below separate-sex toilet rooms. Travel distance is not to exceed 500 feet.

Unisex toilet rooms installed in passenger transportation facilities and airports must have a travel route from separate-sex toilet rooms that does not require passage through security checkpoints.

A clear floor space of not less than 30 inches by 48 inches is required beyond the area of the door swing when a door opens into a unisex toilet or bathing room. Doors providing privacy for unisex toilet and bathing rooms must be securable from within the room. A sign that complies with code requirements is required to designate a unisex toilet or bathing room. Directional signage shall be provided at all separate-sex toilet and bathing rooms indicating the location of the nearest unisex toilet or bathing room.

TOILETS

Toilets will have a normal appearance, but they will sit higher above the floor than a standard toilet. A handicap toilet will rise to a height of between 16 and 20 inches off the finished floor. Eighteen inches is a common height for most handicap toilets. Toilets are required to have a minimum of 30 inches center-to-center between other fixtures and walls. An open space of 21 inches is required in front of the toilet. The same is true for bidets. Urinals require a center-to-center open space of 30 inches. There are many choices in toilet style, they include the following:

- Blowout
- Reverse trap
- Siphon jet
- Siphon vortex
- Siphon wash

SINKS AND LAVATORIES

Visually, handicap sinks, lavatories, and faucets may appear to be standard fixtures, but their method of installation is regulated and the faucets often are unlike a standard faucet. Handicap sinks and lavatories must be positioned to allow a person in a wheelchair to use them easily.

The clearance requirements for a lavatory are numerous. There must be at least 21 inches of clearance in front of the lavatory. This clearance must extend 30 inches from the front edge of the lavatory, or countertop, whichever protrudes the farthest, and to the sides. If you can sit a square box, with a 30" × 30" dimension, in front of the lavatory or countertop, you have adequate clearance for the first requirement. This applies to kitchen sinks and lavatories.

The next requirement calls for the top of the lavatory to be no more than 35 inches from the finished floor. For a kitchen sink, the maximum height is 34 inches. Then, there is knee clearance to consider. The minimum allowable knee clearance requires 29 inches in height and 8 inches in depth. This is measured from the face of the fixture, lavatory or kitchen sink. Toe clearance is another issue. A space nine inches high and nine inches deep is required, as a minimum, for toe space. The last requirement deals with hot water pipes. Any exposed hot water pipes must be insulated, or shielded in some way, to prevent users of the fixture from being burned.

Lavatories in employee and public toilet rooms must be located in the same room as the required water closet.

SINK AND LAVATORY FAUCETS

Handicap faucets frequently have blade handles. The faucets must be located no more than 25 inches from the front edge of the lavatory or counter, whichever is closest to the user. The faucets could use wing handles, single handles, or push buttons to be operated, but the operational force required by the user shall not be more than five pounds.

BATHING UNITS

Handicap bathtubs and showers must meet the requirements of approved fixtures, like any other fixture, but they also are required to have special features and installation methods. The special features are required under the code for approved handicap fixtures. The clear space in front of a bathing unit is required to be a minimum of 1,440 square inches. This is achieved by leaving an open space of 30 inches in front of the unit and forty-eight inches to the sides. If the bathing unit is not accessible from the side, the minimum clearance is increased to an area with a dimension of 48" × 48".

Handicap bathtubs are required to be installed with seats and grab bars. A grab bar, for handicap use, must have a diameter of at least one and one-quarter inch. The diameter may not exceed one and one-half inches. All handicap grab bars are meant to be installed one and one-half inches from walls. The design and strength for these bars are set forth in the building codes.

The seat can be an integral part of the bathtub, or it can be a removable, after-market seat. The grab bars must be at least two feet long. Two of these grab bars are to be mounted on the back wall, one above the other. The bars are to run horizontally. The lower grab bar must be mounted nine inches above the flood-level rim of the tub. The top grab bar must be mounted a minimum of 33 inches, but no more than 36 inches, above the finished floor. The grab bars should be mounted near the seat of the bathing unit.

Additional grab bars are required at each end of the tub. These bars should be mounted horizontally and at the same height as the higher grab bar on the back wall. The bar over the faucet must be at least two feet long. The bar on the other end of the tub may be as short as one foot.

The faucets in these bathing units must be located below the grab bars. The faucets used with a handicap bathtub must be able to operate with a maximum force of five pounds. A personal, handheld shower is required in all handicap bathtubs. The hose for the handheld shower must be at least five feet long.

Two types of showers usually are used for handicap purposes. The first type of shower allows the user to leave a wheelchair and shower while sitting on a seat. The other style of shower stall is meant for the user to roll a wheelchair into the stall and shower while seated in the wheelchair.

If the shower is intended to be used with a shower seat, its dimensions should form a square, with three feet of clearance. The seat should be no more than 16 inches wide and mounted along the side wall. This seat should run the full length of the shower. The height of the seat should be between 17 and 19 inches above the finished floor. Two grab bars should be installed in the shower.

These bars should be located between 33 and 36 inches above the finished floor. The bars are intended to be mounted in an "L" shape. One bar should be 36 inches long and run the length of the seat, mounted horizontally. The other bar should be installed on the side wall of the shower. This bar should be at least 18 inches long.

The faucet for this type of shower must be mounted on the wall across from the seat. The faucet must be at least 38 inches, but not more than 48 inches, above the finished floor. A handheld shower must be installed in the shower. A fixed shower head also can be installed, but there must be a handheld shower, on a hose of at least five feet in length. The faucet must be able to operate with a maximum force of five pounds.

DRINKING UNITS

The distribution of water from a water cooler or drinking fountain must occur at a maximum height of 36 inches above the finished floor. The outlet for drinking water must be located at the front of the unit and the water must flow upward for a minimum distance of four inches. Levers or buttons to control the operation of the drinking unit may be mounted on front of the unit or on the side near the front.

Clearance requirements call for an open space of 30 inches in front of the unit and 48 inches to the sides. Knee and toe clearance is the same as required for sinks and lavatories. If the unit is made so that the drinking spout extends beyond the basic body of the unit, the width clearance may be reduced from 48 inches to 30 inches, as long as knee and toe requirements are met.

STANDARD FIXTURE INSTALLATION REGULATIONS

Standard fixtures also must be installed according to local code regulations. There are space limitations, clearance requirements, and pre-determined, approved methods for installing standard plumbing fixtures. First, let's look at the space and clearance requirements for common fixtures.

STANDARD FIXTURE PLACEMENT

Toilets and bidets require a minimum distance of 15 inches from the center of the fixture's drain to the nearest side wall. These fixtures must have at least 15 inches of clear space between the center of their drains and any obstruction, such as a wall, cabinet, or other fixture. With this rule in mind, a toilet or bidet must be centered in a space of at least 30 inches. Some local codes further require that there be a minimum of 18 inches of clear space in front of these fixtures, and that when toilets are placed in privacy stalls, the stalls must be at least 30 inches wide and 60 inches deep.

Some codes require urinals to be installed with a minimum clear distance of twelve inches from the center of their drains to the nearest obstacle on either side. When urinals are installed side by side, the distance between the centers of their drains must be at least 24 inches. Urinals shall not be substituted for more than 67 percent of the required water closets.

Other local codes may require urinals to have minimum side-wall clearances of at least 15 inches. The center-to-center distance is a minimum of 30 inches. Urinals also must have a minimum clearance of 18 inches in front of them.

These fixtures, as with all fixtures, must be installed level and with good skills. The fixture usually should be set at an equal distance from walls, to avoid a crooked or cocked installation. All fixtures should be designed and installed with proper cleaning in mind.

Bathtubs, showers, vanities, and lavatories should be placed in a manner to avoid violating the clearance requirements for toilets, urinals, and bidets.

SECURING AND SEALING FIXTURES

Some fixtures hang on walls, and others sit on floors. When securing fixtures to walls and floors, there are some rules you must obey. Floor-mounted fixtures, like most residential toilets, should be secured to the floor with the use of a closet flange. The flange is first screwed or bolted to the floor. A wax seal then is placed on the flange, and closet bolts are placed in slots on both sides of the flange. Then, the toilet is set into place.

The closet bolts should be made of brass, or some other material that will resist corrosive action. The closet bolts are tightened until the toilet will not move from side to side

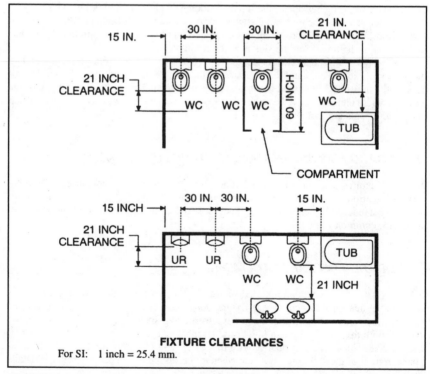

FIXTURE CLEARANCES

For SI: 1 inch = 25.4 mm.

FIGURE 4.3 Bathroom fixtures must have specific clearance from walls, doors or fixtures. (*Courtesy of International Code Council, Inc. and International Plumbing Code 2000.*)

or front to back. In some cases, a flange is not used. When a flange is not used, the toilet should be secured with corrosion-resistant lag bolts.

When toilets or other fixtures are being mounted on a wall, the procedure is a little different. The fixture must be installed on, and supported by, an approved hanger. These hangers usually are packed with the fixture. The hanger must assume the weight placed in and on the fixture, to avoid stress on the fixture itself.

In the case of a wall-hung toilet, the hanger usually has a pattern of bolts extending from the hanger to a point outside the wall. The hanger is concealed in the wall cavity. A watertight joint is made at the point of connection, usually with a gasket ring, and the wall-hung toilet is bolted to the hanger.

With lavatories, the hanger usually is mounted on the outside surface of the finished wall. A piece of wood blocking typically is installed in the wall cavity to allow a solid surface for mounting the bracket. The bracket usually is secured to the blocking with lag bolts. The hanger is put in place and lag bolts are screwed through the bracket and finished wall into the wood blocking. Then the lavatory is hung on the bracket.

The space where the lavatory meets the finished wall must be sealed. This is true of all fixtures coming into contact with walls, floor, or cabinets. The crevice caused by the fixture meeting the finished surface must be sealed to protect against water damage. A caulking compound, such as silicone, usually is used for this purpose. This seal does more than prevent water damage. It eliminates hard-to-clean areas and makes the plumbing easier to keep free of dirt and germs.

Bathtubs must be installed level, and they must be supported properly. The support for most one-piece units is the floor. These units are made to be set into place, leveled, and secured. Other types of tubs, like cast-iron tubs, require more support than the floor will give. They need a ledger or support blocks placed under the rim where the edge of the tub meets the back wall.

The ledger can be a piece of wood, such as a wall stud. The ledger should be about the same length as the tub. This ledger is installed horizontally and level. The ledger should be at a height that will support the tub in a level fashion or with a slight incline, so excess water on the rim of the tub will run back into the tub. The ledger is nailed to wall studs.

If blocks are used, they are cut to a height that will put the bathtub into the proper position. The blocks then are placed at the two ends, and often in the middle of where the tub will sit. The blocks should be installed vertically and nailed to the stud wall.

When the tub is set into place, the rim at the back wall, rests on the blocks or ledger for additional support. This type of tub has feet on the bottom, so the floor supports most of the weight. The edges where the tub meets the walls must be caulked. If shower doors are installed on a bathtub or shower, they must meet safety requirements set forth in the building codes.

Showers today are usually one-piece units, which are meant to sit in their place, be leveled, and secured to the wall. The securing process for one-piece showers and bathtubs usually is accomplished by placing nails or screws through a nailing flange, which is molded as part of the unit, into the stud walls. If only a shower base is being installed, it also must be level and secure. Shower doors must open enough to allow a minimum of 22 inches of unobstructed opening for egress.

Now, let's look at some of the many other regulations involved in installing plumbing fixtures.

THE FACTS ABOUT FIXTURE INSTALLATIONS

When it is time to install fixtures, many rules and regulations must be adhered to. Water supply is one issue. Access is another. Air gaps and overflows are factors. A host of requirements governs the installation of plumbing fixtures. We will start with the fixtures most likely found in residential homes. Then we will look at the fixtures normally associated with commercial applications.

TYPICAL RESIDENTIAL FIXTURE INSTALLATION

Typical residential fixture installations could include everything from hose bibbs to bidets. This section is going to take each fixture that could be considered a typical residential fixture and tell you more about how it must be installed.

With most plumbing fixtures, you have water coming into the fixture and water going out of the fixture. The incoming water lines must be protected against freezing and backsiphonage. Freeze protection usually is accomplished with the placement of the piping. In cold climates, it is advisable to avoid putting pipes in outside walls. Insulation often is applied to water lines to reduce the risk of freezing. Backsiphonage typically is avoided with the use of air gaps and backflow preventers.

Some fixtures, like lavatories and bathtubs, are equipped with overflow routes. These overflow paths must be designed and installed to prevent water from remaining in the overflow after the fixture is drained. They also must be installed in a manner that backsiphonage cannot occur. This usually means nothing more than having the faucet installed so that it is not submerged in water if the fixture floods. By keeping the faucet spout above the high-water mark, you have created an air gap. The path of a fixture's overflow must carry the overflowing water into the trap of the fixture. This should be done by having the overflow path be integrated with the same pipe that drains the fixture.

Bathtubs must be equipped with wastes and overflows. Most codes require these wastes and overflows to have a minimum diameter of one-and-one-half inches. The method for blocking the waste opening must be approved. Common methods for holding water in a tub include the following:

- Lift-and-turn stoppers
- Plunger-style stoppers
- Push-and-pull stoppers
- Rubber stopper

Some fixtures, like handheld showers pose special problems. Since the shower is on a long hose, it could be dropped into a bathtub full of water. If a vacuum was formed in the water pipe while the shower head was submerged, the unsanitary water from the bathtub could be pulled back into the potable water supply. This situation is avoided with the use of an approved backflow preventer.

A drainage connection made with removable connections, such as slip-nuts and washers, must be accessible. If access is not practical, the connections must be soldered, solvent cemented, or screwed so as to form a solid connection. This usually isn't a problems for sinks and lavatories, but it can create some problems with bathtubs. Many builders and home buyers despise having an ugly access panel in the wall where their tub waste is located. To eliminate the need for this type of access, the tub waste can be connected with permanent joints. This could mean soldering a brass tub waste or gluing a plastic one. But, if the tub waste is connected with slip-nuts, an access panel is required.

Washing machines generally receive their incoming water from boiler drains or laundry faucets. There is a high risk of a cross-connection when these devices are used with an automatic clothes washer. This type of connection must be protected against backsiphonage. The drainage from a washing machine must be handled by an indirect waste.

An air break is required and usually accomplished by placing the washer's discharge hose into a two-inch pipe as an indirect waste receptor, such as a laundry tub. The water supply to a bidet also must be protected against backsiphonage.

Dishwashers are another likely source of backsiphonage. These appliances must be equipped with either a back-flow protector or an air gap installed on the water-supply piping. The drainage from dishwashers is handled differently in some codes.

It is common for the code to require the use of an air gap on the drainage of a dishwasher. These air gaps usually are mounted on the countertop or in the rim of the kitchen sink. The air gap forces the waste discharge of the dishwasher through open air and down a separate discharge hose. This eliminates the possibility of backsiphonage or a backup from the drainage system into the dishwasher. Some codes require dishwasher drainage to be trapped separately and vented, or discharged indirectly into a properly trapped and vented fixture. Other codes allow the discharge hose from a dishwasher to enter the drainage system in several ways. It might be individually trapped. It might discharge into a trapped fixture. The discharge hose could be connected to a wye tailpiece in the kitchen sink drainage. Further, it might be connected to the waste connection provided on many garbage disposers.

While we are on the subject of garbage disposers, be advised that garbage disposers require a drain of at least one and one-half inches and must be trapped. It might seem to go without saying, but garbage disposers must have a water source. This doesn't mean you have to pipe a water supply to the disposer; a kitchen faucet provides adequate water supply to satisfy the code.

Floor drains must have a minimum diameter of two inches. Remember, piping run under a floor may never be smaller than two inches in diameter. Floor drains must be trapped, usually must be vented, and must be equipped with removable strainers. It is necessary to install floor drains so that the removable strainer is readily accessible.

Laundry trays are required to have inch-and-a-half drains. These drains should be equipped with crossbars or a strainer. Laundry trays may act as indirect-waste receptors for clothes washers. In the case of a multiple-bowl laundry tray, the use of a continuous waste is acceptable.

Lavatories are required to have drains with a diameter at least one-and-one-quarter inches. The drain must be equipped with some device to prevent foreign objects from entering the drain. These devices could include pop-up assemblies, cross-bars, or strainers.

When installing a shower, it is necessary to secure the pipe serving the shower head with water. This riser usually is secured with a drop-ear ell and screws. It is, however, acceptable to secure the pipe with a pipe clamp.

When we talk of showers here, we are speaking only of showers, not tub-shower combinations. The frequent use of tub-shower combinations confuses many people. A shower has different requirements from those of a tub-shower combination. A shower drain must have a diameter of at least two inches. The reason for this is simple. In a tub-shower combination, a one-and-a-half-inch drain is sufficient because the walls of the bathtub will retain water until the smaller drain can remove it. A shower doesn't have high retaining walls, therefore, a larger drain is needed to clear the shower base of water more quickly. Shower drains must have removable strainers. The strainers should have a diameter of at least three inches.

Code may require all showers to contain a minimum of 900 square inches of shower base. This area must not be less than 30 inches, in any direction. These measurements must be taken at the top of the threshold, and they must be interior measurements. A shower advertised as a 30-inch shower might not meet code requirements. If the measurements are taken from the outside dimensions, the stall will not pass muster. There is one exception to the above ruling. Square showers with a rough-in of 32 inches may be allowed. But the exterior of the base may not measure less than $31\frac{1}{2}$ inches.

In some cases, the minimum interior area of a shower base must be at least 1,024 square inches. When determining the size of the shower base, the measurements should be taken from a height equal to the top of the threshold. The minimum size requirements must be maintained for a vertical height equal to 70 inches above the drain. The only objects allowed to protrude into this space are grab bars, faucets, and shower heads.

The waterproof wall enclosure of a shower, or a tub-shower combination, must extend from the finished floor to a height of no less than six feet. Another criterion for these enclosures is that they must extend at least 70 inches above the height of the drain opening. The enclosure walls must be at the higher of the two determining factors. An example of when this might come into play is a deck-mounted bathing unit. With a tub mounted in an elevated platform, an enclosure that extends six feet above the finished floor might not meet the criteria of being 70 inches above the drain opening.

Though not as common as they once were, built-up shower stalls still are popular in high-end housing. These stalls typically use a concrete base covered with tile. You might never install one of these classic shower bases, but you need to know how, just in case the need arises. These bases often are referred to as shower pans. Cement is poured into the pan to create a base for ceramic tile.

Before these pans can be formed, attention must be paid to the surface to be under the pan. The sub-floor, or other supporting surface, must be smooth and able to accommodate the weight of the shower. When the sub-structure is satisfactory, you are ready to make your shower pan.

Shower pans must be made from a waterproof material. In the old days, these pans were made of lead or copper. Today, they generally re made with coated papers or vinyl materials. These flexible materials make the job much easier. When forming a shower pan, the edges of the pan material must extend at least two inches above the height of the threshold. The lining must not be nailed or perforated at any point less than one inch above the finished threshold. Local code may require the material to extend at least three inches above the threshold. The pan material also must be securely attached to the stud walls.

Some codes go deeper with their shower regulations. The shower threshold must be one inch lower than the other sides of the shower base, but the threshold must never be lower than two inches. The threshold also must never be higher than nine inches. When installed for handicap facilities, the threshold may be eliminated.

Local codes also may require the shower base to slope toward the drain with a minimum pitch of one-quarter of an inch to the foot, but not more than one-half inch per foot. The opening into the shower must be large enough to accept a shower door with minimum dimensions of twenty-two inches.

If PVC sheets are used as a shower base, the plasticized PVC sheets must have a minimum thickness of 0.040 inches. The sheets must be joined by solvent welding in accordance with the manufacturer's installation instructions.

Sheet lead used for pan material is required to weigh not less than 4 pounds per square foot. The lead is to be coated with an asphalt paint or other approved coating. Lead sheets must be insulated from conducting substances other than the connecting drain by 15-pound asphalt felt or its equivalent. Sheet lead shall be joined by burning.

Sheet copper also can be used as a shower pan liner. The copper must not weigh less than 12 ounces per square foot. Insulation must protect the copper from conducting substances other than the connecting drain by 15-pound asphalt felt or its equivalent. When sheet copper is joined, it must be brazed or soldered.

The drains for this type of shower base are new to many young plumbers. Plumbers who have worked under my supervision have attempted to use standard shower drains for these types of bases. You cannot do that, at least not if you don't want the pan to leak. This type of shower base requires a drain that is similar to some floor drains.

The drain must be installed in a way that will not allow water that might collect in the pan to seep around the drain and down the exterior of the pipe. Any water entering the pan must go down the drain. The proper drain will have a flange that sits beneath the pan material. The pan material will be cut to allow water into the drain. Another part of the drain then is placed over the pan material and bolted to the bottom flange. The compression of the top piece and the bottom flange, with the pan material wedged between them, will create a watertight seal. The strainer portion of the drain then will screw into the bottom flange housing. Since the strainer is on a threaded extension, it can be screwed up or down to accommodate the level of the finished shower pan.

Sinks are required to have drains with a minimum diameter of one and one-half inches. Strainers or crossbars are required in the sink drain. If you look, you will see that basket strainers have the basket part, as a strainer, and crossbars below the basket. This provides protection from foreign objects, even when the basket is removed. If a sink is equipped with a garbage disposer, the drain opening in the sink should have a diameter of at least three and one-half inches.

Toilets usually are required to be water-saver models. Older models, which use five gallons per flush, no longer are allowed in many jurisdictions for new installations.

The seat on a residential water closet must be smooth and sized for the type of water closet it is serving. This usually means that the seat will have a round front.

The fill valve or ballcock for toilets must be of the anti-siphon variety. Older ballcocks that are not of the antisiphon style still are sold. Just because these units are available doesn't make them acceptable. Don't use them; you will be putting your license and yourself on the line.

Toilets of the flush-tank type are required to be equipped with overflow tubes. These overflow tubes do double duty as refill conduits. The overflow tube must be large enough to accommodate the maximum water intake entering the water closet at any given time. All parts in a flush tank must be accessible for repair and replacement.

Whirlpool tubs must be installed as recommended by the manufacturer. All whirlpool tubs shall be installed to allow access to the unit's pump. The pump's drain should be pitched to allow the pump to empty its volume of water when the whirlpool is drained. The whirlpool pump should be positioned above the fixture's trap.

All plumbing faucets and valves using both hot and cold water must be piped in a uniform manner. This manner calls for the hot water to be piped to the left side of the faucet or valve. Cold water should be piped to the right side of the faucet or valve. This uniformity reduces the risk of burns from hot water.

Valves or faucets used for showers must be designed to provide protection from scalding. This means that any valve or faucet used in a shower must be pressure-balanced or contain a thermostatic-mixing valve. The temperature control must not allow the water temperature to exceed 110°F in some regions and 120°F in others. This provides safety, especially to the elderly and the very young, against scalding injuries from the shower. Not all codes require these temperature-controlled valves in residential dwellings. The thermostatic-mixing valves must be sized according to the peak demand of fixtures located downstream of the valve. A water heater thermostat must not be used as the temperature-control device for compliance on this issue.

COMMERCIAL FIXTURE APPLICATIONS

Drinking fountains are a common fixture in commercial applications. Restaurants use garbage disposers so big, it can take two plumbers to move them. Gang showers are not uncommon in school gyms and health clubs. Urinals are another common commercial fixture. Then, there are water closets. Water closets are in homes, but the ones installed for commercial applications often differ from residential toilets. Special fixtures and applications exist for some unusual plumbing fixtures, like baptismal pools in churches. This section is going to take you into the commercial field and show you how plumbing needs vary from residential uses to commercial applications.

Let's start with drinking fountains and water coolers. The main fact to remember about water coolers and fountains is this: they are not allowed in toilet facilities. You may not install a water fountain in a room that contains a water closet. If the building for which a plumbing diagram is being designed will serve water, such as a restaurant, or if the building will provide access to bottled water, drinking fountains and water coolers may not be required.

Commercial garbage disposers can be big. These monster grinding machines require a drain with a diameter of no less than two inches. Commercial disposers must have their own drainage piping and trap. As with residential disposers, commercial disposers must have a cold water source. In some jurisdictions, the water source must be of an automatic type. These large disposers may not be connected to a grease interceptor. Commercial dishwashing machines must discharge through an air gap or air break into a standpipe or waste receptor as called for in the local plumbing code.

Garbage can washers are not something you will find in the average home, but they are not uncommon to commercial applications. Because of the nature of this fixture, the water supply to the fixture must be protected against backsiphonage. This can be done with either a backflow preventer or an air gap. The waste pipe from these fixtures must have individual traps. The receptor that collects the residue from the garbage can washer must be equipped with a removable strainer, capable of preventing the entrance of large particles into the sanitary drainage system.

Special fixtures are just that, special. Fixtures that might fall into this category include church baptismal pools, swimming pools, fish ponds, and similar arrangements. The water pipes to any of these special fixtures must be protected against backsiphonage.

Showers for commercial or public use can be very different from those found in a residence. It is not unusual for showers in commercial-grade plumbing to be gang showers. This amounts to one large shower enclosure with many showerheads and shower valves. In gang showers, the shower floor must be properly graded toward the shower drain or drains. The floor must be graded in a way to prevent water generated at one shower station from passing through the floor area of another shower station.

The methods employed to divert water from each shower station to a drain are up to the designer, but it is imperative that water used by one occupant not pass into another bather's space. Zone one requires the gutters of gang showers to have rounded corners. These gutters must have a minimum slope toward the drains of two percent. The drains in the gutter must not be more than eight feet from side walls and not more than 16 feet apart.

Urinals are not a common household item, but they are typical fixtures in public toilet facilities. The amount of water used by a urinal, in a single flush, should be limited to a maximum of one and one-half gallons. Water supplies to urinals must be protected from backflow. Only one urinal may be flushed by a single flush valve. When urinals are used, they must not take the place of more than one-half of the water closets normally required. Public-use urinals are required to have a water trap seal that is visible and unobstructed by strainers.

Floor and wall conditions around urinals are another factor to be considered. These areas are required to be waterproof and smooth. They must be easy to clean, and they may not be made from an absorbent material. These materials may be required around a urinal in several directions. They must extend at least one foot on each side of the urinal. This measurement is taken from the outside edge of the fixture.

The material is required to extend from the finished floor to a point four feet off the finished floor. The floor under a urinal must be made of this same type of material, and the material must extend to a point at least one foot in front of the farthest portion of the urinal.

Commercial-grade water closets can present some of their own variations on residential requirements. The toilets used in public facilities must have elongated bowls. These bowls must be equipped with elongated seats. Further, the seats must be hinged and they must have open or split fronts.

Flush valves are used almost exclusively with commercial-grade fixtures. They are used on water closets, urinals, and some special sinks. If a fixture depends on trap siphonage to empty itself, it must be equipped with a flush-valve or a properly rated flush-tank. These valves or tanks are required for each fixture in use.

Flush valves must be equipped with vacuum breakers that are accessible. Flush valves must be rated as water-conserving valves. These valves must be able to be regulated for water pressure, and they must open and close fully. If water pressure is not sufficient to operate a flush valve, other measures, such as a flush tank, must be incorporated into the design. All manually operated flush tanks should be controlled by an automatic filler, designed to refill the flush tank after each use. The automatic filler will be equipped to cut itself off when the trap seal is replenished and the flush tank is full. If a flush tank is designed to flush automatically, the filler device will be controlled by a timer.

SPECIAL FIXTURES

An entire group of special fixtures usually is found only in facilities providing health care. The requirements for these fixtures are extensive. While you may never have a need to work with these specialized fixtures, you should know the code requirements for them. This section is going to provide you with the information you might need.

Many special fixtures are required to be made of materials providing a higher standard than normal fixture materials. They may be required to endure excessive heat or cold. Many of these special fixtures also are required to be protected against backflow. The fear of backflow extends to the drainage system, as well as the potable water supply. All special fixtures must be of an approved type.

STERILIZERS

Any concealed piping that serves special fixtures and that may require maintenance or inspection must be accessible. All piping for sterilizers must be accessible. Steam piping to a sterilizer should be installed with a gravity system to control condensation and to prevent moisture from entering the sterilizer. Sterilizers must be equipped with a means to control the steam vapors. The drains from sterilizers are to be piped as indirect wastes. Sterilizers are required to have leak detectors, which are designed to expose leaks and to carry unsterile water away from the sterilizer. The interior of sterilizers may not be cleaned with acid or other chemical solutions while the sterilizers are connected to the plumbing system.

CLINICAL SINKS

Clinical sinks sometimes are called bedpan washers. Clinical sinks are required to have an integral trap. The trap seal must be visible and the contents of the sink must be removed by siphonic or blowout action. The trap seal must be replenished automatically, and the sides of the fixture must be cleaned by a flush rim at every flushing of the sink. These special fixtures are required to connect to the DWV system in the same manner as a water closet. Clinical sinks installed in utility rooms are not meant to be a substitute for a service sink. On the other hand, service sinks may never be used to replace a clinical sink. Devices for making or storing ice shall not be placed in a soiled utility room.

VACUUM FLUID-SUCTION SYSTEMS

Vacuum system receptacles are to be built into cabinets or cavities, but they must be visible and readily accessible. Bottle suction systems used for collecting blood and other human fluids must be equipped with overflow prevention devices at each vacuum receptacle. Secondary safety receptacles are recommended as an additional safeguard. Central fluid-suction systems must provide continuous service. If a central suction system requires periodic cleaning or maintenance, it must be installed so that it can continue to operate, even while cleaning or maintenance is being performed. Central systems installed in hospitals must be connected to emergency power facilities. The vent discharge from these systems must be piped separately to the outside air, above the roof of the building.

Waste originating in a fluid-suction system, which is to be drained into the normal drainage piping, must be piped into the drainage system with a direct-connect, trapped arrangement; indirect-waste connection of this type of unit is not allowed. Piping for these fluid-suction systems must be noncorrosive and have a smooth interior surface. The main pipe shall have a diameter of no less than one inch. Branch pipes must not be smaller than one-half inch. All piping is required to have accessible cleanouts and must be sized according to manufacturer's recommendations. The air flow in a central fluid-suction system should not be allowed to exceed 5,000 feet per minute.

SPECIAL VENTS

Institutional plumbing uses different styles of vents for some equipment from what is encountered with normal plumbing. One such vent is called a local vent, one use of which pertains to bedpan washers. A bedpan washer must be connected to at least one vent, with a minimum diameter of two inches, and that vent must extend to the outside air, above the roof of the building.

These local vents are used to vent odors and vapors. Local vents may not tie in with vents from the sanitary plumbing or sterilizer vents. In multistory buildings, a local vent stack can collect the discharge from individual local vents for multiple bedpan washers, located above each other. A two-inch stack can accept up to three bedpan washers. A three-inch stack can handle six units, and a four-inch stack will accommodate up to 12 bedpan washers. These local vent stacks are meant to tie into the sanitary drainage system, and they must be vented and trapped if they serve more than one fixture.

Each local vent must receive water to maintain its trap seal. The water source shall come from the water supply for the bedpan washer being served by the local vent. A minimum of a one-quarter inch tubing shall be run to the local vent, and it must discharge water into the vent each time the bedpan washer is flushed.

Vents serving multiple sterilizers must be connected with inverted wye fittings, and all connections must be accessible. Sterilizer vents are intended to drain to an indirect waste. The minimum diameter of a vent for a bedpan sterilizer shall be one and one-half inches. When serving a utensil sterilizer, the minimum vent size shall be two inches. Vents for pressure-type sterilizers must be at least two and one-half inches in diameter. When serving a pressure instrument sterilizer, a vent stack must be at least two inches in diameter. Up to two sterilizers of this type may be on a two-inch vent. A three-inch stack can handle four units.

WATER SUPPLY

Hospitals are required to have at least two water services. These two water services, however may connect to a single water main. Hot water must be made available to all fixtures, as required by the fixture manufacturer. All water heaters and storage tanks must be of a type approved for the intended use.

Some jurisdictions require the hot-water system to be capable of delivering six and one-half gallons of 125°F water per hour for each bed in a hospital. Some codes further require hospital kitchens to have a hot-water supply of 180°F water equal to four gallons per hour for each bed. Laundry rooms are required to have a supply of 180°F water at a rate of four-and-one-half gallons per hour for each bed. Local code may require hot water storage tanks to have capacities equal to no less than eight percent of the water heating capacity. Some codes continue with their hot water requirements by dictating the use of copper in submerged steam heating coils. If a building is higher than three levels, the hot-water system must be equipped to circulate. Valves are required on the water-distribution piping to fixture groups.

BACKFLOW PREVENTION

Backflow prevention devices must be installed at least six inches above the flood-level rim of the fixture. In the case of handheld showers, the height of installation shall be six inches above the highest point the hose can be used.

In most cases, hospital fixtures will be protected against backflow by the use of vacuum breakers. A boiling-type sterilizer, however, should be protected with an air gap. Vacuum suction systems may be protected with either an air gap or a vacuum breaker.

I know this has been a long chapter, and I apologize for the length, but it was necessary to give you all the pertinent details about fixtures. As you now know, fixtures are not as simple as they first might appear. There are numerous regulations to learn and apply when installing plumbing fixtures. Your local jurisdiction may require additional or different code compliance. As always, check with your local authority before installing plumbing.

CHAPTER 5
WATER HEATERS

Water heaters are only a small portion of a plumbing system, but they are important, and they can be dangerous. Many plumbers take water heaters and the codes pertaining to them lightly. This is a mistake. Water heaters can become lethal if they are not installed properly. I feel very strongly about the safety issues associated with water heaters. My position is based largely on decades of watching plumbers and plumbing contractors take shortcuts with water heaters that could be disastrous.

GENERAL PROVISIONS

Water heaters sometimes are used as a part of a space heating system. When this is the case, the maximum outlet water temperature for the water heater is 140°F, unless a tempering valve is used to maintain an acceptable temperature in the potable water system. It is essential that all potable water in the water heater be maintained throughout the entire system. Potability of water must be maintained at all times. Every water heater is required to be equipped with a drain valve near the bottom. This is true, too, for hot-water storage tanks. All drain valves must conform to ASSE 1005.

The location of water heaters and hot-water storage tanks is important. Code requires both water heaters and hot-water storage tanks to be accessible for observations, maintenance, servicing, and replacement. Every water heater must bear a label of an approved agency.

The temperature of water delivered from a tankless water heater may not exceed 140°F when used for domestic purposes. This portion of the code does not supersede the requirement for protective shower valves, as detailed in the code.

All water heaters shall be third-party-certified, and water heaters must be installed in accordance with manufacturer's requirements. Oil-fired water heaters must meet installation requirements of the *International Mechanical Code*. Electric water heaters must meet the requirements of the *ICC Electrical Code*. Gas-fired water heaters are required to meet the criteria of the *International Fuel Gas Code*.

All storage tanks and water heaters installed for domestic hot water must have the maximum allowable working pressure clearly and indelibly stamped in the metal or marked on a plate welded thereto or otherwise permanently attached. All markings of this type must be in accessible positions outside of the tanks. Inspection or reinspection of these markings must be readily possible.

Number of Bathrooms	1 to 1.5		2 to 2.5			3 to 3.5			
Number of Bedrooms	1	2	3	2	3	4	5	3	4	5	6
First Hour Rating², Gallons	42	54	54	54	67	67	80	67	80	80	80

Notes:
¹The first hour rating is found on the "Energy Guide" label.
²Non-storage and solar water heaters shall be sized to meet the appropriate first hour rating as shown in the table.

FIGURE 5.1 The minimum capacity for water heaters must be is shown above. *(Reprinted from the 2000 Uniform Plumbing Code (UPC) with the permission of the International Association of Plumbing and Mechanical Officials (IAPMO).)*

Every hot-water supply system is required to be fitted with an automatic temperature control. The control must be capable of being adjusted from a minimum temperature to the highest acceptable temperature setting for the intended temperature operating range.

INSTALLING WATER HEATERS

All water heaters are required to be installed in accordance with the manufacturer's recommendations and within the confines of the plumbing code. Water heaters that are fueled by gas or oil must conform to both the plumbing code and the mechanical or gas code. Electric water heaters must conform to the requirements of the plumbing code and the provisions of NFPA 70, as listed in the plumbing code.

A water heater that has an ignition source, when installed in a garage, must be installed on an elevated base that keeps the source of the ignition at least 18 inches above the garage floor.

Rooms are sometimes used as plenums for heating systems. Water heaters using solid, liquid, or gas fuel are not allowed to be installed in rooms that contain air handling machinery when the rooms are being used as plenums. Additionally, fuel-fired water heaters are not allowed to be installed in sleeping rooms, bathrooms, or closets that can be accessed from either bathrooms or sleeping rooms, unless the water heater is equipped with a direct-vent system. There is one exception to this rule. A water heater that has a sealed combustion chamber or is directly vented to the outside can be installed in a sleeping room, bathroom, or closet that is accessible from either of these rooms.

When earthquake loads are applicable, water-heater supports must be designed and installed for the seismic forces in accordance with the *International Building Code.*

When water heaters are installed in attics, special provisions must be made. For example, an attic that houses a water heater must be provided with an opening and unobstructed passageway large enough to allow for the water heater's removal. This should be common sense, but it is also part of the plumbing code. Many measurements come into play when planning the exit route for an attic water heater. They are as follows:

• Minimum height: 30 inches
• Minimum width: 22 inches
• Maximum length: 20 feet

A continuous solid floor is required in the exit area, and the flooring must be at least 24 inches wide. Another requirement calls for a level service area with minimum dimensions of 30 inches deep and 30 inches wide. This service area must be in front of the water heater, or where the service area of the water heater is located. A clear access opening with minimum dimensions of 20 inches by 30 inches, which is large enough to allow removal of the water heater, is required.

Size of Combustion Air Openings or Ducts[1] for Gas- or Liquid-Burning Water Heaters		Btu watts

Btu	watts
1000	293
2000	586
4000	1172
5000	1465
100,000	29,300

Column 1 Buildings of Ordinary Tightness		Column 2 Buildings of Unusually Tight Construction	
Condition	Size of Opening or Duct	Condition	Size of Opening or Duct
Appliance in unconfined[2] space	May rely on infiltration alone.	Appliance in unconfined[2] space: Obtain combustion air from outdoors or from space freely communicating with outdoors.	Provide 2 openings, each having 1 sq. in. (645 mm²) per 5,000 Btu/h input.
Appliance in confined[4] space 1. All air from inside building	Provide two openings into enclosure each having 1sq. in. (645 mm²) per 1,000 Btu/h input freely communicating with other unconfined interior spaces. Minimum 100 sq. in. (0.06 m²) each opening.	Appliance in confined[4] space: Obtain combustion air from outdoors or from space freely communicating with outdoors.	1. Provide two vertical ducts or plenums: 1 sq. in. (645 mm²) per 4,000 Btu/h input each duct or plenum. 2. Provide two horizontal ducts or plenums: 1 sq. in. (645 mm²) per 2,000 Btu/h input each duct or plenum.
2. Part of air from inside building	Provide 2 openings into enclosure[3] from other freely communicating unconfined[2] interior spaces, each having an area of 100 sq. in. (0.06 m²) plus one duct or plenum opening to outdoors having an area of 1 sq. in. (645 mm²) per 5,000 Btu/h input rating.		3. Provide two openings in an exterior wall of the enclosure: each opening 1 sq. in. (645 mm²) per 4,000 Btu/h input. 4. Provide 1 ceiling opening to ventilated attic and 1 vertical duct to attic: each opening 1 sq. in. (645 mm²) per 4,000 Btu/h input.
3. All air from outdoors: Obtain from outdoors or from space freely communicating with outdoors.	Use any of the methods listed for confined space in unusually tight construction as indicated in Column 2.		5. Provide 1 opening in enclosure ceiling to ventilated attic and 1 opening in enclosure floor to ventilated crawl space: each opening 1 sq. in. (645 mm²) per 4,000 Btu/h input.

1 For location of opening, see Section 507.3.
2 As defined in Section 223.0.
3 When the total input rating of appliances in enclosure exceeds 100,000 Btu/h, the area of each opening into the enclosure shall be increased 1 sq. in. (645 mm²) for each 1,000 Btu/h over 100,000 Bth/h.
4 As defined in Section 205.0.

FIGURE 5.2 Gas or liquid-burning water heaters must meet certain standards in the size of combustion air openings or ducts. *(Reprinted from the* 2000 Uniform Plumbing Code (UPC) *with the permission of the International Association of Plumbing and Mechanical Officials (IAPMO).)*

MAKING CONNECTIONS

Making connections to water heaters is not difficult, but the manner in which the connections are made must conform to code requirements. The first consideration is the installation of cutoff valves. A cold-water branch line from a main water supply to a hot-water storage tank or water heater must be provided with a cutoff valve that is accessible on the same floor, located near the equipment and only serving the hot-water storage tank or water heater. The valve must not interfere with or cause a disruption of the cold-water supply to the remainder of the cold-water system.

Any means of connecting a circulating water heater to a tank must provide for proper circulation of water through the water heater. All piping required for the installation of appliances that will draw from the water heater or storage tank must comply with all provisions of the plumbing and mechanical codes.

SAFETY REQUIREMENTS

Plumbers must comply with safety requirements when installing or replacing a water heater or hot-water storage tank. One major concern is the siphoning of water from a water heater or storage tank. An antisiphon device is required to prevent siphoning. A cold-water "dip" tube with a hole at the top or a vacuum-relief valve installed in the cold water supply line above the top of the water heater or storage tank are acceptable means of protection. Some water heaters and storage tanks receive their incoming water from the bottom of the unit. These types of heater and tanks must be supplied with an approved vacuum relief valve that complies with ANSI Z21.22

Energy cutoff valves are required on all water heaters that are controlled automatically. The energy cutoff valve is designed to cut off the supply of heat energy to the water tank before the temperature of the water in the tank exceeds 210°F. The installation of an energy cutoff valve does not relieve the need for a temperature-and-pressure relief valve—both are required.

Every electric water heater must be provided with its own electrical disconnect switch that is in close proximity to the water heater. In the case of gas-fired or oil-fired water heaters, cutoff valves must be installed close to the water heaters to stop the fuel flow when needed.

RELIEF VALVES

Pressure-relief valves and temperature-relief valves, or a temperature-and-pressure relief valve.(the one most commonly used) are required on all water heaters and storage tanks that are operating above atmospheric pressure. The valves must be approved and conform to ANSI Z21.22 ratings. Relief valves must be of a self-closing (levered) type. In no case shall the relief valve be used as a means of controlling thermal expansion.

Relief valves must be installed in the shell of a water heater tank. Any temperature relief valve must be installed so that it is actuated by the water in the top 6 inches of the tank being served by the valve. When separate tanks are used, the valves must be installed on the tank and no valve may be installed between the water heater and the storage tank. It is prohibited to install a cutoff valve or check valve between a relief valve and the water heater or tank being serviced by the relief valve. Never omit the installation of required relief valves. The result of doing so can be catastrophic.

All relief valves, whether temperature, pressure, or a combination of the two, and all energy cutoff devices must bear a label of an approved agency. The valves and devices must have a temperature setting of not more than 210°F and a pressure setting that does not exceed the tank or water heater manufacturer's rated working pressure, or 150 PSI, whichever is less. The relieving capacity of each relief valve must equal or exceed the heat input to the water heater or storage tank.

Since relief valves might create a discharge, the disposal of the discharge must be dealt with. In no case is it allowable for the discharge tube from a relief valve to be connected directly to a drainage system. The discharge tube must be provided in a full-size tube or pipe that is the same size as the outlet of the relief valve. You have two choices on the termination point of the discharge tube. It can be piped outside of a building or it can terminate over an indirect waste receptor inside a building.

When freezing conditions might exist, the discharge tubing or piping for a relief valve must be protected from freezing. This is done by having the tubing discharge through an air gap and into an indirect waste receptor located in a heated space. Local regulations might allow some other form of installation, so check your local code requirements.

Any risk of personal injury or property damage must be avoided when piping a discharge tube from a relief valve. The discharge piping must be installed so that it is readily observable by building occupants. Traps on discharge tubes and pipes are prohibited. All discharge piping must drain by gravity. The tubing must terminate atmospherically not more than 6 inches above the floor, and the end of the discharge tubing or piping may not be threaded. When discharge piping is installed so that it leaves the room or enclosure containing a water heater and relief valve that discharges into an indirect waste receptor, an air gap must be installed before, or at the point of, leaving the room or enclosure. Discharge tubes from relief valves must not discharge into a safety pan. Materials used for discharge piping shall be made to the standards listed in the plumbing code or shall be tested, rated, and approved for such use in accordance with ASME A112.4.1.

Safety pans are required for water heaters and storage tanks that are installed in locations where leakage can cause property damage. Water heaters and storage tanks shall be placed in safety pans that are constructed of galvanized steel or other approved metal materials. The minimum thickness of the metallic pan shall be 24 gauge. Electric water heaters must be installed in pans when leakage might cause property damage, but the pan may be made of a 24-gauge metal pan or a high-impact plastic pan that has a minimum thickness of 0.0625 inches. All piping from safety pan drains must be made with materials approved by the plumbing code.

Safety pans must have a minimum depth of 1.5 inches and be of sufficient size and shape to receive all dripping or condensate from the tank or water heater contained in the pan. A safety pan must drain by an indirect waste. The drainage pipe or tube from the pan must have a minimum diameter of 1 inch or the outlet diameter of the relief valve, whichever is larger.

The drain tube or pipe from a safety pan must run full-size for its entire length and terminate over a suitably located indirect waste receptor or floor drain, or extend to the exterior of the building and terminate not less than 6 inches nor more than 24 inches above the adjacent ground level. Unfired hot-water storage tanks must be insulated so that heat loss is limited to a maximum of 15 British thermal units (Btus) per hour, per square foot of external tank surface area. For purposes of determining this heat loss, the design ambient temperature shall not be higher than 65°F.

VENTING WATER HEATERS

The venting of water heaters that require venting is regulated by the plumbing code. All venting materials must be in compliance with all code requirements. Venting systems might consist of approved chimneys, type B vents, type L vents, or plastic pipe. The recommendations of the equipment manufacturer must be observed in selecting the proper venting material and installation procedure.

Diameter of Connector		Galvanized Sheet
Inches	mm	Gauge No.
5 or less	125 or less	28
Over 5 to 9	Over 125 to 225	26
Over 9 to 12	Over 225 to 300	22
Over 12 to 16	Over 300 to 400	20
Over 16	Over 400	16

FIGURE 5.3 Chimney connectors made of single-wall steel pipe and serving low-heat appliances must be no less than the gauges listed above. *(Reprinted from the* 2000 Uniform Plumbing Code (UPC) *with the permission of the International Association of Plumbing and Mechanical Officials (IAPMO).)*

Vents must be designed and installed to develop a positive flow adequate to convey all products of combustion to the outside atmosphere. Condensing appliances that cool flue gases nearly to the dew point within the appliance, resulting in low-vent gas temperatures, may use plastic venting materials and vent configurations unsuitable for noncondensing appliances. All unused openings in a venting system must be closed or capped to the satisfaction of the local code enforcement officer.

Type B vents are not allowed for use with water heaters that may be converted readily to the use of solid or liquid fuels. Water heaters listed for use with chimneys only may not be vented with type B vents. Manually operated dampers must not be installed in chimneys, vents, or chimney or vent connectors of fuel-burning water heaters. Fixed baffles on the water heater side of draft hoods and draft regulators are not to be considered dampers.

VENT CONNECTORS

Vent connectors used for gas water heaters with draft hoods may be constructed of non-combustible materials having resistance to corrosion not less than that of galvanized sheet steel and be of a thickness not less than that specified in the code. Or, they may be of type B or type L vent material. Single-wall metal vent connectors must be supported securely and joints fastened with sheet metal screws, rivets, or another approved means. Such connectors must not originate in an unoccupied attic or concealed space and shall not pass through any attic, inside wall, floor, or concealed space and shall be located in the same room or space as the fuel-burning water heater.

SUPPORTING VENT SYSTEMS

Combustion products, vents, vent connectors, exhaust ducts from ventilating hoods, chimneys, and chimney connectors must not extend into or through any air duct or plenum, except for when the venting system may pass through a combustion air duct. All vents supported from the ground must rest on a solid masonry or concrete base extending at least 2 inches above adjoining ground level. If the base of a vent is not supported from the ground and is not self-supporting, it must rest on a firm metal or masonry support. All venting systems must be supported adequately for their weight and design. No water heater is allowed to be vented into a fireplace or into a chimney that serves a fireplace.

VENT OFFSET

With minor exceptions, gravity vents must extend in a generally vertical direction with offsets not exceeding 45°. These vents are allowed to have one horizontal offset of not more than 60°. All offsets must be supported properly for their weight and must be installed to maintain proper clearance to prevent physical damage and the separation of joints.

Offsets with angles of more than 45° are considered to be horizontal offsets. Horizontal vent connectors must not be greater than 75 percent of the vertical height of the vent and must comply with all code regulations.

Vent connectors in a gravity-type venting system must have a continuous rise of not less than $\frac{1}{4}$-inch per foot of developed length, measured from the appliance vent collar to the vent. If a single-wall metal vent connector is allowed and installed, it must have a minimum clearance of 6 inches from any combustible material.

TERMINATION

Vents must terminate above the roof surface of the building. The pipe must pass through a flashing and terminate in an approved or listed vent cap that is installed in accordance with the manufacturer's recommendations. There is an exception to this. A direct vent or mechanical draft appliance will be acceptable when installed according to its listing and manufacturer's instructions.

Gravity-type venting systems, with the exception of venting systems that are integral parts of a listed water heater, must terminate at least 5 feet above the highest vent collar being served. Type B gas vents with listed vent caps 12 inches in size or smaller are permitted to be terminated in accordance with the code requirements as long as they are at least 8 feet from a vertical wall or similar obstruction. All other type B vents must terminate not less than 2 feet above the highest point where they pass through the roof and at least 2 feet higher than any portion of a building within 10 feet.

Type L vents shall not terminate less than 2 feet above the roof through which it passes, nor less than 4 feet from any portion of the building that extends at an angle of more than 45° upward from the horizontal. No vent system is allowed to terminate less than 4 feet below or 4 feet horizontally from, nor less than 1 foot above any door, openable window, or gravity air inlet into any building. As usual, there are exceptions.

Vent terminal of direct vent appliances with inputs of 50,000 Btu/h or less shall be located at least 9 inches from an opening through which combustion products could enter a building. Appliances with inputs in excess of 50,000 Btu/h but not exceeding 65,000 Btu/h shall require 12-inch vent termination clearances. The bottom of the vent terminal and the air intake shall be located at least 12 inches above grade.

AREA

The internal cross-sectional area of a venting system must not be less than the area of the vent collar on the water heater, unless the venting system has been designed in accordance with other code requirements. In no case shall the area be less than 7 square inches, unless the venting system is an integral part of a listed water heater.

VENTING MULTIPLE APPLIANCES

It is acceptable to connect multiple oil or listed gas-burning appliances to a common gravity-type venting system, if the appliances have an approved primary safety control capable of shutting off the burners and the venting system is designed in compliance with code requirements.

Multiple appliances connected to a common vent system must be within the same story of the building, unless an engineered system is being used. The inlets for multiple connections must be offset in a way that no inlet is opposite another. Oval vents may be used for multiple appliance venting, but the venting system must be not less than the area of the largest vent connector plus 50 percent of the areas of the additional vent connectors.

EXISTING SYSTEMS

New water heaters installed as replacements must meet code criteria before they can be connected to existing venting systems. The existing system must have been installed lawfully at the time of its installation. Code compliance with the internal area of the venting system must exist. Any connection must be made in a safe manner.

DRAFT HOODS

Draft hoods for water heaters must be located in the same room or space as the combustion air opening for the water heater. The draft hood must be installed in the position for which it is designed and must be located so that the relief opening is not less than 6 inches from any surface other than the water heater being served, measured in a direction 90° to the plane of the relief opening. Exceptions could exist if manufacturer's recommendations vary.

EXISTING MASONRY CHIMNEYS

Existing masonry chimneys with not more than one side exposed to the outside can be used to vent a gas water heater. Some conditions, however, must apply for this to be the case. The local code may require unlined chimneys to be lined with approved materials. Effective cross-sectional area of the chimney must not be more than four times the cross-sectional area of the vent and chimney connectors entering the chimney.

The effective area of the chimney, when connected to multiple connectors, must not be less than the area of the largest vent or connector plus 50 percent of the area of the additional vent or connector.

Automatically controlled gas water heaters connected to a chimney that also serves equipment burning liquid fuel must be equipped with an automatic pilot. A gas water heater connector and a connector from an appliance burning liquid fuel may be connected to the same chimney through separate openings, providing the gas water heater is vented above the other fuel-burning appliance. Both also may be connected through a single opening if joined by a suitable fitting located at the chimney. Multiple connections must not be made at the same horizontal plane of another inlet. Any chimney must be clear of obstructions and cleaned if previously used for venting solid or liquid fuel-burning appliances.

CONNECTORS

Chimney connectors must comply with code requirements as set forth in tables in the local code book. When multiple connections are made, the connector, the manifold, and the chimney must be sized properly. Gravity vents must not be connected to vent systems served by power vents, unless the connection is made on the negative pressure side of the power exhaust. Single-wall metal chimneys require a minimum clearance of 6 inches from combustible materials.

Connectors must be kept as short and as straight as possible. Water heaters are required to be installed as close as possible to the venting system. Connectors must not be longer than 75 percent of the portion of the venting system above the inlet connection unless they are part of an approved engineered system. A connector to a masonry chimney must extend through the wall to the inner face of the liner, but not beyond. The connector must be cemented to the masonry. A thimble may be used to facilitate removal of the connector for cleaning, in which case the thimble shall be cemented in place permanently. Connectors shall not pass through any floor or ceiling.

Draft regulators are required in connectors serving liquid-fuel-burning water heaters, unless the water heater is approved for use without a draft regulator. Draft regulators must be installed in the same room or enclosure as the water heater in such a manner that no difference in pressure between air in the vicinity of the regulator and the combustion air supply will be permitted.

MECHANICAL DRAFT SYSTEMS

It is acceptable to vent water heaters with mechanical draft systems of either forced or induced draft design. Forced draft systems must be designed and installed to be gas-tight or to prevent leakage of combustion products into a building. Connectors vented by natural draft must not be connected to mechanical draft systems operating under positive pressure. Systems using a mechanical draft system must prevent the flow of gas to the main burners when the draft system is not performing so as to satisfy safe operating requirements of the water heater.

Exit terminals of mechanical draft systems must be located not less than 12 inches from any opening through which combustion products could enter the building, nor less than 7 feet above grade when located adjacent to public walkways.

VENTILATING HOODS

Ventilating hoods can be used to vent gas-burning water heaters installed in commercial applications. Dampers are not allowed when automatically operated water heaters are vented through natural draft ventilating hoods. If a power venter is used, the water heater control system must be interlocked so that the water heater will operate only when the power venter is in operation.

SAFETY COMMENTARY

I've been plumbing for more than 25 years. During these years, I've done just about every type of plumbing there is. My career has involved working with all sorts of plumbers and plumbing contractors. As a plumbing contractor for the last 20 years, I've used a lot of other plumbing contractors as subcontractors. My work has even extended into teaching code classes and apprenticeship classes at Central Maine Technical College. I'm telling you this to give you an idea of my background and experience in the industry. This is because I want to tell you something that I believe is extremely important about water heaters.

During my time in the field, I've run into many occasions when plumbing contractors failed to obtain permits for the installation of water heaters. A plumbing permit is required for every water heater installation and replacement. Yet a good number of contractors believe they can get by without a permit. I've heard dozens of contractors say that it takes more time to get a permit and inspection than it does to install a water heater. This can be true, but the point is that a permit and inspection is required by code.

Some code enforcement offices are more active than others in enforcing the need for permits and inspections when water heaters are being replaced. All codes that I know of, however, do require a permit and inspection for the installation or replacement of a water heater. This shouldn't be a big deal, but it is, and it can be a very big deal.

About two years ago, I lost the service contract on more than 180 apartment units because I refused to replace a water heater without a permit and inspection. I didn't like losing the account, but I was not going to violate the plumbing code and risk myself and my business to a lawsuit for any cost. It can be hard for reputable plumbing contractors to compete with bootleggers who are willing to work without permits. The time and money spent on permits and inspections do drive up the cost of a job. Still, you should never install or replace a water heater without adhering to the plumbing code.

Why do I feel so strongly about permits for water heaters? There are several reasons. First, code requires a permit and inspection, so this should be reason enough. Secondly, installing a water heater illegally can open up a huge risk for a lawsuit. If you install a water heater in accordance with code requirements and a problem occurs later, your insurance should cover your losses. This probably would not be the case if you had installed the water heater in violation of the plumbing code. Leaving ethics, morals, and lawsuits out of it, there is still the issue of personal safety.

Believe it or not, I have found water heaters where a plug had been installed in place of a relief valve. Any plumber with even minimal experience knows the risk of this. I've found cutoff valves installed in illegal locations around water heaters. If the proper safety precautions are not taken, a water heater can become a large bomb. I remember watching a video from a manufacturer that showed water heaters blowing up. In one case, the water heater went from a basement location right through the roof of a home. The explosions can be violent.

I won't try to beat you over the head with my vision, but seriously consider the risk you will be taking if you don't install water heaters properly and in accordance with all code provisions. Protect yourself by getting permits and inspections. With this said, let's move to the next chapter and explore water supply and distribution.

CHAPTER 6
WATER SUPPLY
AND DISTRIBUTION

Potable water is essential to life. Water is often taken for granted in today's society, but it shouldn't be. Without good water, we would all die. Part of a plumber's responsibility is to provide safe drinking water. The requirements for making sure water is safe to drink are many. This chapter deals with the installation of potable water systems. It details approved types of piping and installation methods. These guidelines must be followed to ensure the health of our nation.

When we talk about potable water, we are speaking about water that is safe for drinking, cooking, and bathing, among other uses. Any building intended for human occupancy, which has plumbing fixtures installed, must have a potable water supply. If such a building is intended for year-round habitation or as a place where people work, both hot and cold potable water must be made available. All fixtures that are intended for use in bathing, drinking, cooking, food processing, or for creating medically related products must have potable water available, and only potable water. It is permissible to use nonpotable water for flushing toilets and urinals.

Now that you know what must be equipped with potable water, let's see how to get the water to the fixtures.

THE MAIN WATER PIPE

The main water pipe delivering potable water to a building is called a water-service pipe, and it must have a diameter of at least three-quarters of an inch. The pipe must be sized, according to code requirements, to provide adequate water volume and pressure to the fixtures.

Ideally, a water-service pipe should be run from the primary water source to the building in a private trench. By private trench, I mean a trench not used for any purpose, except for the water-service. However, it usually is allowable to place the water-service pipe in the same trench used by a sewer or building drain, when specific installation requirements are followed. The water-service pipe must be separated from the drainage pipe. The bottom of the water-service pipe may not be closer than 12 inches to the drainage pipe at any point.

A shelf must be made in the trench to support the water-service pipe. The shelf must be made solid and stable, at least 12 inches above the drainage pipe. If the water service is not located above the sewer, there must be at least 5 feet of undisturbed or compacted

WATER SERVICE PIPE

MATERIAL	STANDARD
Acrylonitrile butadiene styrene (ABS) plastic pipe	ASTM D 1527; ASTM D 2282
Asbestos–cement pipe	ASTM C 296
Brass pipe	ASTM B 43
Copper or copper-alloy pipe	ASTM B 42; ASTM B 302
Copper or copper-alloy tubing (Type K, WK, L, WL, M or WM)	ASTM B 75; ASTM B 88; ASTM B 251; ASTM B 447
Chlorinated polyvinyl chloride (CPVC) plastic pipe	ASTM D 2846; ASTM F 441; ASTM F 442; CSA B137.6
Ductile iron water pipe	AWWA C151; AWWA C115
Galvanized steel pipe	ASTM A 53
Polybutylene (PB) plastic pipe and tubing	ASTM D 2662; ASTM D 2666; ASTM D 3309; CSA B137.8
Polyethylene (PE) plastic pipe	ASTM D 2239; CSA CAN/CSA-B137.1
Polyethylene (PE) plastic tubing	ASTM D 2737; CSA B137.1
Cross-linked polyethylene (PEX) plastic tubing	ASTM F 876; ASTM F 877; CSA CAN/CSA-B137.5
Cross-linked polyethylene/ aluminum/cross-linked polyethylene (PEX-AL-PEX) pipe	ASTM F 1281; CSA CAN/CSA B137.10
Polyethylene/aluminum/ polyethylene (PE-AL-PE) pipe	ASTM F 1282; CSA CAN/CSA-B137.9
Polyvinyl chloride (PVC) plastic pipe	ASTM D 1785; ASTM D 2241; ASTM D 2672; CSA CAN/CSA-B137.3

FIGURE 6.1 The materials and standards for water-service pipe are listed above. (*Courtesy of International Code Council, Inc. and International Plumbing Code 2000.*)

WATER DISTRIBUTION PIPE

MATERIAL	STANDARD
Brass pipe	ASTM B 43
Chlorinated polyvinyl chloride (CPVC) plastic pipe and tubing	ASTM D 2846; ASTM F 441; ASTM F 442; CSA B137.6
Copper or copper-alloy pipe	ASTM B 42; ASTM B 302
Copper or copper-alloy tubing (Type K, WK, L, WL, M or WM)	ASTM B 75; ASTM B 88; ASTM B 251; ASTM B 447
Cross-linked polyethylene (PEX) plastic tubing	ASTM F 877; CSA CAN/CSA-B137.5
Cross-linked polyethylene/aluminum/cross-linked polyethylene (PEX-AL-PEX) pipe	ASTM F 1281; CSA CAN/CSA-B137.10
Galvanized steel pipe	ASTM A 53
Polybutylene (PB) plastic pipe and tubing	ASTM D 3309; CSA CAN3-B137.8

FIGURE 6.2 Pipe that distributes water must conform to the standards in this table. (*Courtesy of International Code Council, Inc. and International Plumbing Code 2000.*)

earth between the two pipes. It is not acceptable to have a water-service pipe located in an area where pollution is probable. A water service should never run through, above, or under a waste disposal system, such as a septic field.

If a water-service pipe is installed in an area subject to flooding, the pipe must be protected against flooding. Water services also must be protected against freezing. The depth of the water service will depend on the climate of the location. Check with your local code officer to see how deep a water-service pipe must be buried in your area. Care must be taken when backfilling a water-service trench. The backfill material must be free of objects, like sharp rocks, that might damage the pipe.

When a water service enters a building through or under the foundation, the pipe must be protected by a sleeve. This sleeve is usually a pipe with a diameter at least twice that of the water-service pipe. Once through the foundation, the water service may need to be converted to an acceptable water-distribution pipe. As you learned in reading about approved materials, some materials approved for water-service piping are not approved for interior water distribution.

If the water-service pipe is not an acceptable water-distribution material, it must be converted to an approved material, generally within the first five feet of its entry into the building. Once inside a building, the maze of hot and cold water pipes are referred to as water-distribution pipes. Let's see what you need to know about water-distribution systems.

WATER DISTRIBUTION

Sizing a water-distribution system can become complicated. As we move through this chapter, we will leave sizing exercises until last. There are some accepted methods that simplify the sizing of water-distribution pipes. Near the end of the chapter, you will get information about sizing a system. For now, we will concentrate on other regulations.

FIXTURE SUPPLIES

Fixture supplies are the tubes or pipes that rise to the fixture from the fixture branch, which is the pipe coming out of the wall or floor. In zone three, a fixture supply may not have a length of more than 30 inches. The required minimum sizing for a fixture supply is determined by the type of fixture being supplied with water.

PRESSURE-REDUCING VALVES

Pressure-reducing valves are required to be installed on water systems when the water pressure coming to the water-distribution pipes is in excess of 80 pounds per square inch (psi). The only time this regulation generally is waived is when the water-service pipe is bringing water to a device requiring high pressure.

BANGING PIPES

Banging pipes usually are the result of a water hammer. If you don't want complaining customers, avoid water hammers. You can avoid water hammers in several ways. You might install air chambers above each faucet or valve. Water hammer arrestors are available and do a good job in controlling water-hammer. Expansion tanks also can help with water hammer.

Water hammer is most prevalent around such quick-closing valves as ballcocks and washing machine valves. Another way to reduce water hammer is to avoid long, straight runs of pipe. By installing offsets in your water piping, you gradually break up the force of the water. By diminishing the force, you reduce water hammer. All building water supply systems in which quick-acting valves are installed must be provided with devices to absorb the hammer caused by high pressures resulting from the quick closing of the valves. The devices are to be installed as close as possible to the quick-acting valves.

BOOSTER PUMPS

Not all water sources are capable of providing optimum water pressure. When this is the case, a booster pump may be required to increase water pressure. If water pressure fluctuates severely, the water-distribution system must be designed to operate on the minimum water pressure anticipated.

When calculating the water pressure needs of a system, you can use information provided by your code book, which should provide ratings for all common fixtures showing the minimum pressure requirements for each type of fixture. A water-distribution system must be sized to operate satisfactorily under peak demands.

Booster pumps must be equipped with low-water cutoffs. These safety devices are required to prevent the possibility of a vacuum, which might cause backsiphonage.

WATER TANKS

When booster pumps are not a desirable solution, water storage tanks are a possible alternative. Water storage tanks must be protected from contamination. They may not be located under soil or waste pipes. If the tank is of a gravity type, it must be equipped with overflow provisions.

SIZES FOR OVERFLOW PIPES FOR WATER SUPPLY TANKS

MAXIMUM CAPACITY OF WATER SUPPLY LINE TO TANK (gpm)	DIAMETER OF OVERFLOW PIPE (inches)
0 - 50	2
50 - 150	$2^1/_2$
150 - 200	3
200 - 400	4
400 - 700	5
700 - 1,000	6
Over 1,000	8

For SI: 1 inch = 25.4 mm, 1 gallon per minute = 3.785 L/m.

FIGURE 6.3 Overflow pipes for water-supply tanks are required to meet the above standards. *(Courtesy of International Code Council, Inc. and International Plumbing Code 2000.)*

SIZE OF DRAIN PIPES FOR WATER TANKS

TANK CAPACITY (gallons)	DRAIN PIPE (inches)
Up to 750	1
751 to 1,500	$1^1/_2$
1,501 to 3,000	2
3,001 to 5,000	$2^1/_2$
5,001 to 7,500	3
Over 7,500	4

For SI: 1 inch = 25.4 mm, 1 gallon = 3.785 L.

FIGURE 6.4 Water tank drain pipes cannot be smaller than specified here. *(Courtesy of International Code Council, Inc. and International Plumbing Code 2000.)*

Materials	Type of Joints	Horizontal	Vertical
Cast Iron Hub and Spigot	Lead and Oakum	5 feet (1524 mm), except may be 10 feet (3048 mm) where 10 foot (3048 mm) lengths are installed [1, 2, 3]	Base and each floor not to exceed 15 feet (4572 mm)
	Compression Gasket	Every other joint, unless over 4 feet (1219 mm), then support each joint [1,2,3]	Base and each floor not to exceed 15 feet (4572 mm)
Cast Iron Hubless	Shielded Coupling	Every other joint, unless over 4 feet (1249 mm), then support each joint [1,2,3,4]	Base and each floor not to exceed 15 feet (4572 mm)
Copper Tube and Pipe	Soldered, Brazed or Welded	1-1/2 inch (40 mm) and smaller, 6 feet (1829 mm), 2 inch (50 mm) and larger, 10 feet (3048 mm)	Each floor, not to exceed 10 feet (3048 mm) [5]
Steel and Brass Pipe for Water or DWV	Threaded or Welded	3/4 inch (20 mm) and smaller, 10 feet (3048 mm), 1 inch (25 mm) and larger, 12 feet (3658 mm)	Every other floor, not to exceed 25 feet (7620 mm) [5]
Steel, Brass and Tinned Copper Pipe for Gas	Threaded or Welded	1/2 inch (15 mm), 6 feet (1829 mm) 3/4 (20 mm) and 1 inch (25.4 mm), 8 feet (2438 mm), 1-1/4 inch (32 mm) and larger, 10 feet (3048 mm)	1/2 inch (12.7 mm), 6 feet (1829 mm), 3/4 (19 mm) and 1 inch (25.4 mm), 8 feet (2438 mm), 1-1/4 inch (32 mm) and larger, every floor level
Schedule 40 PVC and ABS DWV	Solvent Cemented	All sizes, 4 feet (1219 mm). Allow for expansion every 30 feet (9144 mm) [3, 6]	Base and each floor. Provide mid-story guides. Provide for expansion every 30 feet (9144 mm) [6]
CPVC	Solvent Cemented	1 inch (25 mm) and smaller, 3 feet (914 mm), 1-1/4 inch (32 mm) and larger, 4 feet (1219 mm)	Base and each floor. Provide mid-story guides [6]
Lead	Wiped or Burned	Continuous support	Not to exceed 4 feet (1219 mm)
Copper	Mechanical	In accordance with standards acceptable to the Administrative Authority	
Steel & Brass	Mechanical	In accordance with standards acceptable to the Administrative Authority	
PEX	Metal Insert and Metal Compression	32 inches (800 mm)	Base and each floor. Provide midstory guides

[1] Support adjacent to joint, not to exceed eighteen (18) inches (457 mm).
[2] Brace at not more than forty (40) foot (12192 mm) intervals to prevent horizontal movement.
[3] Support at each horizontal branch connection.
[4] Hangers shall not be placed on the coupling.
[5] Vertical water lines may be supported in accordance with recognized engineering principles with regard to expansion and contraction, when first approved by the Administrative Authority.
[6] See the appropriate IAPMO Installation Standard for expansion and other special requirements.

FIGURE 6.5 This illustration looks at materials, joints and horizontal or vertical use. *(Reprinted from the 2000 Uniform Plumbing Code (UPC) with the permission of the International Association of Plumbing and Mechanical Officials (IAPMO).)*

The water supply to a gravity-style water tank must be controlled automatically through the use of a ballcock or another suitable and approved device. Incoming water should enter the tank by way of an air gap, which should be at least four inches above the overflow. Water tanks also are required to have provisions that allow them to be drained. The drain pipe must have a valve to prevent draining, except when desired.

PRESSURIZED WATER TANKS

Pressurized water tanks, which are used with well systems, are the type most commonly encountered in modern plumbing. All pressurized water tanks should be equipped with a vacuum relief valve at the top. These vacuum relief valves should be rated for proper operation up to maximum temperatures of 200°F and maximum water pressure of 200 psi. The minimum size of the vacuum relief valve shall be one-half inch. There is an exception that waives this requirement for diaphragm/bladder tanks.

It is also necessary to equip these tanks with pressure-relief valves, which must be installed on the supply pipe that feeds the tank or on the tank itself. The relief valve shall discharge when pressure builds to a point to endanger the integrity of the pressure tank. The valve's discharge must be carried by gravity to a safe and approved disposal location. The piping carrying the discharge of a relief valve may not be connected directly to the sanitary drainage system.

SUPPORTING YOUR PIPE

The method used to support your pipes is regulated by the plumbing code. There are requirements for the types of materials you may use and how they may be used. Let's see what they are.

One concern with the type of hangers used is their compatibility with the pipe they are supporting. You must use a hanger that will not have a detrimental effect on your piping. For example, you may not use a galvanized strap hanger to support copper pipe. Generally, the hangers used to support a pipe should be made from the same material as the pipe being supported. For example, copper pipe should be hung with copper hangers. This eliminates the risk of a corrosive action between two different types of materials. Plastic-coated hangers may be used with all types of pipe.

The hangers used to support the pipe must be capable of supporting the pipe at all times. The hanger must be attached to the pipe and to the member holding the hanger in a satisfactory manner. For example, it would not be acceptable to wrap a piece of wire around a pipe and then wrap the wire around the bridging between two floor joists. Hangers should be attached securely to the member supporting it. For example, a hanger should be attached to the pipe and then nailed to a floor joist. The nails used to hold a hanger in place should be made of the same material as the hanger if corrosive action is a possibility.

Both horizontal and vertical pipes require support. The intervals between supports will vary, depending upon the type of pipe being used and whether it is installed vertically or horizontally. The following examples will show you how often you must support the various types of pipes when they are hung horizontally. These examples are the maximum distances allowed between supports in many code jurisdictions:

- ABS—every 4'
- Brass pipe—every 10'
- Cast Iron—every 5'
- Copper—every 6'

HANGER SPACING

PIPING MATERIAL	MAXIMUM HORIZONTAL SPACING (feet)	MAXIMUM VERTICAL SPACING (feet)
ABS pipe	4	10[b]
Aluminum tubing	10	15
Brass pipe	10	10
Cast-iron pipe[a]	5	15
Copper or copper-alloy pipe	12	10
Copper or copper-alloy tubing, $1^1/_4$-inch diameter and smaller	6	10
Copper or copper-alloy tubing, $1^1/_2$-inch diameter and larger	10	10
Cross-linked polyethylene (PEX) pipe	2.67 (32 inches)	10[b]
Cross-linked polyethylene/ aluminum/crosslinked polyethylene (PEX-AL-PEX) pipe	$2^2/_3$ (32 inches)	4
CPVC pipe or tubing, 1 inch or smaller	3	10[b]
CPVC pipe or tubing, $1^1/_4$ inches or larger	4	10[b]
Steel pipe	12	15
Lead pipe	Continuous	4
PB pipe or tubing	2.67 (32 inches)	4
Polyethylene/aluminum/polyethylene (PE-AL-PE) pipe	2.67 (32 inches)	4
PVC pipe	4	10[b]
Stainless steel drainage systems	10	10[b]

For SI: 1 inch = 25.4 mm, 1 foot = 304.8 mm.

a. The maximum horizontal spacing of cast-iron pipe hangers shall be increased to 10 feet where 10-foot lengths of pipe are installed.

b. Midstory guide for sizes 2 inches and smaller.

FIGURE 6.6 Horizontal and vertical spacing of hangers is shown here. *(Courtesy of International Code Council, Inc. and International Plumbing Code 2000)*

- CPVC—every 3'
- Galvanized—every 12'
- PB pipe—every 32"
- PVC—every 4'

Not all code regions are the same. You may find the following support requirements in place in your region:

- ABS—every 4'
- Brass pipe—every 10'
- Cast Iron—every 15'
- Copper—every 10'
- CPVC—every 3'
- Galvanized—every 15'
- PVC—every 4'
- PB pipe—every 4'

WATER CONSERVATION

Water conservation continues to grow as a major concern. When setting the flow rates for various fixtures, water conservation is a factor. The flow rates of many fixtures must be limited to no more than 3 gallons per minute (gpm). In zone 3, these fixtures include:

- Showers
- Lavatories
- Kitchen sinks
- Other sinks

The rating of 3 gpm is based on a water pressure of eighty psi.

When installed in public facilities, lavatories must be equipped with faucets producing no more than one-half gpm. If the lavatory is equipped with a self-closing faucet, it may produce up to one-quarter gpm per use. Water closets are restricted to a use of no more than 3 gallons of water, and urinals must not exceed a usage of 1½ gallons of water with each use.

ANTI-SCALD PRECAUTIONS

It is easy for the very young or the elderly to receive serious burns from plumbing fixtures. In an attempt to reduce accidental burns, it is required that mixed water to gang showers be controlled by thermostatic means or by pressure-balanced valves. All showers, except for showers in residential dwellings in some regions, must be equipped with pressure-balanced valves or thermostatic controls. These temperature-control valves may not allow water with a temperature of more than 120°F to enter the bathing unit. In zone three, the maximum water temperature is 110°F. Some jurisdictions require these safety valves on all showers.

VALVE REGULATIONS

Gate valves and ball valves are examples of full-open valves, as required under valve regulations. These valves do not depend on rubber washers, and when they are opened to their maximum capacity, there is full flow through the pipe. Many locations along the water-distribution installation require the installation of full-open valves, These valves might be required in the following locations:

• On the water-service pipe, before and after the water meter
• On each water-service pipe for each building served
• On discharge pipes of water-supply tanks, near the tank
• On the supply pipe to every water heater, near the heater
• On the main supply pipe to each dwelling

Some code regions require full-open valves as follows:

• On the water-service pipe, near the source connections
• On the main water-distribution pipe, near the water service
• On water supplies to water heaters
• On water supplies to pressurized tanks, such as well-system tanks
• On the building side of every water meter

Full-open valves may be required for use in all water-distribution locations, except as cutoffs for individual fixtures, in the immediate area of the fixtures. Other local regulations may apply to specific building uses; check with your local code officer to confirm where full-open valves may be required in your system. All valves must be installed so that they are accessible.

CUTOFFS

Cutoff valves do not have to be full-open valves. Stop-and-waste valves are an example of cutoff valves that are not full-open valves. Every sillcock must be equipped with an individual cutoff valve. Appliances and mechanical equipment that have water supplies are required to have cutoff valves installed in the service piping. With only a few exceptions, cutoffs are required on all plumbing fixtures. Check with your local code officer for fixtures not requiring cutoff valves. All valves must be accessible.

SOME HARD-LINE FACTS

• All devices used to treat or convey potable water must be protected against all contamination.
• It is not acceptable to install stop-and-waste valves underground.
• If there are two water systems in a building, one potable and one nonpotable, the piping for each system must be marked clearly. The marking can be in the form of a suspended metal tag or a color code. Your local code may require the pipe to be color-coded and tagged. Nonpotable water piping should not be concealed.

APPLICATION FOR BACKFLOW PREVENTERS

DEVICE	DEGREE OF HAZARD[a]	APPLICATION[b]	APPLICABLE STANDARDS
Air gap	High or low hazard	Backsiphonage or backpressure	ASME A112.1.2
Antisiphon-type water closet flush tank ball cock	Low hazard	Backsiphonage only	ASSE 1002 CSA CAN/ B125
Barometric loop	High or low hazard	Backsiphonage only	(See Section 608.13.4)
Reduced pressure principle backflow preventer	High or low hazard	Backpressure or backsiphonage Sizes $3/8''$ - 16"	ASSE 1013 AWWA C511 CSA CAN/CSA-B64.4
Reduced pressure detector assembly backflow preventer	High or low hazard	Backsiphonage or backpressure (Fire sprinkler systems)	ASSE 1047
Double check backflow prevention assembly	Low hazard	Backpressure or backsiphonage Sizes $3/8''$ - 16"	ASSE 1015 AWWA C510
Double check detector assembly backflow preventer	Low hazard	Backpressure or backsiphonage (Fire sprinkler systems) Sizes $1 1/2''$ - 16"	ASSE 1048
Dual-check-valve-type backflow preventer	Low hazard	Backpressure or backsiphonage Sizes $1/4''$ - 1"	ASSE 1024
Backflow preventer with intermediate atmospheric vents	Low hazard	Backpressure or backsiphonage Sizes $1/4''$ - $3/4''$	ASSE 1012 CSA CAN/CSA-B64.3
Dual-check-valve-type backflow preventer for carbonated beverage dispensers/post mix type	Low hazard	Backpressure or backsiphonage Sizes $1/4''$ - $3/8''$	ASSE 1032
Pipe-applied atmospheric-type vacuum breaker	High or low hazard	Backsiphonage only Sizes $1/4''$ - 4"	ASSE 1001 CSA CAN/CSA-B64.1.1
Pressure vacuum breaker assembly	High or low hazard	Backsiphonage only Sizes $1/2''$ - 2"	ASSE 1020
Hose-connection vacuum breaker	High or low hazard	Low head backpressure or backsiphonage Sizes $1/2''$, $3/4''$, 1"	ASSE 1011 CSA CAN/CSA-B64.2
Vacuum breaker wall hydrants, frost-resistant, automatic draining type	High or low hazard	Low head backpressure or backsiphonage Sizes $3/4''$, 1"	ASSE 1019 CSA CAN/CSA-B64.2.2
Laboratory faucet backflow preventer	High or low hazard	Low head backpressure and backsiphonage	ASSE 1035 CSA B64.7
Hose connection backflow preventer	High or low hazard	Low head backpressure, rated working pressure backpressure or backsiphonage Sizes $1/2''$ - 1"	ASSE 1052
Spill-proof vacuum breaker	High or low hazard	Backsiphonage only Sizes $1/4''$ - 2"	ASSE 1056

For SI: 1 inch = 25.4 mm.

a. Low hazard—See Pollution (Section 202).
   High hazard—See Contamination (Section 202).
b. See Backpressure (Section 202).
   See Backpressure, Low Head (Section 202).
   See Backsiphonage (Section 202).

FIGURE 6.7 Various devices used as backflow preventers have different hazard ratings and applications. *(Courtesy of International Code Council, Inc. and International Plumbing Code 2000).*

- Hazardous materials, such as chemicals, may not be placed into a potable water system.
- Piping that has been used for a purpose other than conveying potable water may not be used as a potable water pipe.
- Water used for any purpose should not be returned to the potable water supply; this water should be transported to a drainage system.
- Mechanically extracted outlets must have a height not less than 3 times the thickness of the branch tube wall, and they must be brazed in compliance with the code. Branch tubes must not restrict the flow in the run tube. A dimple/depth stop must be formed in the branch tube to ensure that penetration into the collar is of the correct depth. For inspection purposes, a second dimple must be placed 0.25 inch above the first dimple. The dimples must be aligned with the tube run.

BACKFLOW PREVENTION

Backflow and backsiphonage are genuine health concerns. When a backflow occurs, it can pollute entire water systems. Without backflow and backsiphonage protection, municipal water services could become contaminated. There are many sources that are capable of deteriorating the quality of potable water. Backflow preventers for hose connections must consist of four independent check valves, with each one alternating with an independent atmospheric vent. The system also must have a means of field testing and draining.

Consider this example. A person is using a water hose to spray insecticide on the grounds around a house. The device being used to distribute the insecticide is a bottle-type sprayer, attached to a typical garden hose. The bottle has just been filled for use with a poisonous bug killer. A telephone rings inside the home. The individual lays down the bottle-sprayer and runs into the house to answer the phone.

While the person is in the house, and the bottle-sprayer is connected with the sillcock's valve open, a water main breaks. The backpressure caused by the break in the water main creates a vacuum. The vacuum sucks the poisonous contents of the bottle-sprayer backward, into the potable water system. Now what? How far did the poison go? How much pipe and how many fixtures must be replaced before the water system can be considered safe? The lack of a backflow preventer on the sillcock has created a nightmare. Human health and expensive financial considerations are at stake. A simple, inexpensive backflow preventer could have averted this potential disaster.

All potable water systems must be protected against backsiphonage and backflow with approved devices. Numerous types of devices are available to provide this type of protection. The selection of devices will be governed by the local plumbing inspector. It is necessary to choose the proper device for the use intended.

When more than one backflow-prevention valve is installed on a single premise, and they are installed in one location, each separate valve shall be identified permanently by the permittee in a manner satisfactory to the administrative authority.

Water supply inlets to tanks, vats, sumps, swimming pools, and other receptors must be protected by one of the following means: an approved air gap; a listed vacuum breaker installed on the discharge side of the last valve with the critical level not less than 6 inches or in accordance with its listing; or a backflow suitable for the contamination or pollution, installed in accordance with the requirements for that type of device or assembly as set forth by the code.

An airgap is the most positive form of protection from backflow. However, air gaps are not always feasible. Since air gaps cannot always be used, there are a number of devices available for the protection of potable water systems.

Some backflow preventers are equipped with vents. When these devices are used, the vents must not be installed so that they might become submerged. It is also required that these units be capable of performing their function under continuous pressure.

Some backflow preventers are designed to operate in a manner similar to an air gap. When conditions occur that can cause a backflow, the devices open and create an open air space between the two pipes connected to it. Reduced-pressure backflow preventers perform this action very well. Another type of backflow preventer that performs on a similar basis is an atmospheric-vent backflow preventer.

Vacuum breakers frequently are installed on water heaters, hose bibbs, and sillcocks. Generally they also are installed on the faucet spout of laundry tubs. These devices either mount on a pipe or screw onto a hose connection. Some sillcocks are equipped with factory-installed vacuum breakers. These devices open, when necessary, and break any siphonic action with the introduction of air. Hose bibbs are to be fitted with backflow preventers that are listed as a non-removable type.

In some specialized cases a barometric loop is used to prevent backsiphonage. These loops must extend at least 35 feet high and can be used only as a vacuum breaker. The loops are effective because they rise higher than the point where a vacuum suction can occur. Barometric loops work on the principle that by being 35 feet in height, suction will not be achieved.

Double-check valves are used in some instances to control backflow. When used in this capacity, double-check valves must be equipped with approved vents. for example, this type of protection would be used on a carbonated beverage dispenser.

Backflow preventers must be inspected from time to time. Therefore, backflow preventers must be installed in accessible locations.

Some fixtures require an air gap as protection from backflow. Some of these fixtures are lavatories, sinks, laundry tubs, bathtubs, and drinking fountains. This air gap is accomplished through the design and installation of the faucet or spout serving these fixtures.

Vacuum breakers must be installed at least six inches above the flood-level rim of the fixture. Vacuum breakers, because of the way they are designed to introduce air into the potable water piping, may not be installed where they may suck in toxic vapors or fumes.

Backflow Prevention Devices, Assemblies and Methods					
	Degree of Hazard				
Device, Assembly or Method[1]	Pollution (Low Hazard)		Contamination (High Hazard)		Installation[2, 3]
	Back-Siphon-age	Back-Pressure	Back-Siphon-age	Back-Pressure	
Airgap	X		X		See table in this chapter.
Atmospheric Vacuum Breaker	X		X		Upright position. No valves downstream. Minimum of six (6) inches (152 mm) or listed distance above all downstream piping and flood level rim of receptor.[4, 5]
Spill-Proof Pressure-Type Vacuum Breaker	X		X		Upright position. Minimum of six (6) inches (152 mm) or listed distance above all downstream piping and flood level rim of receptor.[5]
Double Check Valve Backflow Preventer	X	X			Horizontal, unless otherwise listed. Requires one (1) foot (305 mm) minimum clearance at bottom for maintenance. May need platform/ladder for test and repair. Does not discharge water.
Pressure Vacuum Breaker	X		X		Upright position. May have valves downstream. Minimum of twelve (12) inches (305 mm) above all downstream piping and flood level rim of receptor. May discharge water.
Reduced Pressure Principle Backflow Preventer	X	X	X	X	Horizontal unless otherwise listed. Requires one (1) foot (305 mm) minimum clearance at bottom for maintenance. May need platform/ladder for test and repair. May discharge water.

[1] See description of devices and assemblies in this chapter.
[2] Installation in pit or vault requires previous approval by the Administrative Authority.
[3] Refer to general and specific requirements for installation.
[4] Not to be subjected to operating pressure for more than 12 hours in any 24 hour period.
[5] For deck-mounted and equipment-mounted vacuum breakers, see Section 603.4.16.

FIGURE 6.8 There is a variety of devices and methods to help prevent backflow. *(Reprinted from the* 2000 Uniform Plumbing Code (UPC) *with the permission of the International Association of Plumbing and Mechanical Officials (IAPMO).)*

For example, it would not be acceptable to install a vacuum breaker under the exhaust hood of a kitchen range.

When potable water is connected to a boiler for heating purposes, the potable water inlet should be equipped with a vented backflow preventer. If the boiler contains chemicals in its water, the potable water connection should be made with an air gap or a reduced-pressure-principal backflow preventer.

Connections between a potable water supply and an automatic fire sprinkling system should be made with a check valve. If the potable water supply is being connected to a nonpotable water source, the connection should be made through a reduced-pressure-principal backflow preventer.

Lawn sprinklers and irrigation systems must be installed with backflow prevention in mind. Vacuum breakers are a preferred method for backflow prevention, but other types of backflow preventers are allowed.

Cross-connections are prohibited, except where approved protective devices are installed. Water pumps, filters, softeners, tanks and all other devices that handle or treat potable water must be protected from contamination. Pressure-type vacuum breakers must conform to ASSE 1020 for outdoor use.

All pullout spout-type faucets must be in compliance with CSA CAN/CSA-B125 and have an integral vacuum breaker or vent to atmosphere in their design, or shall require a dedicated deck- or wall-mounted vacuum breaker. All faucets with integral atmospheric or spill-proof vacuum breakers must be installed in accordance with the manufacturer's recommendations.

Water heaters and boiler drain valves that are provided with hose connection threads used only for draining the water heater do not require backflow protection. The water connection devices for automatic clothes washers do not require backflow protection when such protection is provided inside the washing machine.

The potable water supply for a carbonator must be protected by either an air gap or a vented backflow preventer installed within the carbonated beverage dispenser. Carbonated beverage dispensing systems without an approved internal air gap or vented backflow preventer must have the water supply protected with a vented backflow preventer for carbonated beverage dispensers.

Potable water supplies for fire-protection systems that are usually under pressure, except in one- and two-family residential sprinkler systems piped using materials approved for potable water distribution, must be protected from backpressure and backsiphonage. This protection can be in the form of a double-check valve assembly, a double-check detector assembly, a reduced pressure backflow preventer, or a reduced pressure detector assembly. Potable water supplies to fire-protection systems that are not usually under pressure must be protected from backflow and must meet the requirements of the appropriate standards.

HOT-WATER INSTALLATIONS

When hot-water pipe is installed, it often is expected to maintain the temperature of its hot water for a distance of up to 100 feet from the fixture it serves. If the distance between the source and the fixture is more than 100 feet, a recirculating system frequently is required. When a recirculating system is not appropriate, other means can be used to maintain water temperature. These means could include insulation or heating tapes. Check with your local code officer for approved alternates to a recirculating system, if necessary.

A circulator pump used on a recirculating line must be equipped with a cutoff switch. The switch may operate manually or automatically.

WATER HEATERS

The standard working pressure for a water heater is 125 psi. The maximum working pressure of a water heater is required to be permanently marked in an accessible location. Every water heater is required to have a drain, located at the lowest possible point on the water heater. Some exceptions may be allowed for very small, under-the-counter water heaters.

All water heaters are required to be insulated. The insulation factors are determined by the heat loss of the tank in an hour's time. These regulations are required of a water heater before it is approved for installation.

Relief valves are mandatory equipment on water heaters. These safety valves are designed to protect against excessive temperature and pressure. The most common type of safety valve will protect against both temperature and pressure. The blowoff rating for these valves must not exceed 210°F and 150 psi. The pressure-relief valve must not have a blowoff rating of more than the maximum working pressure of the water heater it serves—usually 125 psi.

When temperature and pressure relief valves are installed, their sensors should monitor the top six inches of water in the water heater. No valves may be located between the water heater and the temperature and pressure relief valves.

The blowoff from relief valves must be piped down, to protect bystanders in the event of a blowoff. The pipe used for this purpose must be rigid and capable of sustaining temperatures of up to 210°F. The discharge pipe must be the same size as the relief valve's discharge opening, and it must run, undiminished in size, to within six inches of the floor. If a relief valve discharge pipe is piped into the sanitary drainage system, the connection must be through an indirect waste. The end of a discharge pipe may not be threaded, and no valves may be installed in the discharge pipe.

When the discharge from a relief valve might damage property or people, safety pans should be installed. These pans typically have a minimum depth of one-and-one-half inches. Plastic pans commonly are used for electric water heaters, and metal pans are used for fuel-burning heaters. These pans must be large enough to accommodate the discharge flow from the relief valve.

The pan's drain may be piped to the outside of the building or to an indirect waste, where people and property will not be affected. The discharge location should be chosen so that it will be obvious to building occupants when a relief valve discharges. Traps should not be installed on the discharge piping from safety pans.

Water heaters must be equipped with an adjustable temperature control. This control is required to be automatically adjustable from the lowest to the highest temperatures allowed. Some locations restrict the maximum water temperature in residences to 120°F. A switch must be supplied to shut off the power to electric water heaters. When the water heater uses a fuel such as gas, a valve must be available to cut off the fuel source. Both electric and fuel shutoffs must be able to be used, without affecting the remainder of the building's power or fuel. All water heaters requiring venting must be vented in compliance with local code requirements.

PURGING THE SYSTEM OF CONTAMINANTS

When a potable water system has been installed, added to, or repaired, it must be disinfected. For years, this amounted to little more than running water through the system until it appeared clean. This is no longer enough. Under today's requirements, the system must undergo a true cleansing procedure.

The precise requirements for clearing a system of contaminants will be prescribed by the local health department or code enforcement office. Typically, it will require flushing the system with potable water until the water appears clean. This action will be followed by a cleaning with a chlorine solution. The exact requirements for the mixture of chlorine and water will be provided by a local agency. The chlorine mixture will be introduced into the system and usually will be required to remain in the system for between three and 24 hours.

After the chlorine has been in the system for the required time, the system will be flushed with potable water until no trace of chlorine remains. Again, check with your local authorities for the exact specifications for purifying the potable water system.

WORKING WITH WELLS

When you will be working with wells or other private water sources, there are some rules you must follow. This section will apprise you of what you need to know when working with private water systems.

If a building does not have access to a public water source, it must depend on water from a private source. Typically, this source is a well. Under some conditions, however, it could be a cistern, spring, or stream. If surface water is used as a potable water source, it must be tested and approved for use. For that matter, wells generally are required to be tested and approved.

The quality of water from a private source must meet minimum standards as potable water. This is determined through water tests, usually conducted by the local health department or some other local authority.

The quantity of water delivered from a well also must meet certain requirements. The well, or water source, must be capable of supplying enough water for the intended use of the system. For example, zone three rates a single-family home as requiring 75 gallons of water a day for each occupant. Hospitals, on the other hand, require a minimum of between 150 and 250 gallons of water each day for each bed in the hospital.

All private water sources must be protected from contamination. They also must be disinfected before being used. The protection from contamination can include several factors. For example, wells must have water-tight caps. Wells may not be installed in an area where contamination is likely, such as near a septic system. The height of a well casing should extend above the ground. All wells should be located above and upstream from any possible contaminating sources, such as a septic field.

Any well, whether drilled, dug, or driven, generally must not deliver water for potable use from a depth of less than ten feet. There are rules governing the allowable distances between a private water source and possible pollution sources. The following examples show how far a private water source in zone three must be from a few of the possible polluting sources:

- Barnyard- 100'
- Pasture land- 100'
- Septic tank- 25'
- Sewer- 10'
- Underground disposal fields- 50'

Construction Requirements for Wells

Wells must be installed to meet minimum standards. What follows is a description of the minimum requirement for the installation of various types of wells in a common code area.

Dug and Bored Wells

Dug and bored wells are usually relatively shallow. Their casings must be made from waterproof concrete, corrugated metal pipe, galvanized pipe, or tile. These casings must extend to a minimum depth of ten feet below the ground and also extend below the water table. For example, if the well is 16 feet deep and the water table is 13 feet deep, the casing must extend at least 13 feet into the ground.

When wells are dug or bored, there is a large space between the well casing and the undisturbed earth. This space must be filled with a grout material. The grout must encompass the well casing and have a minimum thickness of six inches. This helps prevent surface water from entering the well. It is also necessary for the well casing to rise at least six inches above the well platform.

Then, there is the cover. This type of well is usually large in diameter. The top of the well must be sealed with a watertight cover. Covers must overlap the sides of the well casing and extend downward for a minimum of two inches. Concrete covers are common on this type of well.

Common practice with bored and dug wells is to have the water pipe from the well to the house exit through the side of the casing. This generally is done below ground, below the regional frost line. Where this penetration of the well casing occurs, the hole must be sealed to prevent outside water from entering the well.

If a well is installed in an area subject to flooding, the well casing and cover must be designed and installed to withstand the forces associated with a flood. Grading of the ground around a private water source may be required to divert runoff water from entering the potable water supply.

Drilled and Driven Wells

Drilled and driven wells are different from dug and bored wells. The diameters of these wells are much smaller, and drilled wells are often very deep. The casings for these wells must be made of steel or some other approved material. The casing must extend at least six inches above the well platform.

Grouting is required around the exterior of these casings. The grout material is required to a minimum depth of ten feet or solid contact with rock, whichever comes first. The casing should extend into rock or well beyond the water table level.

Getting the water pipe from a drilled well to a building usually is done through the side of the casing. The usual procedure calls for the use of a pitless adapter, which mounts into the well casing and forms a watertight seal. In any event, the casing must be sealed at any openings that might allow nonpotable water to enter the well.

The cover for a drilled well must be waterproof and usually will allow provisions for electrical wires connected to the pump. These wires must get from the submerged pump to the control box. In all cases, the cover must be designed and installed to prevent the influx of surface water into the well.

Well Pumps

The pumps used with potable water systems must meet minimum standards. This section identifies and explains these standards. Pumps must be approved for use. They must be readily accessible for service, maintenance, or repair. In flood-prone areas, pumps must be designed and installed to resist the potential detrimental effects of a flood. Water pumps are required to be capable of continuous operation.

In some jurisdictions, a pump installed in a home must be installed on an appropriate base. Some pumps are installed on brackets that are connected to pressure-supply tanks. If a pump is installed in a basement, it must be installed at least 18 inches above the base-

DISTANCE FROM SOURCES OF CONTAMINATION TO PRIVATE WATER SUPPLIES AND PUMP SUCTION LINES

SOURCE OF CONTAMINATION	DISTANCE (feet)
Barnyard	100
Farm silo	25
Pasture	100
Pumphouse floor drain of cast iron draining to ground surface	2
Seepage pits	50
Septic tank	25
Sewer	10
Subsurface disposal fields	50
Subsurface pits	50

For SI: 1 foot = 304.8 mm.

FIGURE 6.9 Wells must be located at least a certain distance from possible contamination sites. (*(Courtesy of International Code Council, Inc. and International Plumbing Code 2000.)*

ment floor. It is not acceptable to install the pump in a pit or closer than 18 inches to the finished floor level of a basement floor. This provision is meant to protect the pump from submersion, through basement flooding.

Pump Houses

When shallow-well pumps or two-pipe jet pumps are used, the pumps sometimes are placed in pump houses. These pump houses must be of approved construction.

A building providing shelter to a water pump must be equipped to prevent the pump or related piping from freezing. Such an enclosure also must be provided with adequate drainage facilities to prevent water from rising over the pump and piping.

SIZING POTABLE WATER SYSTEMS

About the only aspect of the potable water system that we have not covered is sizing. Well, this section of the chapter is going to show you how to size potable water piping. But, be advised, this procedure is not always simple and requires concentration. If your mind is not fresh, leave this section for a later reading. If you are ready to learn how to size water pipe, get a pen and paper, and get ready for one of the more complicated aspects of this book.

Some facets of potable water pipe sizing are not very difficult. Many times your code book will provide charts and tables to help you. Some of these graphics will detail precisely what size pipe or tubing is required. Unfortunately, however, code books cannot provide concrete answers for all piping installations.

Many factors affect the sizing of potable water piping. The type of pipe used will have an influence on your findings. Some pipe materials have smaller inside diameters than others. Some pipe materials have a rougher surface or more restrictive fittings than others. Both of these factors will affect the sizing of a water system.

When sizing a potable water system, you must be concerned with the speed of the flowing water, the quantity of water needed, and the restrictive qualities of the pipe being used to convey the water. Most materials approved for potable water piping will provide a flow velocity of five feet per second. The exception is galvanized pipe. Galvanized pipe will provide a speed of eight feet per second.

It might be surprising that galvanized pipe allows a faster flow rate. This occurs because of the wall strength of galvanized pipe. In softer pipes, such as copper, fast-moving water essentially can wear a hole in the pipe. These flow ratings are not carved in stone. I am sure you will find people who will argue for either a higher or a lower rating, but these ratings are in use with current plumbing codes.

Inch	mm				
1/2	15				
3/4	20	**Water Supply Fixture Units (WSFU) and Minimum Fixture Branch Pipe Sizes[3]**			
1	25				
		Minimum Fixture Branch	Private	Public	Assembly[5]
Appliances, Appurtenances or Fixtures[2]		Pipe Size[1,3]			
Bathtub or Combination Bath/Shower (fill)		1/2"	4.0	4.0	
3/4" Bathtub Fill Valve		3/4"	10.0	10.0	
Bidet		1/2"	1.0		
Clotheswasher		1/2"	4.0	4.0	
Dental Unit, cuspidor		1/2"		1.0	
Dishwasher, domestic		1/2"	1.5	1.5	
Drinking Fountain or Watercooler		1/2"	0.5	0.5	0.75
Hose Bibb		1/2"	2.5	2.5	
Hose Bibb, each additional[7]		1/2"	1.0	1.0	
Lavatory		1/2"	1.0	1.0	1.0
Lawn Sprinkler, each head[4]			1.0	1.0	
Mobile Home, each (minimum)			12.0		
Sinks					
Bar		1/2"	1.0	2.0	
Clinic Faucet		1/2"		3.0	
Clinic Flushometer Valve					
with or without faucet		1"		8.0	
Kitchen, domestic		1/2"	1.5	1.5	
Laundry		1/2"	1.5	1.5	
Service or Mop Basin		1/2"	1.5	3.0	
Washup, each set of faucets		1/2"		2.0	
Shower, per head		1/2"	2.0	2.0	
Urinal, 1.0 GPF Flushometer Valve		3/4"	See Footnote 6		
Urinal, greater than 1.0 GPF Flushometer Valve		3/4"	See Footnote 6		
Urinal, flush tank		1/2"	2.0	2.0	3.0
Washfountain, circular spray		3/4"		4.0	
Water Closet, 1.6 GPF Gravity Tank		1/2"	2.5	2.5	3.5
Water Closet, 1.6 GPF Flushometer Tank		1/2"	2.5	2.5	3.5
Water Closet, 1.6 GPF Flushometer Valve		1"	See Footnote 6		
Water Closet, greater than 1.6 GPF Gravity Tank		1/2"	3.0	5.5	7.0
Water Closet, greater than 1.6 GPF Flushometer Valve		1"	See Footnote 6		

Notes:
1. Size of the cold branch pipe, or both the hot and cold branch pipes.
2. Appliances, Appurtenances or Fixtures not included in this Table may be sized by reference to fixtures having a similar flow rate and frequency of use.
3. The listed minimum supply branch pipe sizes for individual fixtures are the nominal (I.D.) pipe size.
4. For fixtures or supply connections likely to impose continuous flow demands, determine the required flow in gallons per minute (GPM) and add it separately to the demand (in GPM) for the distribution system or portions thereof.
5. Assembly [Public Use (See Table 4-1)].
6. When sizing flushometer systems see Section 610.10.
7. Reduced fixture unit loading for additional hose bibbs as used is to be used only when sizing total building demand and for pipe sizing when more than one hose bibb is supplied by a segment of water distributing pipe. The fixture branch to each hose bibb shall be sized on the basis of 2.5 fixture units.

FIGURE 6.10 Fixture units and fixture branch pipe sizes are important in the sizing process. *(Reprinted from the* 2000 Uniform Plumbing Code (UPC) *with the permission of the International Association of Plumbing and Mechanical Officials (IAPMO).)*

When using the three factors previously discussed to determine pipe size, you must use math that you might not have seen since your school days, and you might not have seen it then. Let me give you an example of how a typical formula might look.

- X is equal to the water's rate of flow; in most cases, 5 feet per second.
- Y is equal to the quantity of water in the pipe.
- Z is equal to the inside diameter of the pipe.
- A typical formula might look like this: $Y = XZ$.

Since most plumbers will refuse to do this type of math, most code books offer alternatives. The alternatives are often in the form of tables or charts that show pertinent information about the requirements for pipe sizing. The tables or charts you might find in a code book likely will discuss the following: a pipe's outside diameter, a pipe's inside diameter, a flow rate for the pipe, and a pressure loss in the pipe over a distance of 100 feet. These charts or tables will be dedicated to a particular type of pipe. For example, there would be one table for copper pipe and another table for PB pipe.

The information supplied in a ratings table for PB pipe might look like this:

- Pipe size is ¾"
- Inside pipe diameter is .715
- The flow rate, at 5 feet per second, is 6.26 gpm
- Pressure lost in 100 feet of pipe is 14.98

This type of pipe sizing most often is done by engineers, not plumbers. When sizing a potable water system, the sizing exercise starts at the last fixture and works its way back to the water service.

Commercial jobs, where pipe sizing can get quite complicated, generally are sized by design experts. All a working plumber is required to do is install the proper pipe sizes, in the proper locations and manner. For residential plumbing, where engineers are less likely to have a hand in the design, there is a common method to sizing most jobs. In the average home, a three-quarter-inch pipe is sufficient for the main artery of the water-distribution system. Usually, not more than two fixtures can be served by a half-inch pipe. With this in mind, sizing becomes simple.

Three-quarter-inch pipe usually is run to the water heater and typically is used as a main water-distribution pipe. When nearing the end of a run, the three-quarter-inch pipe is reduced to half-inch pipe, when there are no more than two fixtures remaining to connect. Most water services will have a three-quarter-inch diameter, while those serving homes with numerous fixtures will use a one-inch pipe. This guideline to sizing will work on almost any single-family residence.

The water supplies to fixtures are required to meet minimum standards. These sizes are derived from local code requirements. You simply find the fixture for which you are sizing the supply and check the column heading for the proper size.

Most code requirements seem to agree that there is no definitive way to set a boilerplate formula for establishing potable water pipe sizing. Code officers can require pipe sizing to be performed by a licensed engineer. In most major plumbing systems, the pipe sizing is done by a design professional.

Code books give examples of how a system might be sized. But the examples are not meant as a code requirement. The code requires a water system to be sized properly. Because of the complexity of the process, however, the books do not set firm statistics for the process. Instead, code books give parts of the puzzle, in the form of some minimum standards, but it is up to a professional designer to come up with an approved system.

Where does this leave you? Well, the sizing of a potable water system is one of the most complicated aspects of plumbing. Very few single-family homes are equipped with potable water systems designed by engineers. I already have given you a basic method for sizing small systems. Next, I am going to show you how to use the fixture-unit method of sizing. The fixture-unit method is not very difficult, and it generally is acceptable to code officers. While this method might not be perfect, it is much faster and easier to use than the velocity-method. Other than for additional expense in materials, you can't go wrong by oversizing your pipe. If in doubt about sizing, go to the next larger size. Now, let's see how you might size a single-family residence's potable water system using the fixture-unit method.

Most codes assign a fixture-unit value to common plumbing fixtures. To size with the fixture-unit method, you must establish the number of fixture units to be carried by the pipe. You also must know the working pressure of the water system. Most codes will provide guidelines for these two pieces of information.

For our example, we have a house with the following fixtures: three toilets, three lavatories, one bathtub-shower combination, one shower, one dishwasher, one kitchen sink, one laundry hookup, and two sillcocks. The water pressure serving this house is 50 psi. A one-inch water meter serves the house, and the water-service pipe is 60 feet in length. With this information and the guidelines provided by your local code, you can do a pretty fair job of sizing your potable water system.

The first step is to establish the total number of fixture units on the system. The code regulations will provide this information. In this case, add the total number of fixture-units in the house. You have three toilets, that's nine fixture units. The three lavatories add three fixture units. The tub-shower combination counts as two fixture units, the showerhead over the bathtub doesn't count as an additional fixture. The shower has two fixture units. The dishwasher adds two fixture units and so does the kitchen sink. The laundry hookup counts as two fixture units. Each sillcock is worth three fixture units. This house has a total fixture-unit load of 28.

The first piece of your sizing puzzle is solved. The next step is to determine what size pipe will allow your number of fixture-units.

Our subject house has a water pressure of 50 psi. This pressure rating falls into the category allowed in your local code. First, find the proper water meter size; the one you are looking for is one-inch. You will notice that a one-inch meter and a one-inch water-service pipe are capable of handling 60 fixture units, when the pipe is only running 40 feet. However, when the pipe length is stretched to 80 feet, the fixture load is dropped to 41. At 200 feet, the fixture rating is 25. What is it at 100 feet? At 100 feet, the allowable fixture load is 36. See, this type of sizing is not so hard.

Now, what does this tell us? Well, we know the water-service pipe is 60 feet long. Once inside the house, how far is it to the most remote fixture? In this case, the farthest fixture is 40 feet from the water-service entrance. This gives us a developed length of 100 feet, 60 feet for the water-service and 40 feet for the interior pipe. We know that for 100 feet of pipe, under the conditions in this example, we are allowed 36 fixture units. The house only has 28 fixture units; so our pipe sizing is fine.

What would happen if the water meter was a three-quarter-inch meter, instead of a one-inch meter? With a three-quarter-inch meter and a one-inch water-service and main distribution pipe, we could have 33 fixture units. This still would be a suitable arrangement, since we only have 28 fixture-units. Could we use a three-quarter-inch water-service and water-distribution pipe with the three-quarter-inch meter? No, we couldn't. With all sizes set at three-quarters of an inch, the maximum number of fixture units allowed is 17.

In this example, the piping is over-sized. If you want to be safe, however, use this type of procedure. If you are required to provide a riser diagram showing the minimum pipe sizing allowed, you will have to do a little more work. Once inside a building, water-distribution pipes usually extend for some distance, supplying many fixtures with water. As the distribution pipe continues on it journey, it reduces the fixture load as it goes.

WATER DISTRIBUTION SYSTEM DESIGN CRITERIA
REQUIRED CAPACITIES AT FIXTURE SUPPLY PIPE OUTLETS

FIXTURE SUPPLY OUTLET SERVING	FLOW RATE[a] (gpm)	FLOW PRESSURE (psi)
Bathtub	4	8
Bidet	2	4
Combination fixture	4	8
Dishwasher, residential	2.75	8
Drinking fountain	0.75	8
Laundry tray	4	8
Lavatory	2	8
Shower	3	8
Shower, temperature controlled	3	20
Sillcock, hose bibb	5	8
Sink, residential	2.5	8
Sink, service	3	8
Urinal, valve	15	15
Water closet, blow out, flushometer valve	35	25
Water closet, flushometer tank	1.6	15
Water closet, siphonic, flushometer valve	25	15
Water closet, tank, close coupled	3	8
Water closet, tank, one piece	6	20

For SI: 1 psi = 6.895 kPa, 1 gallon per minute (gpm) = 3.785 L/m.

a. For additional requirements for flow rates and quantities, see Section 604.4.

FIGURE 6.11 Capacities of fixture pipe outlets must meet tie in with water distribution design. *(Courtesy of International Code Council, Inc. and International Plumbing Code 2000.)*

For example, assume that the distribution pipe serves a full bathroom group within ten feet of the water service. After this group is served with water, the fixture-unit load on the remainder of water-distribution piping is reduced by six fixture units. As the pipe serves other fixtures, the fixture-unit load continues to decrease. So, it is feasible for the water-distribution pipe to become smaller as it goes along.

Let's take our same sample house and see how we could use smaller pipe. Okay, we know we need a one-inch water-service pipe. Once inside the foundation, the water service becomes the water-distribution pipe. The water heater is located five feet from the cold-water distribution pipe. The one-inch pipe will extend over the water heater and supply it with cold water. Then, a hot-water distribution pipe will originate at the water heater. Now you have two water-distribution pipes to size.

MAXIMUM FLOW RATES AND CONSUMPTION
FOR PLUMBING FIXTURES AND FIXTURE FITTINGS

PLUMBING FIXTURE OR FIXTURE FITTING	MAXIMUM FLOW RATE OR QUANTITY[b]
Water closet	1.6 gallons per flushing cycle
Urinal	1.0 gallon per flushing cycle
Shower head[a]	2.5 gpm at 60 psi
Lavatory, private	2.2 gpm at 60 psi
Lavatory (other than metering), public	0.5 gpm at 60 psi
Lavatory, public (metering)	0.25 gallon per metering cycle
Sink faucet	2.2 gpm at 60 psi

For SI: 1 gallon = 3.785 L, 1 gallon per minute = 3.785 L/m, 1 psi = 6.895 kPa.

a. A hand-held shower spray is a shower head.

b. Consumption tolerances shall be determined from referenced standards.

FIGURE 6.12 There are limits for flow and consumption for fixtures and their fittings. *(Courtesy of International Code Council, Inc. and International Plumbing Code 2000.)*

When sizing the hot- and cold-water pipes, you could make adjustments for fixture-unit values on some fixtures. For example, a bathtub is rated as two fixture units. Since the bathtub rating is inclusive of both hot and cold water, obviously the demand for just the cold water pipe is less than that shown in our table. I will not break the fixture-units down into fractions or reduced amounts to keep this simple. I will show you the example as if a bathtub requires two fixture units of hot water and two fixture units of cold water. However, you could reduce the amounts listed in the codes about 25 percent to obtain the rating for each hot- and cold-water pipe. For example, the bathtub, when being sized for only cold water, could take on a value of one-and-one-half fixture units.

Now then, let's get on with the exercise. We are at the water heater. We ran a one-inch cold-water pipe overhead and dropped a three-quarter-inch pipe into the water heater. What size pipe do we bring up for the hot water? First, count the number of fixtures that use hot water and assign them a fixture-unit value. The fixtures using hot water are all fixtures, except the toilets and sillcocks. The total count for hot-water fixture-units is lucky number 13. From the water heater, our most remote hot-water fixture is 33 feet away.

What size pipe should we bring up from the water heater? By looking at your local code book, find a distance and fixture-unit count that will work in this case. You would look under the 40-feet column, since our distance is less than 40 feet. When you look in the column, the first fixture-unit number you see is nine; this won't work. The next number is 27; this one will work, because it is greater than the 13 fixture units we need. Looking across the table in your local code book, you will see that the minimum size is three-quarter-inch pipe.

Okay, now we start our hot-water run with three-quarter pipe. As our hot water pipe goes along the 33-foot stretch, it provides water to various fixtures. When the total fixture-count remaining to be served drops to fewer than nine fixture units, we can reduce the pipe to half-inch pipe. We also can run our fixture branches off the main in half-inch pipe. We can do this because the highest fixture-unit rating on any of our hot-water fix-

SIZE OF PIPE IDENTIFICATION

PIPE DIAMETER (inches)	LENGTH OF BACKGROUND COLOR FIELD (inches)	SIZE OF LETTERS (inches)
$3/4$ to $1^1/4$	8	0.5
$1^1/2$ to 2	8	0.75
$2^1/2$ to 6	12	1.25
8 to 10	24	2.5
over 10	32	3.5

For SI: 1 inch = 25.4 mm.

FIGURE 6.13   Pipe with different diameters will have varying background color field and letter size. *(Courtesy of International Code Council, Inc. and International Plumbing Code 2000.)*

tures is two. Even with a pipe run of 200 feet we can use half-inch pipe for up to four fixture units. Is this sizing starting to ring a bell? Remember the sizing guideline I gave you earlier? These sizing examples are making the guidelines method ring true.

With the hot-water sizing done, let's look at the remainder of the cold water. We have less than forty feet to our farthest cold-water fixture. We branch off near the water-heater drop for a sillcock, and a full bathroom group is within seven feet of our water-heater drop. The sillcock branch can be half-inch pipe. The pipe going under the full bathroom group probably could be reduced to three-quarter pipe, but it would be best to run it as a one-inch pipe. However, after serving the bathroom group and the sillcock, how may fixture units are left? There are only 19 fixture units left. We now can reduce to three-quarter pipe, and when the demand drops to below nine fixture units, we can reduce to half-inch pipe. All of our fixture branches can be run with half-inch pipe.

This is one way to size a potable water system that works, without driving you crazy. There may be some argument to the sizes I gave in these examples. The argument would be that some of the pipe is over-sized, but, as I said earlier, when in doubt, go bigger. In reality, the cold-water pipe in the last example probably could have been reduced to three-quarter-inch pipe where the transition was made from water-service to water-distribution pipe. It almost certainly could have been reduced to three-quarter pipe after the water-heater drop. Local codes will have their own interpretation of pipe sizing, but this method usually will serve you well.

Always refer to your local code book for specific sizing requirements and practices.

SOME MORE FACTS TO KEEP YOU OUT OF TROUBLE

Here are some more facts to keep you out of trouble when you are working with potable water systems:

1. When working with a solvent-cement joint, you are not required to use a primer when all of the following conditions apply: the cement being used is third-party certified as conforming to ASTM F493; the cement being used is yellow; the cement is used only for joining half-inch to 2-inch diameter CPVC pipe and fittings, and the CPVC pipe and fittings are manufactured in accordance with ASTM D2846.

MINIMUM SIZES OF FIXTURE WATER SUPPLY PIPES

FIXTURE	MINIMUM PIPE SIZE (inch)
Bathtubs (60″ × 32″ and smaller)[a]	$1/2$
Bathtubs (larger than 60″ × 32″)	$1/2$
Bidet	$3/8$
Combination sink and tray	$1/2$
Dishwasher, domestic[a]	$1/2$
Drinking fountain	$3/8$
Hose bibbs	$1/2$
Kitchen sink[a]	$1/2$
Laundry, 1, 2 or 3 compartments[a]	$1/2$
Lavatory	$3/8$
Shower, single head[a]	$1/2$
Sinks, flushing rim	$3/4$
Sinks, service	$1/2$
Urinal, flush tank	$1/2$
Urinal, flush valve	$3/4$
Wall hydrant	$1/2$
Water closet, flush tank	$3/8$
Water closet, flush valve	1
Water closet, flushometer tank	$3/8$
Water closet, one piece[a]	$1/2$

For SI: 1 inch = 25.4 mm, 1 foot = 304.8 mm, 1 psi = 6.895 kPa.

a. Where the developed length of the distribution line is 60 feet or less, and the available pressure at the meter is a minimum of 35 psi, the minimum size of an individual distribution line supplied from a manifold and installed as part of a parallel water distribution system shall be one nominal tube size smaller than the sizes indicated.

FIGURE 6.14 Do not use less than the minimum sizes for fixture water supply pipes. *(Courtesy of International Code Council, Inc. and International Plumbing Code 2000.)*

2. Cross-linked polyethylene plastic requiring joints between tubing or fittings must comply with the plumbing code.

3. PEX tubing must be marked appropriately to identify the uses for which the material is approved.

4. PEX fittings must be made of metal and must be secured with metal compression fittings.

MANIFOLD SIZING

NOMINAL SIZE INTERNAL DIAMETER (inches)	MAXIMUM DEMAND (gpm)	
	Velocity at 4 feet per second	Velocity at 8 feet per second
$^1/_2$	2	5
$^3/_4$	6	11
1	10	20
$1^1/_4$	15	31
$1^1/_2$	22	44

For SI: 1 inch = 25.4 mm, 1 gallon per minute = 3.785 L/m, 1 foot per second
= 0.305 m/s.

FIGURE 6.15   These are the recommended sizes for manifolds. *(Courtesy of International Code Council, Inc. and International Plumbing Code 2000.)*

5. PEX tubing may not be used for water heater connections within the first 18 inches of the piping connected to the water heater.

6. Flared pipe ends must be made with a tool designed for that operation.

7. Mechanical joints must be made in accordance with the local plumbing code and the manufacturer's recommendations. Metallic lock rings and insert fittings as described in ASTM F1807 are required for the installation.

8. Tempered water is required to be delivered from accessible hand-washing facilities.

9. Maintaining energy efficiency must conform to the *International Energy Conservation Code.*

10. Vacuum breakers for hose connections in health care or laboratory areas shall not be less than 6 feet above the floor.

11. Pipe identification must include the contents of the piping system and an arrow indicating the direction of flow. Any hazardous piping systems must contain information that addresses the nature of the hazard. Pipe identification must be repeated at maximum intervals of 25 feet and at each point where the piping passes through a wall, floor, or roof. All lettering must be readily observed within the room or space the piping is located. Any coloring of pipe identification must be discernable and consistent throughout the building. Identification labeling must be sized in compliance with code requirements.

12. Pressure-type vacuum breakers shall not be installed in locations where spillage could cause damage to the structure.

13. The discharge from a reverse osmosis system must enter a drainage system through an air gap or an air-gap device.

14. Construction, installation, alterations, and repair of solar systems, equipment, and appliances intended to utilize solar energy for space heating or cooling, domestic hot water heating, swimming pool heating or process heating shall be in accordance with the *International Mechanical Code.*

15. Flexible corrugated connectors made of copper or stainless steel are limited in their length. Water heater connectors must not be more than 24 inches long. Fixture connectors are limited to a maximum length of 30 inches. Connectors for washing machines must not exceed 72 inches in length. Flex connectors for dishwashers and icemakers must not be longer than 120 inches.

16. Female PVC screwed fittings for water piping must be used with plastic male fittings and plastic male threads only.

PIPE FITTINGS

MATERIAL	STANDARD
Acrylonitrile butadiene styrene (ABS) plastic	ASTM D 2468
Cast iron	ASME B16.4; ASME B16.12
Chlorinated polyvinyl chloride (CPVC) plastic	ASTM F 437; ASTM F 438; ASTM F 439
Copper or copper alloy	ASME B16.15; ASME B16.18; ASME B16.22; ASME B16.23; ASME B16.26; ASME B16.29; ASME B16.32
Gray iron and ductile iron	AWWA C110; AWWA C153
Malleable iron	ASME B16.3
Metal Insert Fittings Utilizing a Copper Crimp Ring SDR9 (PEX) Tubing	ASTM F 1807
Polyethylene (PE) plastic	ASTM D 2609
Polyvinyl chloride (PVC) plastic	ASTM D 2464; ASTM D 2466; ASTM D 2467; CSA CAN/CSA-B137.2
Steel	ASME B16.9; ASME B16.11; ASME B16.28

FIGURE 6.16 This table lists material for pipe fittings and the standards they must meet. (*Courtesy of International Code Council, Inc. and International Plumbing Code 2000.*)

MINIMUM REQUIRED AIR GAPS

FIXTURE	MINIMUM AIR GAP	
	Away from a wall[a] (inches)	Close to a wall (inches)
Lavatories and other fixtures with effective opening not greater than $1/2$ inch in diameter	1	$1^{1}/_{2}$
Sink, laundry trays, gooseneck back faucets and other fixtures with effective openings not greater than $^3/_4$ inch in diameter	1.5	2.5
Over-rim bath fillers and other fixtures with effective openings not greater than 1 inch in diameter	2	3
Drinking water fountains, single orifice not greater than $^7/_{16}$ inch in diameter or multiple orifices with a total area of 0.150 square inch (area of circle $^7/_{16}$ inch in diameter)	1	$1^{1}/_{2}$
Effective openings greater than 1 inch	Two times the diameter of the effective opening	Three times the diameter of the effective opening

For SI: 1 inch = 25.4 mm.

a. Applicable where walls or obstructions are spaced from the nearest inside edge of the spout opening a distance greater than three times the diameter of the effective opening for a single wall, or a distance greater than four times the diameter of the effective opening for two intersecting walls.

FIGURE 6.17 These are the minimum air-gap listings for fixtures. (*Courtesy of International Code Council, Inc. and International Plumbing Code 2000.*)

Minimum Airgaps for Water Distribution[4]		
Fixtures	When not affected by side walls[1] Inches (mm)	When affected by side walls[2] Inches (mm)
Effective openings[3] not greater than one-half (1/2) inch (12.7 mm) in diameter	1 (25.4)	1-1/2 (38)
Effective openings[3] not greater than three-quarters (3/4) inch (20 mm) in diameter	1-1/2 (38)	2-1/4 (57)
Effective openings[3] not greater than one (1) inch (25 mm) in diameter	2 (51)	3 (76)
Effective openings[3] greater than one (1) inch (25 mm) in diameter	Two (2) times diameter of effective opening	Three (3) times diameter of effective opening

[1] Side walls, ribs or similar obstructions do not affect airgaps when spaced from the inside edge of the spout opening a distance greater than three times the diameter of the effective opening for a single wall, or a distance greater than four times the effective opening for two intersecting walls.

[2] Vertical walls, ribs or similar obstructions extending from the water surface to or above the horizontal plane of the spout opening other than specified in Note 1 above. The effect of three or more such vertical walls or ribs has not been determined. In such cases, the airgap shall be measured from the top of the wall.

[3] The effective opening shall be the minimum cross-sectional area at the seat of the control valve or the supply pipe or tubing which feeds the device or outlet. If two or more lines supply one outlet, the effective opening shall be the sum of the cross-sectional areas of the individual supply lines or the area of the single outlet, whichever is smaller.

[4] Airgaps less than one (1) inch (25.4 mm) shall only be approved as a permanent part of a listed assembly that has been tested under actual backflow conditions with vacuums of 0 to 25 inches (635 mm) of mercury.

FIGURE 6.18 For proper water distribution, these air-gap minimums should be observed. (*Reprinted from the* 2000 Uniform Plumbing Code (UPC) *with the permission of the International Association of Plumbing and Mechanical Officials (IAPMO).)*

Minimum Required Air Chamber Dimensions

Nominal Pipe Diameter	Length of Pipe (ft.)	Flow Pressure P.S.I.G.	Velocity in Ft. Per. Sec.	Required Vol. in Cubic Inch	Air Chamber Phys. Size in Inches
1/2" (15 mm)	25	30	10	8	3/4" x 15"
1/2" (15 mm)	100	60	10	60	1" x 69.5"
3/4" (20 mm)	50	60	5	13	1" x 5"
3/4" (20 mm)	200	30	10	108	1.25" x 72.5"
1" (25 mm)	100	60	5	19	1.25" x 12.7 "
1" (25 mm)	50	30	10	40	1.25" x 27"
1-1/4" (32 mm)	50	60	10	110	1.25" x 54"
1-1/2" (40 mm)	200	30	5	90	2" x 27"
1-1/2" (40 mm)	50	60	10	170	2" x 50.5"
2" (50 mm)	100	30	10	329	3" x 44.5"
2" (50 mm)	25	60	10	150	2.5" x 31"
2" (50 mm)	200	60	5	300	3" x 40.5"

FIGURE 6.19 Air chambers must meet certain minimum standards. *(Courtesy of International Code Council, Inc. and International Plumbing Code 2000.)*

CHAPTER 7
SANITARY
DRAINAGE SYSTEMS

Drainage systems intimidate many people. When these people look at their code books, they see charts and math requirements that make them nervous. Their fear is largely unjustified. For the inexperienced, the fundamentals of building a suitable drainage system can appear formidable. With a basic understanding of plumbing, however, the process becomes much less complicated.

This chapter is going to take you, step-by-step, through the procedures of making a working drainage system. You are going to learn the criteria for sizing pipe. You will be shown which types of fittings can be used in various applications. During the process, you will be given instructions for the proper installation of a drainage system.

PIPE SIZING

Sizing pipe for a drainage system is not difficult. There are a few benchmark numbers you must know, but you don't have to memorize them. Your code book will have charts and tables that provide the benchmarks. All you must know is how to interpret and use the information provided.

The size of a drainage pipe is determined by using various factors, the first of which is the drainage load. This refers to the volume of drainage the pipe will be responsible for carrying. When you refer to your code book, you will find ratings that assign a fixture-unit value to various plumbing fixtures. For example, a residential toilet has a fixture-unit value of four. A bathtub's fixture-unit value is two.

By using the ratings given in your code book, you can quickly assess the drainage load for the system you are designing. Since plumbing fixtures require traps, you also must determine what size traps are required for particular fixtures. Again, you don't need a math degree to accomplish this task.

In fact, your code book will tell you what trap sizes are required for most common plumbing fixtures.

For example, by referring to the ratings in your code book, you will find that a bathtub requires an inch-and-a-half trap. A lavatory may be trapped with an inch-and-a-quarter trap. The list will go on to describe the trap needs for all common plumbing fixtures. Trap sizes will not be provided for toilets, since toilets have integral traps.

Inch	mm
1-1/4	32
1-1/2	40
2	50
2-1/2	65
3	80

Drainage Fixture Unit Values (DFU)

Plumbing Appliance, Appurtenance or Fixture	Min. Size Trap and Trap Arm[7]	Private	Public	Assembly[8]
Bathtub or Combination Bath/Shower	1-1/2"	2.0	2.0	
Bidet	1-1/4"	1.0		
Bidet	1-1/2"	2.0		
Clothes Washer, domestic, standpipe[5]	2"	3.0	3.0	3.0
Dental Unit, cuspidor	1-1/4"		1.0	1.0
Dishwasher, domestic, with independent drain	1-1/2"[2]	2.0	2.0	2.0
Drinking Fountain or Watercooler (per head)	1-1/4"	0.5	0.5	1.0
Food-waste-grinder, commercial	2"		3.0	3.0
Floor Drain, emergency	2"		0.0	0.0
Floor Drain (for additional sizes see Section 702)	2"	2.0	2.0	2.0
Shower single head trap	2"	2.0	2.0	2.0
Multi-head, each additional	2"	1.0	1.0	1.0
Lavatory, single	1-1/4"	1.0	1.0	1.0
Lavatory in sets of two or three	1-1/2"	2.0	2.0	2.0
Washfountain	1-1/2"		2.0	2.0
Washfountain	2"		3.0	3.0
Mobile Home, trap	3"	12.0		
Receptor, indirect waste[1,3]	1-1/2"		See footnote 1,3	
Receptor, indirect waste[1,4]	2"		See footnote 1,4	
Receptor, indirect waste[1]	3"		See footnote 1	
Sinks				
Bar	1-1/2"	1.0		
Bar	1-1/2"[2]		2.0	2.0
Clinical	3"		6.0	6.0
Commercial with food waste	1-1/2"[2]		3.0	3.0
Special Purpose	1-1/2"	2.0	3.0	3.0
Special Purpose	2"	3.0	4.0	4.0
Special Purpose	3"		6.0	6.0
Kitchen, domestic	1-1/2"[2]	2.0	2.0	
(with or without food-waste-grinder and/or dishwasher)				
Laundry	1-1/2"	2.0	2.0	2.0
(with or without discharge from a clothes washer)				
Service or Mop Basin	2"		3.0	3.0
Service or Mop Basin	3"		3.0	3.0
Service, flushing rim	3"		6.0	6.0
Wash, each set of faucets			2.0	2.0
Urinal, integral trap 1.0 GPF[2]	2"	2.0	2.0	5.0
Urinal, integral trap greater than 1.0 GPF	2"	2.0	2.0	6.0
Urinal, exposed trap	1-1/2"[2]	2.0	2.0	5.0
Water Closet, 1.6 GPF Gravity Tank[6]	3"	3.0	4.0	6.0
Water Closet, 1.6 GPF Flushometer Tank[6]	3"	3.0	4.0	6.0
Water Closet, 1.6 GPF Flushometer Valve[6]	3"	3.0	4.0	6.0
Water Closet, greater than 1.6 GPF Gravity Tank[6]	3"	4.0	6.0	8.0
Water Closet, greater than 1.6 GPF Flushometer Valve[6]	3"	4.0	6.0	8.0

1. Indirect waste receptors shall be sized based on the total drainage capacity of the fixtures that drain therein to, in accordance with Table 7-4.
2. Provide a 2" (51 mm) minimum drain.
3. For refrigerators, coffee urns, water stations, and similar low demands.
4. For commercial sinks, dishwashers, and similar moderate or heavy demands.
5. Buildings having a clothes washing area with clothes washers in a battery of three (3) or more clothes washers shall be rated at six (6) fixture units each for purposes of sizing common horizontal and vertical drainage piping.
6. Water closets shall be computed as six (6) fixture units when determining septic tank sizes based on Appendix K of this Code.
7. Trap sizes shall not be increased to the point where the fixture discharge may be inadequate to maintain their self-scouring properties.
8. Assembly [Public Use (See Table 4-1)].

FIGURE 7.1 The chart shows drainage fixture unit values. (*Reprinted from the* 2000 *Uniform Plumbing Code (UPC)* with the permission of the International Association of Plumbing and Mechanical Officials (IAPMO).)

When necessary, you can determine a fixture's drainage unit value by the size of the fixture's trap. An inch-and-a-quarter trap, the smallest trap allowed, will carry a fixture-unit rating of one. An inch-and-a-half trap will have a fixture-unit rating of two. A two-inch trap will have a rating of three fixture units. A three-inch trap will have a fixture-unit rating of five, and a four-inch trap will have a fixture-unit rating of six. This information can be found in your code book and may be applied for a fixture not specifically listed with a rating in your code book.

Determining the fixture-unit value of a pump does require a little math, but it's simple. By taking the flow rate, in gallons per minute (gpm), assign two fixture units for every

DRAINAGE FIXTURE UNITS FOR FIXTURES AND GROUPS

FIXTURE TYPE	DRAINAGE FIXTURE UNIT VALUE AS LOAD FACTORS	MINIMUM SIZE OF TRAP (inches)
Automatic clothes washers, commercial[a]	3	2
Automatic clothes washers, residential	2	2
Bathroom group as defined in Section 202 (1.6 gpf water closet)[f]	5	—
Bathroom group as defined in Section 202 (water closet flushing greater than 1.6 gpf)[f]	6	—
Bathtub[b] (with or without overhead shower or whirlpool attachments)	2	$1^{1}/_{2}$
Bidet	1	$1^{1}/_{4}$
Combination sink and tray	2	$1^{1}/_{2}$
Dental lavatory	1	$1^{1}/_{4}$
Dental unit or cuspidor	1	$1^{1}/_{4}$
Dishwashing machine,[c] domestic	2	$1^{1}/_{2}$
Drinking fountain	$^{1}/_{2}$	$1^{1}/_{4}$
Emergency floor drain	0	2
Floor drains	2	2
Kitchen sink, domestic	2	$1^{1}/_{2}$
Kitchen sink, domestic with food waste grinder and/or dishwasher	2	$1^{1}/_{2}$
Laundry tray (1 or 2 compartments)	2	$1^{1}/_{2}$
Lavatory	1	$1^{1}/_{4}$
Shower	2	$1^{1}/_{2}$
Sink	2	$1^{1}/_{2}$
Urinal	4	Footnote d
Urinal, 1 gallon per flush or less	2[e]	Footnote d
Wash sink (circular or multiple) each set of faucets	2	$1^{1}/_{2}$
Water closet, flushometer tank, public or private	4[e]	Footnote d
Water closet, private (1.6 gpf)	3[e]	Footnote d
Water closet, private (flushing greater than 1.6 gpf)	4[e]	Footnote d
Water closet, public (1.6 gpf)	4[e]	Footnote d
Water closet, public (flushing greater than 1.6 gpf)	6[e]	Footnote d

For SI: 1 inch = 25.4 mm, 1 gallon = 3.785 L.
a. For traps larger than 3 inches, use Table 709.2.
b. A showerhead over a bathtub or whirlpool bathtub attachments does not increase the drainage fixture unit value.
c. See Sections 709.2 through 709.4 for methods of computing unit value of fixtures not listed in Table 709.1 or for rating of devices with intermittent flows.
d. Trap size shall be consistent with the fixture outlet size.
e. For the purpose of computing loads on building drains and sewers, water closets or urinals shall not be rated at a lower drainage fixture unit unless the lower values are confirmed by testing.
f. For fixtures added to a dwelling unit bathroom group, add the DFU value of those additional fixtures to the bathroom group fixture count.

FIGURE 7.2 Above are common drainage fixture units for frequently used plumbing jobs. *(Courtesy of International Code Council, Inc. and International Plumbing Code 2000.)*

gpm of flow. For example, a pump with a flow rate of thirty gpm would have a fixture-unit rating of sixty. Some code jurisdictions are more generous. For example, you might find that your local code will allow one fixture-unit to be assigned for every seven and a half gpm. With the same pump, producing thirty gpm, the liberal code fixture-unit rating would be four. That's quite a difference from the ratings in more conservative code regions.

Other considerations when sizing drainage pipe is the type of drain you are sizing and the amount of fall the pipe will have. For example, the sizing for a sewer will be done a little differently from the sizing for a vertical stack. A pipe with a quarter-inch fall will be rated differently from the same pipe with a eighth-inch fall.

SIZING BUILDING DRAINS AND SEWERS

Building drains and sewers use the same criteria in determining the proper pipe size. The two components you must know to size these types of pipes are the total number of drainage fixture units entering the pipe and the amount of fall placed on the pipe. The

DRAINAGE FIXTURE UNITS FOR FIXTURE DRAINS OR TRAPS

FIXTURE DRAIN OR TRAP SIZE (inches)	DRAINAGE FIXTURE UNIT VALUE
$1\frac{1}{4}$	1
$1\frac{1}{2}$	2
2	3
$2\frac{1}{2}$	4
3	5
4	6

For SI: 1 inch = 25.4 mm.

FIGURE 7.3 The illustration shows sizes for fixture units. *(Courtesy of International Code Council, Inc. and International Plumbing Code 2000.)*

MINIMUM CAPACITY OF SEWAGE PUMP OR SEWAGE EJECTOR

DIAMETER OF THE DISCHARGE PIPE (inches)	CAPACITY OF PUMP OR EJECTOR (gpm)
2	21
$2\frac{1}{2}$	30
3	46

For SI: 1 inch = 25.4 mm, 1 gpm = 3.785 L/m.

FIGURE 7.4 Discharge pipe diameter is compared to the capacity of a pump or ejector. *(Courtesy of International Code Council, Inc. and International Plumbing Code 2000.)*

amount of fall is based on how much the pipe drops in each foot it travels. A normal grade is generally one-quarter inch to the foot, but the fall could be more or less. Drainage fixture-unit values for continuous and semicontinuous flow into a drainage system is to be computed on the basis that 1 gpm of flow is equal to 2 fixture units.

When you refer to your code book you will find information, probably a table, to aid you is sizing building drains and sewers. Let's take a look at how a building drain for a typical house would be sized.

SIZING EXAMPLE

Our sample house has two and one-half bathrooms, a kitchen, and a laundry room. To size the building drain for this house, we must determine the total fixture-unit load that may be placed on the building drain. To do this, we start by listing all of the plumbing fixtures producing a drainage load. In this house we have the following fixtures:

- One bathtub
- One shower
- Three toilet
- Three lavatories
- One kitchen sink
- One dishwasher

Minimum Horizontal Distance Required From Building Sewer

Buildings or structures[1]	2 feet (610 mm)
Property line adjoining private property	Clear[2]
Water supply wells	50 feet[3] (15240 mm)
Streams	50 feet (15240 mm)
On-site domestic water service line	1 foot[4] (305 mm)
Public water main	10 feet[5,6] (3048 mm)

Note:

1. Including porches and steps, whether covered or uncovered, breezeways, roofed porte-cocheres, roofed patios, carports, covered walks, covered driveways, and similar structures or appurtenances.

2. See also Section 313.3.

3. All drainage piping shall clear domestic water supply wells by at least fifty (50) feet (15240 mm). This distance may be reduced to not less than twenty-five (25) feet (7620 mm) when the drainage piping is constructed of materials approved for use within a building.

4. See Section 720.0.

5. For parallel construction.

6. For crossings, approval by the Health Department or Administrative Authority shall be required.

FIGURE 7.5 Sewers are required to be a minimum distance away from buildings, water supplies and property lines. (*Reprinted from the 2000 Uniform Plumbing Code (UPC) with the permission of the International Association of Plumbing and Mechanical Officials (IAPMO).*)

BUILDING DRAINS AND SEWERS

DIAMETER OF PIPE (inches)	MAXIMUM NUMBER OF DRAINAGE FIXTURE UNITS CONNECTED TO ANY PORTION OF THE BUILDING DRAIN OR THE BUILDING SEWER, INCLUDING BRANCHES OF THE BUILDING DRAIN[a]			
	Slope per foot			
	1/16 inch	1/8 inch	1/4 inch	1/2 inch
1 1/4	—	—	1	1
1 1/2	—	—	3	3
2	—	—	21	26
2 1/2	—	—	24	31
3	—	36	42	50
4	—	180	216	250
5	—	390	480	575
6	—	700	840	1,000
8	1,400	1,600	1,920	2,300
10	2,500	2,900	3,500	4,200
12	3,900	4,600	5,600	6,700
15	7,000	8,300	10,000	12,000

For SI: 1 inch = 25.4 mm, 1 inch per foot = 0.0833 mm/m.

a. The minimum size of any building drain serving a water closet shall be 3 inches.

FIGURE 7.6 There is a limit to the number of drainage fixture units connected with any part of a building drain or sewer. (*Courtesy of International Code Council, Inc. and International Plumbing Code 2000.*)

HORIZONTAL FIXTURE BRANCHES AND STACKS[a]

DIAMETER OF PIPE (inches)	MAXIMUM NUMBER OF DRAINAGE FIXTURE UNITS (dfu)			
	Total for a horizontal branch	Stacks[b]		
		Total discharge into one branch interval	Total for stack of three branch intervals or less	Total for stack greater than three branch intervals
1$\frac{1}{2}$	3	2	4	8
2	6	6	10	24
2$\frac{1}{2}$	12	9	20	42
3	20	20	48	72
4	160	90	240	500
5	360	200	540	1,100
6	620	350	960	1,900
8	1,400	600	2,200	3,600
10	2,500	1,000	3,800	5,600
12	2,900	1,500	6,000	8,400
15	7,000	Footnote c	Footnote c	Footnote c

For SI: 1 inch = 25.4 mm.

a. Does not include branches of the building drain. Refer to Table 710.1(1).

b. Stacks shall be sized based on the total accumulated connected load at each story or branch interval. As the total accumulated connected load decreases, stacks are permitted to be reduced in size. Stack diameters shall not be reduced to less than one-half of the diameter of the largest stack size required.

c. Sizing load based on design criteria.

FIGURE 7.7 There are limits on the number of horizontal fixture branches and stacks linked to drainage fixture units. (Courtesy of International Code Council, Inc. and International Plumbing Code 2000.)

7.7

BUILDING SEWER PIPE

MATERIAL	STANDARD
Acrylonitrile butadiene styrene (ABS) plastic pipe	ASTM D 2661; ASTM D 2751; ASTM F 628
Asbestos-cement pipe	ASTM C 428
Cast-iron pipe	ASTM A 74; ASTM A 888; CISPI 301
Coextruded composite ABS DWV sch 40 IPS pipe (solid)	ASTM F 1488
Coextruded composite ABS DWV sch 40 IPS pipe (cellular core)	ASTM F 1488
Coextruded composite PVC DWV sch 40 IPS pipe (solid)	ASTM F 1488
Coextruded composite PVC DWV sch 40 IPS pipe (cellular core)	ASTM F 1488
Coextruded composite PVC IPS DR - PS DWV, PS140, PS200	ASTM F 1488
Coextruded composite ABS sewer and drain DR - PS in PS35, PS50, PS100, PS140, PS200	ASTM F 1488
Coextruded composite PVC sewer and rain DR - PS in PS35, PS50, PS100, PS140, PS200	ASTM F 1488
Concrete pipe	ASTM C 14; ASTM C 76; CSA A257.1; CSA CAN/CSA A257.2
Copper or copper-alloy tubing (Type K or L)	ASTM B 75; ASTM B 88; ASTM B 251
Polyvinyl chloride (PVC) plastic pipe (Type DWV, SDR26, SDR35, SDR41, PS50 or PS100)	ASTM D 2665; ASTM D 2949; ASTM D 3034; ASTM F 891; CSA B182.2; CSA CAN/CSA-B182.4
Stainless steel drainage systems, Type 316L	ASME/ANSI A112.3.1
Vitrified clay pipe	ASTM C 4; ASTM C 700

FIGURE 7.8 There are many kinds of building sewer pipe. *(Courtesy of International Code Council, Inc. and International Plumbing Code 2000.)*

UNDERGROUND BUILDING DRAINAGE AND VENT PIPE

MATERIAL	STANDARD
Acrylonitrile butadiene styrene (ABS) plastic pipe	ASTM D 2661; ASTM F 628; CSA B181.1
Asbestos-cement pipe	ASTM C 428
Cast-iron pipe	ASTM A 74; CISPI 301; ASTM A 888
Coextruded composite ABS DWV sch 40 IPS pipe (solid)	ASTM F 1488
Coextruded composite ABS DWV sch 40 IPS pipe (cellular core)	ASTM F 1488
Coextruded composite PVC DWV sch 40 IPS pipe (solid)	ASTM F 1488
Coextruded composite PVC DWV sch 40 IPS pipe (cellular core)	ASTM F 1488
Coextruded composite PVC IPS - DR, PS140, PS200 DWV	ASTM F 1488
Copper or copper alloy tubing (Type K, L, M or DWV)	ASTM B 75; ASTM B 88; ASTM B 251; ASTM B 306
Polyolefin pipe	CSA CAN/CSA-B181.2
Polyvinyl chloride (PVC) plastic pipe (Type DWV)	ASTM D 2665; ASTM D 2949; ASTM F 891; CSA CAN/CSA-B181.2
Stainless steel drainage systems, Type 316L	ASME/ANSI A112.3.1

FIGURE 7.9 This is drainage and vent pipe material for use underground. *(Courtesy of International Code Council, Inc. and International Plumbing Code 2000.)*

• One clothes washer

• One laundry tub

By using the chart in my local code book, I can determine the number of drainage fixture units assigned to each of these fixtures. When I add all the fixture units, the total load of 28 fixture units established. It is always best to allow a little extra in your fixture-unit load, so your pipe will be in no danger of becoming overloaded. The next step is to look at the chart in the code book to determine the sizing of our building drain.

Our building drain will be installed with a quarter-inch fall. By looking at the chart in the code book, I see that we can use a 3-inch pipe for our building drain, based on the number of fixture units, but notice the footnote below the chart. The note indicates that a 3-inch pipe may not carry the discharge of more than two toilets, and our test house has three toilets. This means we will have to move up to a 4-inch pipe.

Suppose our test house only had two toilets; what would the outcome be then? If we eliminate one of the toilets, our fixture load drops to 24. According to the table, we could use a 2½-inch pipe, but we know our building drain must be at least a 3-inch pipe to connect to the toilets. A fixture's drain may enter a pipe the same size as the fixture drain or a pipe that is larger, but it may never be reduced to a smaller size, except with a 4 x 3-inch closet bend.

So, with two toilets, our sample house could have a building drain and sewer with a 3-inch diameter. But, should we run a 3-inch pipe or a 4-inch pipe? In a highly competitive bidding situation, 3-inch pipe probably would win the coin toss. It would be less expensive to install a 3-inch drain and you would be more likely to win the bid on the job. When feasible, however, it would be better to use a four-inch drain. This allows the homeowner to add another toilet in the future. If you install a 3-inch sewer, the homeowner would not be able to add a toilet without replacing the sewer with 4-inch pipe.

HORIZONTAL BRANCHES

Horizontal branches are the pipes branching off from a stack to accept the discharge from fixture drains. These horizontal branches usually leave the stack as a horizontal pipe, but they may turn to a vertical position, while retaining the name of horizontal branch.

The procedure for sizing a horizontal branch is similar to the one used to size a building drain or sewer, but the ratings are different. Your code book will contain the benchmarks for your sizing efforts, but let me give you some examples.

The number of fixture units allowed on a horizontal branch is determined by pipe size and pitch. All the following examples are based on a pitch of one-quarter inch to the foot. A 2-inch pipe can accommodate up to six fixture units in most code regions. Some jurisdictions allow up to eight fixture units. A 3-inch pipe can handle 20 fixture units, but not more than two toilets. A 3-inch pipe is allowed up to 35 units and up to three toilets in some jurisdictions. An inch-and-a-half pipe will carry either two or three fixture units, depending upon your local code requirements. When the additional fixture unit is allowed, it may not be from sinks, dishwashers, or urinals. A 4-inch pipe will take up to 160 fixture units in most code areas. You might find a jurisdiction that will allow up to 216 units.

STACK SIZING

Stack sizing is not too different from the other sizing exercises we have studied. When you size a stack, you must base your decision on the total number of fixture units carried by the stack and the amount of discharge into branch intervals. This may sound complicated, but it isn't.

Tables in your local code book can help you with sizing pipes. You will notice that there are three columns. The first is for pipe size, the second represents the discharge of a branch interval, and the last column shows the ratings for the total fixture-unit load on a stack. This table is based on a stack with no more than three branch intervals.

Sizing the stack requires you to first determine the fixture load entering the stack at each branch interval. Let me give you an example of how this type of sizing works. In our example, we will size a stack that has two branch intervals. The lower branch has a half-bath and a kitchen on it. Using the ratings from common code regions, the total fixture-unit count for this branch is six. This is determined by the table providing fixture unit ratings for various fixtures.

The second stack has a full bathroom group on it. The total fixture unit count on this branch is six, if you use a bathroom group rating, or seven, if you count each fixture

Maximum Unit Loading and Maximum Length of Drainage and Vent Piping

Size of Pipe, inches (mm)	1-1/4 (32)	1-1/2 (40)	2 (50)	2-1/2 (65)	3 (80)	4 (100)	5 (125)	6 (150)	8 (200)	10 (250)	12 (300)
Maximum Units											
Drainage Piping[1]											
Vertical	1	2[2]	16[3]	32[3]	48[4]	256	600	1380	3600	5600	8400
Horizontal	1	1	8[3]	14[3]	35[4]	216[5]	428[5]	720[5]	2640[5]	4680[5]	8200[5]
Maximum Length											
Drainage Piping											
Vertical, feet (m)	45 (14)	65 (20)	85 (26)	148 (45)	212 (65)	300 (91)	390 (119)	510 (155)	750 (228)		
Horizontal (Unlimited)											
Vent Piping (See note)											
Horizontal and Vertical											
Maximum Units	1	8[3]	24	48	84	256	600	1380	3600		
Maximum Lengths, feet (m)	45 (14)	60 (18)	120 (37)	180 (55)	212 (65)	300 (91)	390 (119)	510 (155)	750 (228)		

1 Excluding trap arm.
2 Except sinks, urinals and dishwashers.
3 Except six-unit traps or water closets.
4 Only four (4) water closets or six-unit traps allowed on any vertical pipe or stack; and not to exceed three (3) water closets or six-unit traps on any horizontal branch or drain.
5 Based on one-fourth (1/4) inch per foot (20.9 mm/m) slope. For one-eighth (1/8) inch per foot (10.4 mm/m) slope, multiply horizontal fixture units by a factor of 0.8.

Note: The diameter of an individual vent shall not be less than one and one-fourth (1-1/4) inches (31.8 mm) nor less than one-half (1/2) the diameter of the drain to which it is connected. Fixture unit load values for drainage and vent piping shall be computed from Tables 7-3 and 7-4. Not to exceed one-third (1/3) of the total permitted length of any vent may be installed in a horizontal position. When vents are increased one (1) pipe size for their entire length, the maximum length limitations specified in this table do not apply.

FIGURE 7.10 The chart shows the relationship between maximum unit loading, and drainage and vent piping length. (Reprinted from the 2000 Uniform Plumbing Code (UPC) with the permission of the International Association of Plumbing and Mechanical Officials (IAPMO).)

SLOPE OF HORIZONTAL DRAINAGE PIPE

SIZE (inches)	MINIMUM SLOPE (inch per foot)
$2^{1}/_{2}$ or less	$^{1}/_{4}$
3 to 6	$^{1}/_{8}$
8 or larger	$^{1}/_{16}$

For SI: 1 inch = 25.4 mm, 1 inch per foot = 0.0833 mm/m.

FIGURE 7.11 This points out the size and slope of drainage pipe. *(Courtesy of International Code Council, Inc. and International Plumbing Code 2000.)*

individually. I would use the larger of the two numbers, for a total of seven.

When you look at the table in your code book, you will look at the horizontal listings for a 3-inch pipe. You know the stack must have a minimum size of 3 inches to accommodate the toilets. As you look across the table, you will see that each 3-inch branch may carry up to 20 fixture units. Well, your first branch has six fixture units and the second branch has seven fixture units, so both branches are within their limits.

When you combine the total fixture units from both branches you have a total of thirteen fixture units. Continuing to look across the table, you see that the stack can accommodate up to 48 fixture units. Obviously, a 3-inch stack is adequate for your needs. If the fixture-unit loads had exceeded the numbers in either of the columns, the pipe size would have had to have be increased.

When sizing a stack, it is possible that the developed length of the stack will be composed of different sizes of pipe. For example, at the top of the stack, the pipe size may be 3 inches, but at the bottom of the stack the pipe size may be 4 inches. This is because as you get to the lower portion of the stack, the total fixture units placed on the stack are greater.

Remember to check your local code requirements, since they may be different from the ones with which I or other plumbers work.

PIPE INSTALLATIONS

Once the pipe is sized properly, it is ready for installation. There are a few regulations pertaining to pipe installation of which you need to be aware.

Grading Your Pipe

You must install horizontal drainage piping so that it falls toward the waste-disposal site. A typical grade for drainage pipe is ¼ inch of fall per foot. This means the lower end of a 20-foot piece of pipe would be 5 inches lower than the upper end when properly installed. While the quarter-inch-to-the-foot grade is typical, it is not the only acceptable grade for all pipes.

If you are working with pipe that has a diameter of 2½ inches, or less, the minimum grade for the pipe is ¼ inch to the foot. Pipes with diameters between 3 and 6 inches are allowed a minimum grade of ⅛ inch to the foot. Some code zones require that special permission be granted prior to installing pipe with a one-eighth-of-an-inch-to-the-foot grade. Pipes with diameters of 8 inches or more may be allowed to be installed with an acceptable grade of ¹⁄₁₆ inch to the foot.

PIPE FITTINGS

MATERIAL	STANDARD
Acrylonitrile butadiene styrene (ABS) plastic	ASTM D 3311; CSA B181.1; ASTM D 2661
Cast iron	ASME B16.4; ASME B16.12; ASTM A 74; ASTM A 888; CISPI 301
Coextruded composite ABS DWV sch 40 IPS pipe (solid or cellular core)	ASTM D 2661; ASTM D 3311; ASTM F 628
Coextruded composite PVC DWV sch 40 pipe IPS-DR, PS140, PS200 (solid or cellular core)	ASTM D 2665; ASTM D 3311; ASTM F 891
Coextruded composite ABS sewer and drain DR-PS in PS35, PS50, PS100, PS140, PS200	ASTM D 2751
Coextruded composite PVC sewer and drain DR-PS in PS35, PS50, PS100, PS140, PS200	ASTM D 3034
Copper or copper alloy	ASME B16.15; ASME B16.18; ASME B16.22; ASME B16.23; ASME B16.26; ASME B16.29; ASME B16.32
Glass	ASTM C 1053
Gray iron and ductile iron	AWWA C110
Malleable iron	ASME B16.3
Polyvinyl chloride (PVC) plastic	ASTM D 3311; ASTM D 2665
Stainless steel drainage systems, Types 304 and 316L	ASME/ANSI A112.3.1
Steel	ASME B16.9; ASME B16.11; ASME B16.28

FIGURE 7.12 Above are the standard units for pipe-fitting material. *(Courtesy of International Code Council, Inc. and International Plumbing Code 2000.)*

Joints

There are a number of requirements for making joints between different types of pipes and fittings. Of course, all connections must be made in compliance with the local plumbing code. Some aspects of joining pipes are very specific, and those are what we are going to talk about here.

Mechanical joints on drainage pipes must be made with elastomeric seal. They may be used only on underground piping and must comply with manufacturer's recommendations. Solvent-weld joints are required to be made with pipe where the ends of the pipe surfaces are clean and free from moisture. Joints must be made while the cement is wet. These joints can be made above or below grade.

Schedule 80, or heavier, pipe can be threaded with dies specifically designed for plastic piping. Some approved thread lubricant or tape must be applied to the male threads only.

Asbestos-cement pipe is joined by a sleeve coupling of the same composition as the pipe, sealed with an elastomeric ring.

Brazed joints must be made on clean surfaces with an approved flux. The filler used for brazing must be an approved material. The same is true for welded joints.

Caulked joints are not used much anymore, but they are still used. Joints for hub and spigot pipe must be firmly packed with oakum or hemp. Molten lead is poured in one operation to a depth of not less than one inch. The lead must not recede more than 0.125 inch below the rim of the hub and shall be caulked tight. The lead joint cannot be painted, varnished, or otherwise coated until after the joint is tested and approved.

When a lead joint is made, the joint is to be made in one pouring and caulked tight. Acid-resistant rope and acid-proof cement are permitted.

Compression gasket joints must be compressed when the pipe is fully inserted. Joints between concrete pipe and fittings and mechanical joints on drainage pipe are to be made with elastomeric seal. It is not permissible to install mechanical joints in above-grade systems, unless otherwise approved. When making joints with stainless steel drainage systems and other systems a mechanical joint must be used. An O ring used with a stainless steel drainage system must be used with an elastomeric seal.

Supporting Your Pipe

How you support your pipes also is regulated by the plumbing code. There are requirements for the type of materials you may use and how they may be used. Let's see what they are.

One concern with the type of hangers is their compatibility with the pipe they are supporting. You must use a hanger that will not have a detrimental effect on your piping. For example, you may not use galvanized strap hanger to support copper pipe. Generally, the hangers used to support a pipe should be made from the same material as the pipe being supported. For example, copper pipe should be hung with copper hangers. This eliminates the risk of a corrosive action between two different types of materials. A plastic, or plastic-coated hanger may be used with all types of pipe. The exception to this rule might be when the piping is carrying a liquid with a temperature that might affect or melt the plastic hanger.

The hangers used to support pipe must be capable of supporting the pipe at all times. The hanger must be attached to the pipe and to the member holding the hanger in a satisfactory manner. For example, it would not be acceptable to wrap a piece of wire around a pipe and then wrap the wire around the bridging between two floor joists. Hangers should be securely attached to the member supporting it. For example, a hanger should be attached to the pipe and then nailed to a floor joist. The nails used to hold a hanger in place should be made of the same material as the hanger, if corrosive action is a possibility.

Both horizontal and vertical pipes require support. The intervals between supports will vary, depending upon the type of pipe being used and whether it is installed vertically or horizontally. The following examples will show you how often you must support the various types of pipes when they are hung horizontally, these examples are the maximum distances allowed between supports for zone 3:

- ABS—every 4'
- Cast iron—every 5'
- Galvanized—every 1'
- PVC—every 4'
- DWV copper—every 10'

When these same types of pipes are installed vertically, they must be supported at no less than the following intervals:

- ABS—every 4'
- Cast iron—every 15'
- Galvanized—every 15'
- PVC—every 4'
- DWV copper—every 10'

When installing cast-iron stacks, the base of each stack must be supported because of the weight of cast iron pipe. Pipe with flexible couplings, bands, or unions, must be installed and supported to prevent these flexible connections from moving. In pipes larger than 4 inches, all flexible couplings must be supported to prevent the force of the pipe's flow from loosening the connection at changes in direction.

Caulking Ferrules			
Pipe size (inches)	Inside diameter (inches)	Length (inches)	Minimum weight each Lb. Oz.
2	2-1/4	4-1/2	1 0
3	3-1/4	4-1/2	1 12
4	4-1/4	4-1/2	2 8

Caulking Ferrules (Metric)			
Pipe size (mm)	Inside diameter (mm)	Length (mm)	Minimum weight each (kg)
50	57	114	0.454
80	83	114	0.790
100	108	114	1.132

Soldering Bushings			
Pipe size (inches)	Minimum weight each Lb. Oz.	Pipe size (inches)	Minimum weight each Lb. Oz.
1-1/4	0 6	2-1/2	1 6
1-1/2	0 8	3	2 0
2	0 14	4	3 8

Soldering Bushings (Metric)			
Pipe size (mm)	Minimum weight each (kg)	Pipe size (mm)	Minimum weight each (kg)
32	0.168	65	0.622
40	0.224	80	0.908
50	0.392	100	1.586

FIGURE 7.13 Above are descriptions of various kinds of ferrules and bushings. *(Reprinted from the 2000 Uniform Plumbing Code (UPC) with the permission of the International Association of Plumbing and Mechanical Officials (IAPMO).)*

PIPE-SIZE REDUCTION

As mentioned earlier, you may not reduce the size of a drainage pipe as it heads for the waste-disposal site. The pipe size may be enlarged, but it may not be reduced. There is one exception to this rule. Reducing closet bends, such as a 4 x 3 closet bend, are allowed.

OTHER FACTS TO REMEMBER

A drainage pipe installed underground must have a minimum diameter of 2 inches. When you are installing a horizontal branch fitting near the base of a stack, keep the branch fitting away from the point where the vertical stack turns to a horizontal run. The branch fitting should be installed at least 30 inches back on a three-inch pipe and 40 inches back on a 4-inch pipe. By multiplying the size of the pipe by a factor of ten, you can determine how far back the branch fitting should be installed.

All drainage piping must be protected from the effects of flooding. When leaving a stub of pipe to connect with fixtures planned for the future, the stub must not be more than

2 feet in length and it must be capped. Some exceptions are possible on the prescribed length of a pipe stub. If you have a need for a longer stub, consult your local code officer. Clean-out extensions are not affected by the two-foot rule.

Multiple buildings situated on the same lot may not share a common building sewer that connects to a public sewer. Horizontal branches that connect to the bases of stacks must connect at a point not less than ten pipe diameters downstream from the stack. Unless otherwise determined, horizontal branches must connect to horizontal stack offsets at a point located not less than ten pipe diameters downstream from the upper stack.

The offset must be vented when horizontal branches connect to stacks within 2 feet above or below a vertical stack offset when the offset is located more than four branch intervals below the top of a stack. Vents for vertical offsets are not required when the stack and its offset is sized as a building drain. Horizontal branches may not connect to a horizontal stack offset or within 2 feet above or below the offset when the offset is located more than four branch intervals below the top of the stack.

A vent is required for a stack with a horizontal offset that is located more than four branch intervals below the top of a stack. Where a vertical offset occurs in a soil or waste stack below the lowest horizontal branch, change in the diameter of the stack because of the offset is not required. If a horizontal offset occurs in a soil or waste stack below the lowest horizontal branch, the required diameter of the offset and the stack below it must be sized as a building drain.

FITTINGS

Fittings are also a part of the drainage system. Knowing when, where, and how to use the proper fittings is mandatory to the installation of a drainage system. Fittings are used to make branches and to change direction. The use of fittings to change direction is where we will start. When you wish to change direction with a pipe, you may have it change from a horizontal run to a vertical rise. You may be going from a vertical position to a horizontal one, or you might only want to offset the pipe in a horizontal run. Each of these three categories require the use of different fittings. Let's take each circumstance and examine the fittings allowed.

OFFSETS IN HORIZONTAL PIPING

When you want to change the direction of a horizontal pipe, you must use fittings approved for that purpose. You have six choices to choose from in zone 3. Those choices are:

- Sixteenth bend
- Sixth bend
- Combination wye and eighth bend
- Eighth bend
- Long-sweep fittings
- Wye

Any of these fittings generally are approved for changing direction with horizontal piping, but as always, it is best to check with your local code officer for current regulations.

GOING FROM HORIZONTAL TO VERTICAL

You have a wider range of choice in selecting a fitting for going from a horizontal position to a vertical position. Nine possible candidates are available for this type of change in

ABOVE-GROUND DRAINAGE AND VENT PIPE

MATERIAL	STANDARD
Acrylonitrile butadiene styrene (ABS) plastic pipe	ASTM D 2661; ASTM F 628; CSA B181.1
Brass pipe	ASTM B 43
Cast-iron pipe	ASTM A 74; CISPI 301; ASTM A 888
Coextruded composite ABS DWV sch 40 IPS pipe (solid)	ASTM F 1488
Coextruded composite ABS DWV sch 40 IPS pipe (cellular core)	ASTM F 1488
Coextruded composite PVC DWV sch 40 IPS pipe (solid)	ASTM F 1488
Coextruded composite PVC DWV sch 40 IPS pipe (cellular core)	ASTM F 1488
Coextruded composite PVC IPS - DR, PS140, PS200 DWV	ASTM F 1488
Copper or copper-alloy pipe	ASTM B 42; ASTM B 302
Copper or copper-alloy tubing (Type K, L, M or DWV)	ASTM B 75; ASTM B 88; ASTM B 251; ASTM B 306
Galvanized steel pipe	ASTM A 53
Glass pipe	ASTM C 1053
Polyolefin pipe	CSA CAN/CSA-B181.2
Polyvinyl chloride (PVC) plastic pipe (Type DWV)	ASTM D 2665; ASTM D 2949; ASTM F 891; CSA CAN/CSA-B181.2; ASTM F 1488
Stainless steel drainage systems, types 304 and 316L	ASME/ANSI A112.3.1

FIGURE 7.14 This explains the material and standards for different types of above-ground drainage and vent pipe. *(Courtesy of International Code Council, Inc. and International Plumbing Code 2000.)*

direction, when working in zone three. The choices are:

- Sixteenth bend
- Sixth bend
- Combination wye and eighth bend
- Quarter bend
- Sanitary tee

- Eighth bend
- Long-sweep fittings
- Wye
- Short-sweep fittings

You may not use a double sanitary tee in a back-to-back situation if the fixtures being served are of a blowout or pump type. Double sanitary tees must not be used to receive the waste of back-to-back water closets. For example, you could not use a double sanitary tee to receive the discharge of two washing machines if they were back to back. The sanitary tee's throat is not deep enough to keep drainage from feeding back and forth between the fittings. In a case like this, use a double combination wye and eighth bend. The combination fitting has a much longer throat and will prohibit waste water from transferring across the fitting, to the other fixture.

VERTICAL TO HORIZONTAL CHANGES IN DIRECTION

Seven fittings are allowed to change direction from vertical to horizontal. These fittings are:

- Sixteenth bend
- Eighth bend
- Sixth bend
- Long-sweep fittings
- Combination wye and eighth bend
- Wye
- Short sweep fittings that are 3-inch or larger

Some code zones prohibit a fixture outlet connection within eight feet of a vertical to horizontal change in direction of a stack, if the stack serves a suds-producing fixture. A suds-producing fixture could be a laundry fixture, a dishwasher, a bathing unit, or a kitchen sink. This rule does not apply to single-family homes and stacks in buildings with less than three stories.

CHAPTER 8
INDIRECT
AND SPECIAL WASTES

Indirect-waste requirements can pertain to several types of plumbing fixtures and equipment. These might include a clothes washer drain, a condensate line, a sink drain, or the blowoff pipe from a relief valve, just to name a few. These indirect wastes are piped in this manner to prevent the possibility of contaminated matter backing up the drain, into a potable water or food source, among other things.

Most indirect-waste receptors are trapped. If the drain from the fixture is more than two feet long, the indirect-waste receptor must be trapped. This trap rule, however, applies to fixtures such as sinks, not to an item such as a blowoff pipe from a relief valve. In some areas, a drain that is less than five feet long does not have to be trapped. If a floor drain is located within an area subject to freezing, the waste line serving the floor drain shall not be trapped and shall indirectly discharge into a waste receptor located outside of the area subject to freezing.

The safest method of indirect-waste disposal is using an air gap. When an air gap is used, the drain from the fixture terminates above the indirect-waste receptor, with open-air space between the waste receptor and the drain. This prevents any backup or backsiphonage.

Some fixtures, depending on local code requirements, may be piped with an air break, rather than an air gap. With an air break, the drain may extend below the flood-level rim and terminate just above the trap's seal. The risk to an air break is the possibility of a backup. Since the drain is run below the flood-level rim of the waste receptor, the waste receptor could overflow and back up into the drain. This could create contamination, but in cases where contamination is likely, an air gap will be required. Check with your local code office before using an air break.

Standpipes, such as those used for washing machines, are a form of indirect-waste receptors. A standpipe used for this purpose in most jurisdictions must extend at least 18 inches above the trap's seal, but they may not extend more than 30 inches above the trap seal. If a clear-water waste receptor is located in a floor, some codes require the lip of the receptor to extend at least 2 inches above the floor. This eliminates the waste receptor from being used as a floor drain.

Choosing the proper size for a waste receptor generally is based on the receptor's ability to handle the discharge from a drain, without excessive splashing. If you are concerned about sizing a particular waste receptor, talk with your local code officer for a ruling.

Buildings used for food preparation, storage, and similar activities are required to have their fixtures and equipment discharge drainage through an air gap. Dishwashers and open culinary sinks sometimes are excepted. Some code regions require that a discharge pipe terminate at least 2 inches above the receptor. Zone 1 requires the distance to be a minimum of 1 inch. You may find that your local code requires the air gap distance to be a minr imum of twice the size of the pipe discharging the waste. For example, a ½-inch discharge pipe would require a 1-inch air gap.

Floor drains within walk-in refrigerators or freezers in food service and food establishments shall be connected indirectly to the sanitary drainage system with an air gap. There is an exception. Where protected against backflow by a back-water valve, such floor drains shall be connected indirectly to the sanitary drainage system by means or an airbreak or an air gap. Waste receptors shall be permitted in the form of a hub or pipe extending not less than 1 inch above a water-impervious floor and are not required to have a strainer.

Most codes prohibit the installation of an indirect-waste receptor in any room containing toilet facilities. An exception, however, is the installation of a receptor for a clothes washer, when the clothes washer is installed in the same room. Indirect-waste receptors are not allowed to be installed in closets and other unvented areas. Indirect-waste receptors must be accessible. Code generally requires all receptors to be equipped with a means of preventing solids with diameters of one-half inch or larger from entering the drainage system. These straining devices must be removable to allow for cleaning.

When you are dealing with extreme water temperatures in waste water, such as with a commercial dishwasher, the dishwasher drain must be piped to an indirect waste. The indirect waste will be connected to the sanitary plumbing system, but the dishwasher drain may not connect to the sanitary system directly if the waste water temperature exceeds 140°F. Steam pipes may not be connected directly to a sanitary drainage system. Local regulations may require the use of special piping, sumps, or condensers to accept high-temperature water. The direct connection of any dishwasher to the sanitary drainage system likely will be prohibited.

Clear-water waste, from a potable source, must be piped to an indirect-waste receptor using an air gap. Sterilizers and swimming pools are two examples of when this rule would be used. Clear water from nonpotable sources, such as a drip from a piece of equipment, must be piped to an indirect-waste receptor. Some jurisdictions allow an airbreak in place of an air gap. Other code regions require any waste entering the sanitary drainage system from an air conditioner to do so through an indirect-waste receptor.

Where waste water from swimming pools, backwash from filters, and water from pool deck drains discharge to the building drainage system, the discharge must be through an indirect-waste pipe by means of an air gap.

SPECIAL WASTES

Special wastes are those wastes that might harm a plumbing system or waste-disposal system. Possible locations for special-waste piping include photographic labs, hospitals, or buildings where chemicals or other potentially dangerous wastes are dispersed. Small, personal-type photo darkrooms generally do not fall under the scrutiny of these regulations. Buildings that are considered to have a need for special-waste plumbing often are required to have two plumbing systems—one system for normal sanitary discharge and a separate system for the special wastes. Before many special wastes are allowed to enter a sanitary drainage system, the wastes must be neutralized, diluted, or otherwise treated.

Depending upon the nature of the special wastes, special materials may be required. When you venture into the plumbing of special wastes, it is always best to consult the local code officer before proceeding with your work.

CHAPTER 9
VENTS

Most people don't think much about vents when they consider the plumbing in their home or office, but vents play a vital role in the scheme of sanitary plumbing. Many plumbers underestimate the importance of vents. The sizing and installation of vents often causes more confusion than the same tasks applied to drains. This chapter will teach you the role and importance of vents. It also will instruct you in the proper methods of sizing and installing vents.

Whether you are working with simple individual vents or complex island vents, this chapter will improve your understanding and installation of them. Why do we need vents? Vents perform three easily identified functions. The most obvious function of a vent is its capacity to carry sewer gas out of a building and into the open air. A less obvious, but equally important aspect of the vent is its ability to protect the seal in the trap it serves. The third characteristic of the vent is its ability to enable drains to drain faster and better. Let's look more closely at each of these factors.

TRANSPORTATION OF SEWER GAS

Vents transport sewer gas through a building, without exposing occupants of the building to the gas, to an open-air space. Why is this important? Sewer gas can cause health problems. The effect of sewer gas on individuals will vary, but it should be avoided by all individuals. In addition to health problems caused by sewer gas, explosions are also possible when sewer gas is concentrated in a poorly ventilated area. Yes, sewer gas can create an explosion, when it is concentrated, confined, and ignited. As you can see, just from looking at this single purpose of vents, they are an important element of a plumbing system.

PROTECTING TRAP SEALS

Another job plumbing vents perform is the protection of trap seals. The water sitting in a fixture's trap blocks the path of sewer gas trying to enter the plumbing fixture. Without a trap seal, sewer gas could rise through the drainage pipe and enter a building through a plumbing fixture. As mentioned moments ago, this could result in health problems and the risk of explosion. Good trap seals are essential to sanitary plumbing systems.

MAXIMUM DISTANCE OF FIXTURE TRAP FROM VENT

SIZE OF TRAP (inches)	SIZE OF FIXTURE DRAIN (inches)	SLOPE (inch per foot)	DISTANCE FROM TRAP (feet)
$1^1/_4$	$1^1/_4$	$^1/_4$	$3^1/_2$
$1^1/_4$	$1^1/_2$	$^1/_4$	5
$1^1/_2$	$1^1/_2$	$^1/_4$	5
$1^1/_2$	2	$^1/_4$	6
2	2	$^1/_4$	6
3	3	$^1/_8$	10
4	4	$^1/_8$	12

For SI: 1 inch = 25.4 mm, 1 foot = 304.8 mm, 1 inch per foot = 0.0833 mm/m.

FIGURE 9.1 Fixture traps may not be too far from the vent they affect. *(Courtesy of International Code Council, Inc. and International Plumbing Code 2000.)*

COMMON VENT SIZES

PIPE SIZE (inches)	MAXIMUM DISCHARGE FROM UPPER FIXTURE DRAIN (dfu)
$1^1/_2$	1
2	4
$2^1/_2$ to 3	6

For SI: 1 inch = 25.4 mm.

FIGURE 9.2 There are three common vent sizes.*(Courtesy of International Code Council, Inc. and International Plumbing Code 2000.)*

Vents protect trap seals. How do they do it? Vents regulate the atmospheric pressure applied to trap seals. It is possible for pressures to rise in unvented traps to a point where the trap contents actually expel into the fixture it serves. This is not a common problem, but if it occurs, the plumbing fixture could become contaminated.

A more likely problem is when the pressure on a trap seal is reduced and becomes something of a vacuum. When this happens, the water creating the trap seal is sucked out of the trap and down the drain. When the water is taken from the trap, there is no trap seal. The trap will remain unsealed until water is replaced in the trap. Without water in it, a trap is useless. Vents prevent these extreme atmospheric pressure changes, therefore, protecting the trap seal.

Air-admittance valves must be sized in accordance with the standard for the size of the vent to which the valve is connected. The design of a vent system can be created with an approved computer program design method. Capacity requirements for a vent system must be based on the air capacity requirements of the drainage system under a peak load condition.

TINY TORNADOS

Have you ever drained your sink or bathtub and watched the tiny water tornados? When you see the fast swirling action of water being pulled down a drain, it usually indicates that the drain is well vented. If water is sluggish and moves out of the fixture like a lazy river, the vent for the fixture, if there is one, is not performing at its best.

Vents help fixtures to drain faster. The air allowed from the vent keeps the water moving at a more rapid pace. This not only entertains us with tiny tornados, but it aids in the prevention of clogged pipes. It is possible for drains to drain too quickly, removing the liquids and leaving hair, grease, and other potential pipe-blockers present. Such problems, however, should not occur if a pipe is properly graded and does not contain extreme vertical drops into improper fittings.

DO ALL PLUMBING FIXTURES HAVE VENTS?

Most local plumbing codes require all fixture traps to be vented, but there are exceptions. In some jurisdictions, combination waste-and-vent systems are used. In a combination waste-and-vent system, vertical vents are rare. Larger drainage pipes are used instead. The larger diameter of the drain allows air to circulate in the pipe, eliminating the need for a vent, as far as satisfactory drainage is concerned. I have worked with both types of systems, predominately vented systems, and in my opinion, vented systems perform much better than combination waste-and-vent systems.

Combination waste-and-vent systems do not have vents on each fixture, so how is the trap seal protected? Trap seals in a combination waste-and-vent system are protected through the use of antisyphon traps or drum traps. Vented systems usually use P traps. By using an antisiphon or drum trap, the trap is not susceptible to backsiphonage. Since these traps are larger, deeper, and made so that the water in the trap is not replaced by fresh water with each use of the fixture, they are not required to be vented, subject to local code requirements. Most jurisdictions prohibit the use of drum traps and require traps to be vented. Before you install your plumbing, check with the local code officer for the facts pertinent to your location. Fittings for vent piping must be compatible with the piping used.

SIZE OF COMBINATION DRAIN AND VENT PIPE

DIAMETER PIPE (inches)	MAXIMUM NUMBER OF DRAINAGE FIXTURE UNITS (dfu)	
	Connecting to a horizontal branch or stack	Connecting to a building drain or building subdrain
2	3	4
$2\frac{1}{2}$	6	26
3	12	31
4	20	50
5	160	250
6	360	575

For SI: 1 inch = 25.4 mm.

FIGURE 9.3 Combination drain and vent pipes have a maximum allowable size. *(Courtesy of International Code Council, Inc. and International Plumbing Code 2000.)*

INDIVIDUAL VENTS

Individual vents are, as the name implies, vents that serve individual fixtures. These vents vent only one fixture, but they may connect into another vent that will extend to the open air. Individual vents do not have to extend from the fixture being served to the outside air, without joining another part of the venting system, but they must vent to open air space.

Sizing an individual vent is easy. The vent must be at least one-half the size of the drain it serves, but it may not have a diameter of less than 1¼ inches. For example, a vent for a 3-inch drain could, in most cases, have a diameter of an inch and a half. A vent for an inch-and-a-half drain may not have a diameter of less than 1¼ inches.

RELIEF VENTS

Relief vents are used in conjunction with other vents to provide additional air to the drainage system when the primary vent is too far from the fixture. Relief vents must be at least one-half the size of the pipe they are venting.

For example, if a relief vent is venting a 3-inch pipe, the relief vent must have a diameter of 1½ inches or larger. Use the sizing tables in your local code book to establish minimum size requirements. Relief vents may be used to vent more than one fixture.

When relief vents are required on stacks of more than 10 branch intervals, the lower end of each relief vent shall connect to the soil or waste stack through a wye below the horizontal branch serving the floor, and the upper end shall connect to the vent stack through a wye not less than 3 feet above the floor.

CIRCUIT VENTS

Circuit vents are used with a battery of plumbing fixtures. Circuit vents usually are installed just before the last fixture of the battery. The circuit vent then is extended upward to the open air or tied into another vent that extends to the outside. Circuit vents can tie into stack vents or vent stacks. When sizing a circuit vent, you must account for its developed length. In any event, the diameter of a circuit vent must be at least one-half the size of the drain it is serving.

VENT SIZING USING DEVELOPED LENGTH

What effect does the length of the vent have on the vent's size? The developed length, the total liner footage of pipe making up the vent, is used in conjunction with factors provided in code books to determine vent sizes. To size circuit vents, branch vents, and individual vents for horizontal drains, you must use this method of sizing.

The criteria needed for sizing a vent based on developed length are: the grade of the drainage pipe, the size of the drainage pipe, the developed length of the vent, and the factors allowed by local code requirements. Knowing this information, you will use the sizing tables in your local code book to establish pipe sizing.

BRANCH VENTS

Branch vents extend horizontally and connect multiple vents together. Branch vents are sized with the developed-length method. A branch vent or individual vent that is the same

SIZE AND DEVELOPED LENGTH OF STACK VENTS AND VENT STACKS

DIAMETER OF SOIL OR WASTE STACK (inches)	TOTAL FIXTURE UNITS BEING VENTED (dfu)	MAXIMUM DEVELOPED LENGTH OF VENT (feet)[a] DIAMETER OF VENT (inches)										
		1¼	1½	2	2½	3	4	5	6	8	10	12
1¼	2	30										
1½	8	50	150									
1½	10	30	100									
2	12	30	75	200								
2	20	26	50	150								
2½	42		30	100	300							
3	10		42	150	360	1,040						
3	21		32	110	270	810						
3	53		27	94	230	680						
3	102		25	86	210	620						
4	43			35	85	250	980					
4	140			27	65	200	750					
4	320			23	55	170	640					
4	540			21	50	150	580					
5	190				28	82	320	990				
5	490				21	63	250	760				
5	940				18	53	210	670				
5	1,400				16	49	190	590				
6	500					33	130	400	1,000			
6	1,100					26	100	310	780			
6	2,000					22	84	260	660			
6	2,900					20	77	240	600			
8	1,800						31	95	240	940		
8	3,400						24	73	190	720		
8	5,600						20	62	160	610		
8	7,600						18	56	140	560		
10	4,000							31	78	310	960	
10	7,200							24	60	240	740	
10	11,000							20	51	200	630	
10	15,000							18	46	180	570	
12	7,300								31	120	380	940
12	13,000								24	94	300	720
12	20,000								20	79	250	610
12	26,000								18	72	230	500
15	15,000									40	130	310
15	25,000									31	96	240
15	38,000									26	81	200
15	50,000									24	74	180

FIGURE 9.4 The minimum required diameter of stack vents and vent stacks are determined from developed length and the numbers of drainage fixture units attached. (Courtesy of International Code Council, Inc. and International Plumbing Code 2000.)

size as the drain it serves is unlimited in the developed length it may obtain. Be advised, not all local codes use the same sizing charts, so check your local code before you trust your sizing.

VENT STACKS

A vent stack is a pipe used only for the purpose of venting. Vent stacks extend upward from the drainage piping and extend to the open air outside a building. Vent stacks are used as connection points for other vents, such as branch vents. A vent stack is a primary vent that accepts the connection of other vents and vents an entire system. Vent stacks run vertically and are sized a little differently.

The basic procedure for sizing a vent stack is similar to that used with branch vents, but there are some differences. You must know the size of the soil stack, the number of fixture units carried by the soil stack, and the developed length of your vent stack. With this information and the regulations of your local plumbing code, you can size your vent stack. The same sizing method is used when computing the size of stack vents.

STACK VENTS

Stack vents are really two pipes in one. The lower portion of the pipe is a soil pipe, and the upper portion is a vent. This is the type of primary vent most often found in residential plumbing. Stack vents are sized with the same methods used on vent stacks. Offsets are permitted in the stack vent and shall be located at least 6 inches above the flood level of the highest fixture.

COMMON VENTS

Common vents are single vents that vent multiple traps. Common vents are allowed only when the fixtures being served by the single vent are on the same floor level. Some jurisdictions require the drainage of fixtures being vented with a common vent to enter the drainage system at the same level. Usually, not more than two traps can share a common vent, but there is an exception in some regions. In some areas, you may vent the traps of up to three lavatories with a single common vent. Common vents are sized with the same technique applied to individual vents.

ISLAND VENTS

Island vents are unusual-looking vents. Island vents are allowed for use with sinks and lavatories. The primary use for these vents is with the trap of a kitchen sink when it is placed in an island cabinet. The vent must rise as high as possible under the cabinet before it takes a U-turn and heads back downward. Since this piping does not rise above the flood-level rim of the fixture, it must be considered a drain. Fittings approved for drainage must be used in making an island vent. The vent portion of an island vent must be equipped with a cleanout. The vent may not tie into a regular vent until it rises at least 6 inches above the flood-level rim of the fixture.

WET VENTS

Wet vents are pipes that serve as a vent for one fixture and a drain for another. Only the fixtures within a bathroom group may connect to a wet-vented horizontal branch drain. Additional fixtures must discharge downstream of the wet vent. Wet vents, once you know how to use them, can save you a lot of money and time. By effectively using wet vents, you can reduce the amount of pipe, fittings, and labor required to vent a bathroom group or two. Dry vents connected to wet vents must be sized based on the largest required diameter of pipe within the wet vent system served by the dry vent.

The sizing of wet vents is based on fixture units. The size of the pipe is determined by how many fixture units it may be required to carry. A 3-inch wet vent can handle twelve fixture units. A 2-inch wet vent is rated for four fixture units, and an 1½-inch wet vent is allowed only one fixture unit. It is acceptable to wet vent two bathroom groups, six fixtures, with a single vent, but the bathroom groups must be on the same floor level.

Depending upon local regulations, the horizontal branch connecting to the drainage stack may have to enter at a level equal to, or below, the water-closet drain. The branch, however, may connect to the drainage at the closet bend. When wet venting two bathroom groups, the wet vent must have a minimum diameter of 2 inches. Kitchen sinks and washing machines may not be drained into a 2-inch combination waste-and-vent. Water closets and urinals are restricted on vertical combination waste-and-vent systems.

WET VENT SIZE

WET VENT PIPE SIZE (inches)	DRAINAGE FIXTURE UNIT LOAD (dfu)
$1^1/_2$	1
2	4
$2^1/_2$	6
3	12

For SI: 1 inch = 25.4 mm.

FIGURE 9.5 Wet vents are required to have a minimum size, based on fixture-unit discharge. *(Courtesy of International Code Council, Inc. and International Plumbing Code 2000.)*

WASTE STACK VENT SIZE

STACK SIZE (inches)	MAXIMUM NUMBER OF DRAINAGE FIXTURE UNITS (dfu)	
	Total discharge into one branch interval	Total discharge for stack
$1^1/_2$	1	2
2	2	4
$2^1/_2$	No limit	8
3	No limit	24
4	No limit	50
5	No limit	75
6	No limit	100

For SI: 1 inch = 25.4 mm.

FIGURE 9.6 The waste stack must be based on total discharge to the stack and discharge within a branch interval. *(Courtesy of International Code Council, Inc. and International Plumbing Code 2000.)*

If wet venting is allowed on different floor levels in your region, the wet vents must have at least a 2-inch diameter. Water closets that are not located on the highest floor must be back vented. If, however, the wet vent is connected directly to the closet bend, with a 45° bend, the toilet being connected to is not required to be back vented, even if it is on a lower floor.

Wet venting in some regions may be limited to vertical piping. These vertical pipes are restricted to receiving only the waste from fixtures with fixture-unit ratings of two or less, and that serve to vent no more than four fixtures. Wet vents must be one pipe-size larger than usually required, but they must never be smaller than 2 inches in diameter.

CROWN VENTS

A crown vent extends upward from a trap or trap arm. Crown-vented traps are not allowed. Crown vents usually are used on trap arms, but even then, they are not common. The vent must be on the trap arm, and it must be behind the trap by a distance equal to

SIZE AND LENGTH OF SUMP VENTS

DISCHARGE CAPACITY OF PUMP (gpm)	MAXIMUM DEVELOPED LENGTH OF VENT (feet)[a]						
	Diameter of vent (inches)						
	1¼	1½	2	2½	3	4	
10	No limit[b]	No limit	No limit	No limit	No limit	No limit	
20	270	No limit	No limit	No limit	No limit	No limit	
40	72	160	No limit	No limit	No limit	No limit	
60	31	75	270	No limit	No limit	No limit	
80	16	41	150	380	No limit	No limit	
100	10[c]	25	97	250	No limit	No limit	
150	Not permitted	10[c]	44	110	370	No limit	
200	Not permitted	Not permitted	20	60	210	No limit	
250	Not permitted	Not permitted	10	36	132	No limit	
300	Not permitted	Not permitted	10[c]	22	88	380	
400	Not permitted	Not permitted	Not permitted	10[c]	44	210	
500	Not permitted	Not permitted	Not permitted	Not permitted	24	130	

For SI: 1 inch = 25.4 mm, 1 foot = 304.8 mm, 1 gallons per minute = 3.785 L/m.

a. Developed length plus an appropriate allowance for entrance losses and friction due to fittings, changes in direction and diameter. Suggested allowances shall be obtained from NBS Monograph 31 or other approved sources. An allowance of 50 percent of the developed length shall be assumed if a more precise value is not available.

b. Actual values greater than 500 feet.

c. Less than 10 feet.

FIGURE 9.7 Sump vents must adhere to size and length requirements. (*Courtesy of International Code Council, Inc. and International Plumbing Code 2000.*)

twice the pipe size. For example, on a 1½-inch trap, the crown vent would have to be 3 inches behind the trap on the trap arm.

VENTS FOR SUMPS AND SEWER PUMPS

When sumps and sewer pumps are used to store and remove sanitary waste, the sump must be vented. If you will be installing a pneumatic sewer ejector, you will need to run the sump vent to outside air, without tying it into the venting system for the standard sanitary plumbing system. If your sump will be equipped with a regular sewer pump, you may tie the vent from the sump back into the main venting system for the other sanitary plumbing.

Additional rulings apply in some regions. You may find that sump vents must not be smaller than an 1¼-inch pipe. The size requirements for sump vents are determined by the discharge of the pump. For example, a sewer pump capable of producing 20 gallons a minute could have its sump vented for an unlimited distance with a 1½-inch pipe. If the pump was capable of producing 60 gallons per minute, a 1½-inch pipe could not have a developed length of more than 75 feet.

In most cases, a 2-inch vent is used on sumps, and the distance allowed for developed length is not a problem. If your pump will pump more than 100 gallons per minute, however, you had better take the time to do some math. Your code book will provide you with the factors you need to size your vent, and the sizing is easy. You look for the maximum discharge capacity of your pump and match it with a vent that allows the developed length you need.

This concludes the general description and sizing techniques for various vents. Next, we are going to look at regulations dealing with the methods of installation for vents.

VENT-INSTALLATION REQUIREMENTS

Since there are so many types of vents and their role to the plumbing system is so important, many regulations affect the installation of vents. What follows are specifics for installing various vents.

Any building equipped with plumbing also must be equipped with a main vent. The size of this vent must be no less than one-half the size of the building drain. This vent must run undiminished in size and as directly as possible from the building drain to the open air or to a vent header that extends to the open air. Any plumbing system that receives the discharge from a water closet must have either a main vent stack or stack vent. This vent must originate at a 3-inch drainage pipe and extend upward until it penetrates the roof of the building and meets outside air. The vent must have for a minimum diameter of 3 inches. Some codes, however, allow the main stack in detached buildings, where the only plumbing is a washing machine or laundry tub, to have a diameter of 1½-inches.

Main vents that are vent stacks must connect to building drains or to the bases of drainage stacks in compliance with the plumbing code. A main vent that is a stack vent shall be an extension of the drainage stack. When a vent stack connects to a building drain, the connection is to be located downstream of the drainage stack and within a distance of ten times the diameter of the drainage stack.

Multiple branch vents that exceed 40 feet in developed length must be increased by nominal size for the entire developed length of the vent pipe. When a vent penetrates a roof, it must be flashed or sealed to prevent water from leaking past the pipe and through the roof. Metal flashings with rubber collars usually are used, but more modern flashings are made from plastic, rather than metal.

The vent must extend above the roof to a certain height, which may fluctuate between geographical locations. Average vent extensions are between 12 and 24 inches, but check with your local regulations to determine the minimum height in your area.

When vents terminate in the open air, the proximity of their location to windows, doors, or other ventilating openings must be considered. If a vent were placed too close to a window, sewer gas might be drawn into the building when the window was open. Vents should be kept 10 feet from any window, door, opening, or ventilation device. If the vent cannot be kept at least ten feet from the opening, the vent should extend at least two feet above the opening. Depending upon your local region, the vent may be required to extend at least three feet above the opening.

If the roof penetrated by a vent is used for activities other than just weather protection, such as a patio, the vent must extend seven feet above the roof in zone 3. Zone 2 requires these vents to rise at least 5 feet above the roof. In cold climates, vents must be protected from freezing. Condensation can collect on the inside of vent pipes, and in cold climates this condensation might turn to ice. As the ice mass grows, the vent becomes blocked and useless.

This type of protection usually is accomplished by increasing the size of the vent pipe. This ruling usually applies only in areas where temperatures are expected to be below 0 °F. Some codes require vents in this category to have a minimum diameter of 3 inches. If this requires an increase in pipe size, the increase must be made at least one foot below the roof. In the case of side-wall vents, the change must be made at least one foot inside the wall. In some regions, all vents must have diameters of at least 2 inches, but never less than the normally required vent size. Any change in pipe size also must take place at least 12 inches before the vent penetrates into open air, and the vent must extend to a height of ten inches.

There might be occasions when it is better to terminate a plumbing vent out a wall, rather than through a roof. Some jurisdictions don't allow this practice, but others do. Some regions prohibit side-wall vents from terminating under any building's overhang. Side-wall vents must be protected against birds and rodents with a wire mesh or similar cover. Side-wall vents must not extend closer than ten feet to the property boundary of the building lot. If the building is equipped with soffit vents, side-wall vents may terminate under the soffit vents. This rule is in effect to prevent sewer gas from being sucked into the attic of the home.

Some codes require buildings having soil stacks with more than five branch intervals to be equipped with a vent stack. Others require a vent stack with buildings having at least ten stories above the building drain. The vent stack usually will run up near the soil stack. The vent stack must connect into the building drain at or below the lowest branch interval. The vent stack must be sized according to tables in your local code book. The vent stack may be required to be connected within ten times its pipe size on the downward side of the soil stack. This means that a 3-inch vent stack must be within 30 inches of the soil stack, on the downward side of the building drain.

Check your local code to see if stack vents must be connected to the drainage stack at intervals of every five stories. If so, the connection must be made with a relief yoke vent. The yoke vent must be at least as large as either the vent stack or soil stack, whichever is smaller. This connection must be made with a wye fitting at least 42 inches off the floor.

In large plumbing jobs, where there are numerous branch intervals, it may be necessary to vent offsets in the soil stack. The offset usually must be more than 45° to warrant an offset vent. It is common for offset vents to be required when the soil stack offsets and has five, or more, branch intervals above it.

Just as drains are installed with a downward pitch, vents also must be installed with a consistent grade. Vents should be graded to allow any water entering the vent pipe to drain into the drainage system. A typical grade for vent piping is a quarter-of-an-inch-to-the- foot.

MINIMUM DIAMETER AND MAXIMUM LENGTH OF INDIVIDUAL BRANCH FIXTURE VENTS AND INDIVIDUAL FIXTURE HEADER VENTS FOR SMOOTH PIPES

DIAMETER OF VENT PIPE (inches)	INDIVIDUAL VENT AIRFLOW RATE (cubic feet per minute)																			
	Maximum developed length of vent (feet)																			
	1	2	3	4	5	6	7	8	9	10	11	12	13	14	15	16	17	18	19	20
$1/2$	95	25	13	8	5	4	3	2	1	1	1	1	1	1	1	1	1	1	1	1
$3/4$	100	88	47	30	20	15	10	9	7	6	5	4	3	3	3	2	2	2	2	1
1	—	—	100	94	65	48	37	29	24	20	17	14	12	11	9	8	7	7	6	6
$1^1/_4$	—	—	—	—	—	—	—	100	87	73	62	53	46	40	36	32	29	26	23	21
$1^1/_2$	—	—	—	—	—	—	—	—	—	—	—	100	96	84	75	67	60	54	49	45
2	—	—	—	—	—	—	—	—	—	—	—	—	—	—	—	—	—	—	—	100

For SI: 1 inch = 25.4 mm, 1 foot = 304.8 mm, 1 cfm = 0.4719 L/s.

FIGURE 9.8 Branch-fixture and fixture-header vent requirements often come under local control. (*Courtesy of International Code Council, Inc. and International Plumbing Code 2000.*)

Dry vents must be installed in a manner to prevent clogging and blockages. You may not lay a fitting on its side and use a quarter bend to turn the vent up vertically. Dry vents should leave the drainage pipe in a vertical position. An easy way to remember this is that if you need an elbow to get the vent up from the drainage, you are doing it wrong.

Most vents can be tied into other vents, such as a vent stack or stack vent. The connection for the tie-in, however, must be at least six inches above the flood-level rim of the highest fixture served by the vent.

Some regions allow the use of circuit vents to vent fixtures in a battery. The drain serving the battery must be operating at one-half of its fixture-unit rating. If the application is on a lower-floor battery with a minimum of three fixtures, relief vents are required. You also must pay attention to the fixtures draining above these lower-floor batteries.

When a fixture with a rating of four or less and a maximum drain size of two inches is above the battery, every vertical branch must have a continuous vent. If a fixture has a rating exceeding four, all fixtures in the battery must be vented individually. Circuit-vented batteries may not receive the drainage from fixtures on a higher level.

Circuit vents should rise vertically from the drainage. The vent, however, can be taken off the drainage horizontally if the vent is washed by a fixture with a rating of no more than four fixture units.

The washing cannot come from a water closet. The pipe being washed must be at least as large as the horizontal drainage pipe it is venting.

Circuit vents may, at times, be used to vent up to eight fixtures utilizing a common horizontal drain. Circuit vents must be dry vents and they should connect to the horizontal drain in front of the last fixture on the branch. The horizontal drain being circuit-vented must not have a grade of more than one inch per foot. Some code requirements interpret the horizontal section of drainage being circuit-vented as a vent. If a circuit vent is venting a drain with more than four water closets attached, a relief vent must be installed in conjunction with the circuit vent.

Vent placement in relation to the trap it serves is important and regulated. The maximum allowable distance between a trap and its vent will depend on the size of the fixture drain and trap.

All vents, except those for fixtures with integral traps, should connect above the trap seal. A sanitary-tee fitting should be used when going from a vertical stack vent to a trap. Other fittings, with a longer turn, such as a combination wye-and-eighth bend, will place the trap in more danger of backsiphonage. I know this goes against the common sense of a smoother flow of water, but the sanitary tee reduces the risk of a vacuum.

All individual, branch, and circuit vents are required to connect to a vent stack, stack vent, air-admittance valve or extend to open air. Vents for future-use rough-ins must be not less than one-half the size of the drain to be served.

Rough-in vents must be labeled as vents and must either be connected to the vent system or extend to open air.

SUPPORTING YOUR PIPE

Vent pipes must be supported. Vents may not be used to support antennas, flag poles, and similar items. Depending upon the type of material you are using, and whether the pipe is installed horizontally or vertically, the spacing between hangers will vary. Both horizontal and vertical pipes require support. The regulations in the plumbing code apply to the maximum distance between hangers.

Some interceptors, such as those used as a settling tank that discharges through a horizontal indirect-waste receptor, are not required to be vented in certain regions. The interceptor receiving the discharge from the unvented interceptor, however, must be properly vented and trapped.

Traps for sinks that are a part of a piece of equipment, such as a soda fountain, are not required to be vented when venting is impossible. These drains, however, must drain through an indirect-waste receptor to an approved receptor.

Depending upon your region, you may find that all soil stacks that receive the waste of at least two vented branches must be equipped with a stack vent or a main stack vent. Except when approved, fixture drainage may not be allowed to enter a stack at a point above a vent connection. Side-inlet closet-bends are allowed to accept the connection of fixtures that are vented. These connections, however, may not be used to vent a bathroom, unless the connection is washed by a fixture. All fixtures dumping into a stack below a higher fixture must be vented, except when special approval is granted for a variance. Stack vents and vent stacks must connect to a common vent header prior to vent termination.

Up to two fixtures, set back to back or side by side, within the allowable distance between the traps and their vents may be connected to a common horizontal branch that is vented by a common vertical vent. The horizontal branch, however, must be one pipe-size larger than usual. When applying this rule, the following ratings apply: shower drains, three-inch floor drains, four-inch floor drains, pedestal urinals, and water closets with fixture-unit ratings of four fixture-units shall be considered to have three-inch drains.

Some fixture groups are allowed to be stack vented, without individual back vents. These fixture groups must be in one-story buildings or must be located on the top floor of the building, with some special provisions. Fixtures located on the top floor must connect independently to the soil stack, and the bathing units and water closets must enter the stack at the same level.

This same stack-venting procedure can be adapted to work with fixtures on lower floors. The stack being stack-vented must enter the main soil stack though a vertical eighth-bend and wye combination. The drainage must enter above the eighth-bend. A two-inch vent must be installed on the fixture group. This vent must be six inches above the flood-level rim of the highest fixture in the group.

Some fixtures are allowed to be served by a horizontal waste that is within a certain distance to a vent. When piped in this manner, bathtubs and showers both are required to have two-inch P-traps. These drains must run with a minimum grade of one-quarter inch per foot.

A single drinking fountain can be rated as a lavatory for this type of piping. On this type of system, fixture drains for lavatories may not exceed one-and-one-quarter inches, and sink drains cannot be larger than one and one-half inches, in diameter.

In multistory situations, it is possible to drain up to three fixtures into a soil stack above the highest water closet or bathtub connection, without reventing. To do this, certain requirements must be met:

• Minimum stack size of 3 inches is required.

• Approved fixture-unit load on stack is met.

• All lower fixtures must be vented properly.

• All individually unvented fixtures are within allowable distances to the main vent.

• Fixture openings shall not exceed the size of their traps.

• All code requirements must be met and approved.

WORKING WITH A COMBINATION WASTE-AND-VENT SYSTEM

Most jurisdictions limit the extent of what fixtures can be served by combination waste-and-vent systems. In many locations, it is a code violation to include a toilet on a combination system, but Maine, for example, will allow toilets on a combination waste-and-vent system. Since combination waste-and-vent systems can get you into a sticky situation, you should consult your local code officer before using such a system. I will, however, explain how this system works, in general.

The type of fixtures you are allowed to connect to a combination waste-and-vent system may be limited. In some areas, the only fixtures allowed on the combination system are floor drains, standpipes, sinks, and lavatories. Other areas will allow showers, bathtubs, and even toilets to be installed with the combo system. You will have to check your local regulations to see how they affect your choice in types of plumbing systems.

It is intended that the combination waste-and-vent system will be composed mainly of horizontal piping. Generally, the only vertical piping is the vertical risers to lavatories, sinks, and standpipes. These vertical pipes usually may not exceed 8 feet in length. This type of system relies on an oversized drain pipe to provide air circulation for drainage. The pipe often is required to be twice the size required for a drain vented normally. The combination system typically must have at least one vent, which should connect to a horizontal drain pipe.

A dry vent is required to be connected at any point within the system or the system shall connect to a horizontal drain that is vented according to the plumbing code. Combination drain-and-vent systems connecting to building drains receiving only the discharge from a stack or stacks shall be provided with a dry vent. The vent connection to the combo system must extend at least 6 inches vertically above the flood-level rim of the highest fixture being vented before offsetting horizontally.

Any vertical vent must rise to a point at least six inches above the highest fixture being served before it may take a horizontal turn. In a combination system, the pipes are rated for fewer fixture units. A 3-inch pipe connecting to a branch or stack may be allowed to carry only 12 fixture units. A 4-inch pipe, under the same conditions, could be restricted to conveying 20 fixture units. Similarly, a 2-inch pipe might only handle three fixture units, and a 1½-inch-inch pipe may not be allowed. The ratings for these pipes can increase when the pipes connect to a building drain.

Stack vents are allowed, but not always in the usual way. All fixtures on a combo system may be required to enter the stack vent individually, as opposed to on a branch, as usually would be the case. A stack vent used in a combo system generally must be a straight vertical vent, without offsets. The stack vent usually cannot even be offset vertically; it simply cannot be offset. This rule is different in some locations, so check with your local plumbing inspection to see if you are affected by the no-offset rule.

Since stack vents are common, and often required, in a combination system, you must know how to size these pipes. The sizing generally is done based on the number of fixture units entering the stack. I will give you an example of how a stack vent for a combo system might be sized in zone 3.

Since not all pipes run in conjunction with a combination waste-and-vent system have to apply to the combo rules, you might have a 1½-inch pipe entering a stack. The 1½-inch pipe could be used only if it had an individual vent. The stack vent also might be a 1½-inch pipe.

First, let's look at the maximum number of fixture units (fu) allowed on a stack:

- 1½ " stack = 2 fu
- 2" stack = 4 fu
- 3" stack = 24 fu
- 4" stack = 50 fu
- 5" stack = 75 fu
- 6" stack = 100 fu

When you are concerned about the size of a drain dumping into the stack, you must contend with only two pipe sizes. All pipe sizes larger than 2 inches may dump an unlimited number of fixture units into the stack. An 1½-inch pipe may run one fixture unit into the stack, and a 2-inch pipe may deliver two fixture units to the stack. Sizing your stack is as simple as finding your fixture-unit load on the chart in your local code book. Compare your fixture-unit load to the chart and select a pipe size rated for that load.

Again, I want to remind you that combination waste-and-vent systems vary a great deal in what is, and is not allowed. To show that contrast, let me describe how the regulations in Maine differ from the ones already described.

At the time of this writing, Maine has regulations that allow a much different style of combination waste-and-vent plumbing. The Maine regulations allow water closets, showers, and bathtubs to be included on the combo system. These regulations are under consideration for change, but at the moment, they are still in effect.

This alternate form of plumbing is applicable only to buildings with two stories or less. The regulations require that the only fixture located in a basement is a clothes washer. Not only can a water closet be used on this system, up to three water closets are allowed to enter a single 3-inch stack from a branch. Wait, it gets better. You are allowed to pipe up to two of these branches, each carrying the discharge from up to three toilets, to the stack at the same location. If a toilet is installed in the building, there must be at least one 3-inch vent.

Vertical rises from the building drain are allowed to extend ten feet. A branch may contain up to eight water closets, if vented by a circuit vent. Up to three fixtures, not counting floor drains, on the same floor level or a level not exceeding 30 inches apart, may be tied into the building drain at a distance greater than 10 feet. Basically, this means that an entire bathroom group can be piped to the building drain, without limitation on the distance.

When it comes to sizing on the Maine system, there are some big differences. With a grade of a quarter-of-an-inch to the foot, a 1½-inch pipe can carry up to five fixture units; a 2-inch pipe is allowed up to twelve fixture units. These numbers are much higher than the ones given earlier. As you can see, there are some major differences to be found in the rulings for combination waste-and-vent systems. Maine currently operates on a plumbing code that is different from the ones used in most areas.

CHAPTER 10
TRAPS, CLEANOUTS AND INTERCEPTORS

We have covered most of the regulations you will need to know about drains and vents. This chapter will round out your knowledge. Here you will learn about traps. Traps have been mentioned before, and you have learned the importance of vents to trap seals, but here, you will learn more about traps.

Cleanouts are a necessary part of the drainage system. This chapter will tell you what types of cleanouts you can use and when and where they must be used. Back-water valves will be explained. Grease receptors, or grease traps as they often are called, will also be explored. By the end of this chapter, you should be prepared to tackle just about any DWV job.

CLEANOUTS

What are cleanouts, and why are they needed? Cleanouts are a means of access to the interior of drainage pipes. They are needed so blockages in drains can be cleared. Without cleanouts, it is much more difficult to snake a drain. In general, the more cleanouts you have, the better. Plumbing codes establish minimums for the number of cleanouts required and their placement. Let's look at how these regulations apply to you.

WHERE ARE CLEANOUTS REQUIRED?

Cleanouts are required in many places in a plumbing system. Let's start with sewers. All sewers must have cleanouts. The distances between these cleanouts vary from region to region. Generally, cleanouts will be required where the building drain meets the building sewer. The cleanouts may be installed inside the foundation or outside.

Cleanouts		
Size of Pipe (inches)	Size of Cleanout (inches)	Threads per inch
1-1/2	1-1/2	11-1/2
2	1-1/2	11-1/2
2-1/2	2-1/2	8
3	2-1/2	8
4 & larger	3-1/2	8

FIGURE 10.1 This illustration matches pipe and cleanout size, and threading. *(Reprinted from the 2000 Uniform Plumbing Code (UPC) with the permission of the International Association of Plumbing and Mechanical Officials (IAPMO).)*

Cleanouts (Metric)

Size of Pipe (mm)	Size of Cleanout (mm)	Threads per 25.4 mm
40	38	11-1/2
50	38	11-1/2
65	64	8
80	64	8
100 & larger	89	8

FIGURE 10.2 This is a metric version of figure 10.1. *(Reprinted from the* 2000 Uniform Plumbing Code (UPC) *with the permission of the International Association of Plumbing and Mechanical Officials (IAPMO).)*

The cleanout opening, however, must extend upward to the finished floor level or the finished grade outside.

Some jurisdictions prefer that the cleanouts at the junction of building drains and sewers be located outside. If the cleanout is installed inside, it may be required to extend above the flood-level rim of the fixtures served by the horizontal drain. When this is not feasible, allowances may be made. The requirement for a junction clean-out may be waived if a cleanout of at least a 3-inch diameter is within 10 feet of the junction.

Once the sewer is begun, cleanouts should be installed every 100 feet. Some regions require cleanouts at an interval distance of 75 feet for 4-inch and larger pipe and 50 feet for pipe smaller than 4-inch. Cleanouts also are required in sewers when the pipe changes direction. Cleanouts usually are required every time a sewer turns more than 45°. In some cases, a cleanout is required whenever the change in direction is more than 135°.

The cleanouts installed in a sewer must be accessible. This generally means that a standpipe will rise from the sewer to just below ground level. At that point, a cleanout fitting and plug is installed on the standpipe. This allows the sewer to be snaked out from ground level, with little to no digging required.

For building drains and horizontal branches, cleanout location will depend upon pipe size, but cleanouts usually are required every 50 feet. For pipes with diameters of 4 inches or less, cleanouts must be installed every 50 feet. Larger drains may have their cleanouts spaced at 100-foot intervals. Cleanouts also are required on these pipes with a change in direction. The degree of change that triggers the need for a cleanout is usually anything in excess of 45°. Cleanouts must be installed at the end of all horizontal drain runs. Some jurisdictions do not require cleanouts at 50-foot intervals, only 100-foot intervals.

As with most rules, there are some exceptions to the rules, such as:

- If a drain is less than five feet long and is not draining sinks or urinals, a cleanout is not required.
- A change in direction from a vertical drain with a fifth-bend does not require a cleanout.
- Cleanouts are not required on pipes, other than building drains and their horizontal branches, that are above the first-floor level.

P traps and water closets often are allowed to act as cleanouts. When these devices are approved for cleanout purposes, the usually required cleanout fitting and plug at the end of a horizontal pipe run may be eliminated. Not all jurisdictions will accept P traps and toilets as cleanouts; check your local requirements before omitting standard cleanouts.

Cleanouts must be installed in a way that the cleanout opening is accessible and allows adequate room for drain cleaning. The cleanout must be installed to go with the flow. This means that when the cleanout plug is removed, a drain-cleaning device should be able to enter the fitting and the flow of the drainage pipe without trouble.

Cleanouts frequently are required at the base of every stack. This is good procedure at any time, but it is not required by all codes. The height of this cleanout should not exceed four feet. Many plumbers install test tees at these locations to plug their stacks for pressure testing. The test tee doubles as a cleanout.

When the pipes holding cleanouts will be concealed, the cleanout must be made accessible. For example, if a stack will be concealed by a finished wall, provisions must be made for access to the cleanout. This access could take the form of an access door, or the cleanout simply could extend past the finished wall covering. If the cleanout is serving a pipe concealed by a floor, the cleanout must be brought up to floor level and made accessible. This ruling applies not only to cleanouts installed beneath concrete floors, but also to cleanout installed in crawl spaces, with very little room to work.

WHAT ELSE DO I NEED TO KNOW ABOUT CLEANOUTS?

There is still more to learn about cleanouts. Size is one of the lessons to be learned. Cleanouts are required to be the same size as the pipe they are serving, unless the pipe is larger than 4 inches. If you are installing a two-inch pipe, you must install 2-inch cleanouts. When a P trap is allowed for a cleanout, however, it may be smaller than the drain. An example would be an 1¼-inch trap on a 1½-inch drain. Remember though, not all code enforcement officers will allow P traps as cleanouts, and they may require the P trap to be the same size as the drain, if the trap is allowed as a cleanout. Once the pipe size exceeds 4 inches, the cleanouts used should have a minimum size of 4 inches.

Cleanouts must provide adequate clearance for drain cleaning. The clearance required for pipes with diameters of three inches, or more, is 18 inches. Smaller pipes require a minimum clearance of twelve inches in front of their cleanouts. Many plumbers fail to remember this regulation. It is common to find cleanouts pointing toward floor joists or too close to walls. You will save yourself time and money by committing these clearance distances to memory.

A cleanout is installed in a floor may be required to have a minimum height clearance of 18 inches and a minimum horizontal clearance of 30 inches. No under-floor cleanout is allowed to be placed more than 20 feet from an access opening.

Cleanout plugs and caps must be lubricated with water-insoluble, non-hardening material or tape. Only listed thread tape or lubricants and sealants specifically intended for use with plastics shall be used on plastic threads. Conventional pipe thread compounds, putty, linseed oil base products and unknown lubricants and sealants must not be used on plastic threads.

ACCEPTABLE TYPES OF CLEANOUTS

Cleanout plugs and plates must be removed easily. Access to the interior of the pipe should be available without undue effort or time. Cleanouts can take on many appearances. The "U" bend of a P trap can be considered a cleanout, depending upon local interpretation. A rubber cap, held onto the pipe by a stainless steel clamp, can serve as a cleanout. The standard female adapter and plug is a fine cleanout. Test-tees will work as cleanouts. Special cleanouts, designed to allow rodding of a drain in either direction, are acceptable.

Cleanouts with plate-style access covers shall be fitted with corrosion-resisting fasteners. Plugs used for cleanouts are to be constructed of plastic or brass. Countersunk heads are required where raised heads might pose a tripping hazard. Plastic cleanout plugs must conform to code requirements. Brass cleanout plugs can be used only with metallic drain, waste, and vent piping.

VERY BIG CLEANOUTS

The ultimate cleanout is a manhole. You can think of manholes as very big cleanouts. When a pipe's diameter exceeds a certain size, usually either 8 or 10 inches, manholes replace cleanouts. Manholes typically are required every 400 feet. In addition, they are required at all changes in direction, elevation, grade, and size. Manholes shall be protected against flooding and must be equipped with covers to prevent the escape of gases. Connections with manholes often are required to be made with flexible compression joints. These connections must not be made closer than one foot to the manhole and not farther than t3 feet away from the manhole.

TRAPS

Traps are required on drainage-type plumbing fixtures. No fixture is allowed to be double trapped, and traps serving automatic clothes washers or laundry tubs must not discharge into a kitchen sink. With some fixtures, such as toilets, traps are not apparent because they are an integral part of the fixture. The following regulations do not apply to integral traps. Integral traps, being a part of a fixture, are governed by regulations controlling the use of approved fixtures. Drawn brass tubing traps are not allowed for use with urinals. We already have talked about trap seals, so now, let's learn more about traps.

P Traps

P traps are the ones most frequently used in modern plumbing systems. These traps are self-cleaning and frequently have removable U bends, which may act as cleanouts, pending local approval. Fixture traps must be self-scouring and are not allowed to have interior partitions. An exception to interior partitions comes into play with integral traps and traps that are constructed of an approved material that is resistant to corrosion and degradation. P traps must be vented properly. Without adequate venting, the trap seal can be removed by backpressure. Slip joints must be made with an approved elastomeric gasket and shall be installed only on the trap inlet, trap outlet, and within the trap seal.

S traps

S traps were very common when most plumbing drains came up through the floor, instead of out a wall. Many S traps are still in operation, but they no longer are allowed in new installations. S traps are subject to losing their trap seal through self-siphoning.

Drum Traps

Drum traps usually are not allowed in new installations, without special permission from the code officer. The only occasion when drum traps still are used frequently is when they are installed with a combination waste-and-vent system. It is acceptable to use drum traps when they are used as solids interceptors and when they serve chemical waste systems.

Bell Traps

Bell traps are not allowed for use in new installations.

Horizontal Distance of Trap Arms
(Except for water closets and similar fixtures)*

Trap Arm Inches	Distance Trap to Vent Feet Inches		Trap Arm mm	Distance Trap to Vent mm
1-1/4	2	6	32	762
1-1/2	3	6	40	1067
2	5	0	50	1524
3	6	0	80	1829
4 & larger	10	0	100 & larger	3048

Slope one-fourth (1/4) inch per foot (20.9 mm/m)
*The developed length between the trap of a water closet or similar fixture (measured from the top of the closet ring [closet flange] to the inner edge of the vent) and its vent shall not exceed six (6) feet (1829 mm).

FIGURE 10.3 Trap arms must follow this formula for distance. *(Reprinted from the* 2000 Uniform Plumbing Code (UPC) *with the permission of the International Association of Plumbing and Mechanical Officials (IAPMO).)*

House Traps

House traps no longer are allowed; they represent a double-trapping of all fixtures. Local codes may allow house traps under certain circumstances. House traps once were installed where the building drain joined with the sewer. Most house traps were installed inside the structure, but a fair number were installed outside, underground. Their purpose was to prevent sewer gas from coming out of the sewer and into the plumbing system. House traps, however, make drain cleaning very difficult and they create a double-trapping situation, which is not allowed. This regulation, like most regulations, is subject to amendment and variance by the local code official.

Crown-vented Traps

Crown-vented traps are not allowed in new installations. These traps have a vent rising from the top of the trap. As you learned earlier, crown venting must be done at the trap arm, not the trap.

Other Traps

Traps that depend on moving parts or interior partitions are not allowed in new installations.

DOES EVERY FIXTURE REQUIRE AN INDIVIDUAL TRAP?

Basically, every fixture requires an individual trap, but there are exceptions. One such exception is the use of a continuous waste to connect the drains from multiple sink bowls to a common trap. This is done frequently with kitchen sinks.

Some restrictions involve the use of continuous wastes Let's take a kitchen sink as an example. When you have a double-bowl sink, it is all right to use a continuous waste, as long as the drains from each bowl are no more than 30 inches apart and neither bowl is more than 6 inches deeper than the other bowl. Some jurisdictions require that all sinks connected to a continuous waste must be of equal depth. Exceptions to this rule do exist. What if your sink has three bowls? Three-compartment sinks may be connected with a continuous waste. You may use a single trap to collect the drainage from up to three separate sinks or lavatories, as long as the sinks or lavatories are next to each other and in the same room. The trap, however, must be in a location central to all sinks or lavatories.

TRAP SIZES

Trap sizes are determined by the local code. A trap may not be larger than the drain pipe into which it discharges.

TAILPIECE LENGTH

The tailpiece between a fixture drain and the fixture's trap may not exceed 24 inches.

STANDPIPE HEIGHT

A standpipe must extend at least 18 inches, but not more than 30 inches, above its trap. The standpipe should not extend more than 4 feet from the trap. Some local codes require that a standpipe not exceed a height of more than 2 feet above the trap. Plumbers installing laundry standpipes often forget this regulation. When setting your fitting height in the drainage pipe, keep in mind the height limitations on your standpipe. Otherwise, your take-off fitting might be too low, or too high, to allow your standpipe receptor to be placed at the desired height. Traps for kitchen sinks may not receive the discharge from a laundry tub or clothes washer.

PROPER TRAP INSTALLATION

There is more to proper trap installation than location and trap selection. Traps must be installed level, in order for the trap seal to function properly. An average trap seal will consist of two inches of water. Some large traps may have a seal of four inches, and where evaporation is a problem, deep-sealing traps may have a deeper water seal. The positioning of the trap is critical for the proper seal. If the trap is cocked, the water seal will not be uniform, and may contribute to self-siphoning.

When a trap is installed below grade and must be connected from above grade, the trap must be housed in a box, of sorts. An example of such a situation would be a trap for a tub's waste. When installing a bathtub on a concrete floor, the trap is located below the floor. Since the trap cannot be reasonably installed until after the floor is poured, access must be made for the connection. This access, frequently called a tub box or trap box, must provide protection against water, insect, and rodent infiltration.

WHEN IS A TRAP NOT A TRAP?

One type of trap we have not yet discussed is a grease trap. The reason we haven't talked about grease traps is that grease traps are not really traps; they are grease interceptors. There is a big difference between a trap and an interceptor. Grease traps must conform to PDI G101 and must be installed in accordance with the manufacturer's instructions.

Grease traps must be equipped with devices to control the rate of water flow so it does not exceed the rated flow of the trap. A flow-control device must be vented. The vent cannot terminate less than six inches above the flood-rim level or installed in accordance with the manufacturer's instructions.

The vented flow-control device shall be located such that no system vent shall be between the flow control and the grease trap inlet. The vent or air inlet of the flow-control device shall connect with the sanitary drainage vent system as elsewhere required by the code or shall terminate through the roof of the building and shall not terminate to the free atmosphere inside the building.

Traps are meant to prevent sewer gas from entering a building. Traps do not restrict what goes down the drain, only what comes up the drain. Of course, traps do prevent objects larger than the trap from entering the drain, but this is not their primary objective.

Interceptors, on the other hand, are designed to control what goes down a drain. Interceptors are used to keep harmful substances from entering the sanitary drainage systems. Separators, because they separate the materials entering them, retaining some and allowing others to continue into the drainage system, also are required in some circumstances. Interceptors are used to control grease, sand, oil, and other materials.

Interceptors and separators are required when conditions provide opportunity for harmful or unwanted materials to enter a sanitary drainage system. An interceptor is required when oil, grease, sand, or other harmful substances might enter a drainage system. For example, a restaurant is required to be equipped with a grease interceptor because of the large amount of grease present in commercial food establishments. An oil separator would be required for a building where automotive repairs are made. Interceptors and separators must be designed for each individual situation. There is no general method of choosing the proper interceptor or separator, without expert design.

Some guidelines are provided in plumbing codes for interceptors and separators. The capacity of a grease interceptor is based on two factors: grease retention and flow rate. These determinations typically are made by a professional designer. The size of a receptor or separator is also usually determined by a design expert.

A grease trap or grease interceptor is required to receive grease-laden waste from fixtures and equipment that are located in food preparation areas, such as restaurants, hotel kitchens, and hospitals.

Where food-waste grinders connect to grease traps or grease interceptors, the grease interceptor must be sized and rated for the discharge of the food-waste grinder. Grease traps and interceptors are not required in private living quarters and individual dwelling units.

Interceptors for sand and other heavy solids must be readily accessible for cleaning. These units must contain a water seal of not less than 6 inches. Some codes require a minimum water depth of only 2 inches. When an interceptor is used in a laundry, a water seal is not required. Laundry receptors, used to catch lint, string, and other objects, usually are made of wire, and they must be removed easily for cleaning. Their purpose is to prevent solids with a diameter of a half of an inch or more, from entering the drainage system.

CAPACITY OF GREASE TRAPS

TOTAL FLOW-THROUGH RATING (gpm)	GREASE RETENTION CAPACITY (pounds)
4	8
6	12
7	14
9	18
10	20
12	24
14	28
15	30
18	36
20	40
25	50
35	70
50	100

For SI: 1 gallons per minute = 3.785 L/m, 1 pound = 0.454 kg.

FIGURE 10.4 Grease traps need to have the retention capacity listed above when combined with flow-through. *(Courtesy of International Code Council, Inc. and International Plumbing Code 2000.)*

Oil separators are required at repair garages, gasoline stations with grease racks, grease pits or work racks, car-washing facilities with engine or undercarriage cleaning capability, and at factories where oily and flammable liquid wastes are produced. The separators must keep oil-bearing, grease-bearing, and flammable wastes from entering the building drainage system or other point of disposal.

Other types of separators are used for various plants, factories, and processing sites. The purpose of all separators is to keep unwanted objects and substances from entering

Grease Traps

Total Number of Fixtures Connected	Required Rate of Flow per Minute, Gallons	Grease Retention Capacity, Pounds
1	20	40
2	25	50
3	35	70
4	50	100

FIGURE 10.5 This type of trap needs to meet the requirements listed above. . *(Reprinted from the 2000 Uniform Plumbing Code (UPC) with the permission of the International Association of Plumbing and Mechanical Officials (IAPMO).)*

Grease Traps (Metric)

Total Number of Fixtures Connected	Required Rate of Flow per Minute, Liters	Grease Retention Capacity, kg
1	76	18
2	95	22
3	132	31
4	189	45

FIGURE 10.6 This is the metric version of figure 10.5. *(Reprinted from the 2000 Uniform Plumbing Code (UPC) with the permission of the International Association of Plumbing and Mechanical Officials (IAPMO).)*

the drainage system. Vents are required if it is suspected that these devices will be subject to the loss of a trap seal. All interceptors and separators must be readily accessible for cleaning, maintenance, and repairs.

BACK-WATER VALVES

Back-water valves are essentially check valves installed in drains and sewers to prevent the backing up of waste and water in the drain or sewer. Back-water valves are required to be readily accessible and installed any time a drainage system is likely to encounter backups from the sewer.

The intent behind back-water valves is to prevent sewers from backing up into individual drainage systems. Buildings that have plumbing fixtures below the level of the street, where a main sewer is installed, are candidates for back-water valves.

This concludes our section on traps, cleanouts, interceptors, and other drainage-related regulations. While this is a short chapter, it is an important one. You might not have a need for installing manholes or back-water valves every day, but as a plumber, you frequently will work with traps and cleanouts.

CHAPTER 11

STORM DRAINAGE

Storm-water drainage piping is designed to control and convey excess groundwater to a suitable location. A suitable location might be a catch basin, storm sewer, or a pond. But, storm-water drainage may never be piped into a sanitary sewer or plumbing system.

When you wish to size a storm-water drainage system, you must have some benchmark information with which to work. One consideration is the amount of pitch a horizontal pipe will have. Another piece of the puzzle is the number of square feet of surface area your system will be required to drain. You also will need data about the rainfall rates in your area.

When you use your code book to size a storm-water system, you should have access to all the key elements required to size the job, except local rainfall amounts. You should be able to obtain rainfall figures from your state or county offices. Your code book should provide you with a table to use in making your sizing calculations.

SIZING A HORIZONTAL STORM DRAIN OR SEWER

The first step to take when sizing a storm drain or sewer is to establish your known criteria. How much pitch will your pipe have on it? Your code book should offer choices for a pipe pitch.

What else do you need to know? You must know what the rainfall rate is for the area where you will be installing the storm-water system. There should be a table in your code book that lists many regions and their rates of rainfall. You also must know the surface area that your system will be responsible for handling. The surface area must include both roof and parking areas.

When you are working with a standard table, like the ones found in most code books, you must convert the information to suit your local conditions. For example, if a standardized table is based on one inch of rainfall an hour and your location has 2.4 inches of rainfall per hour, you must convert the table, but this is not difficult.

When I want to convert a table based on a 1-inch rainfall to meet my local needs, all I have to do is divide the drainage area in the table by my rainfall amount. For example, if my standard chart shows an area of 10,000 square feet requiring a 4-inch pipe, I can change the table by dividing my rainfall amount, 2.4, into the surface area of 10,000 square feet.

SIZE OF VERTICAL CONDUCTORS AND LEADERS

DIAMETER OF LEADER (inches)[a]	HORIZONTALLY PROJECTED ROOF AREA (square feet)											
	Rainfall rate (inches per hour)											
	1	2	3	4	5	6	7	8	9	10	11	12
2	2,880	1,440	960	720	575	480	410	360	320	290	260	240
3	8,800	4,400	2,930	2,200	1,760	1,470	1,260	1,100	980	880	800	730
4	18,400	9,200	6,130	4,600	3,680	3,070	2,630	2,300	2,045	1,840	1,675	1,530
5	34,600	17,300	11,530	8,650	6,920	5,765	4,945	4,325	3,845	3,460	3,145	2,880
6	54,000	27,000	17,995	13,500	10,800	9,000	7,715	6,750	6,000	5,400	4,910	4,500
8	116,000	58,000	38,660	29,000	23,200	19,315	16,570	14,500	12,890	11,600	10,545	9,660

For SI: 1 inch = 25.4 mm, 1 square foot = 0.0929 m^2.

a. Sizes indicated are the diameter of circular piping. This table is applicable to piping of other shapes provided the cross-sectional shape fully encloses a circle of the diameter indicated in this table.

FIGURE 11.1 This chart illustrates the relationship among a roof, its drains and rain. (*Courtesy of International Code Council, Inc. and International Plumbing Code 2000.*)

SIZE OF HORIZONTAL STORM DRAINAGE PIPING

SIZE OF HORIZONTAL PIPING (inches)	HORIZONTALLY PROJECTED ROOF AREA (square feet)					
	Rainfall rate (inches per hour)					
	1	2	3	4	5	6
$1/8$ unit vertical in 12 units horizontal (1-percent slope)						
3	3,288	1,644	1,096	822	657	548
4	7,520	3,760	2,506	1,800	1,504	1,253
5	13,360	6,680	4,453	3,340	2,672	2,227
6	21,400	10,700	7,133	5,350	4,280	3,566
8	46,000	23,000	15,330	11,500	9,200	7,600
10	82,800	41,400	27,600	20,700	16,580	13,800
12	133,200	66,600	44,400	33,300	26,650	22,200
15	218,000	109,000	72,800	59,500	47,600	39,650
$1/4$ unit vertical in 12 units horizontal (2-percent slope)						
3	4,640	2,320	1,546	1,160	928	773
4	10,600	5,300	3,533	2,650	2,120	1,766
5	18,880	9,440	6,293	4,720	3,776	3,146
6	30,200	15,100	10,066	7,550	6,040	5,033
8	65,200	32,600	21,733	16,300	13,040	10,866
10	116,800	58,400	38,950	29,200	23,350	19,450
12	188,000	94,000	62,600	47,000	37,600	31,350
15	336,000	168,000	112,000	84,000	67,250	56,000
$1/2$ unit vertical in 12 units horizontal (4-percent slope)						
3	6,576	3,288	2,295	1,644	1,310	1,096
4	15,040	7,520	5,010	3,760	3,010	2,500
5	26,720	13,360	8,900	6,680	5,320	4,450
6	42,800	21,400	13,700	10,700	8,580	7,140
8	92,000	46,000	30,650	23,000	18,400	15,320
10	171,600	85,800	55,200	41,400	33,150	27,600
12	266,400	133,200	88,800	66,600	53,200	44,400
15	476,000	238,000	158,800	119,000	95,300	79,250

For SI: 1 inch = 25.4 mm, 1 square foot = 0.0929 m².

FIGURE 11.2 The chart examines horizontal storm drainage piping for a roof. (*Courtesy of International Code Council, Inc. and International Plumbing Code 2000.*)

11.3

Sizing Roof Drains, Leaders, and Vertical Rainwater Piping

Size of Drain, Leader or Pipe, Inches	Flow, gpm	Maximum Allowable Horizontal Projected Roof Areas Square Feet at Various Rainfall Rates					
		1"/hr	2"/hr	3"/hr	4"/hr	5"/hr	6"/hr
2	23	2176	1088	725	544	435	363
3	67	6440	3220	2147	1610	1288	1073
4	144	13,840	6920	4613	3460	2768	2307
5	261	25,120	12,560	8373	6280	5024	4187
6	424	40,800	20,400	13,600	10,200	8160	6800
8	913	88,000	44,000	29,333	22,000	17,600	14,667

Sizing Roof Drains, Leaders, and Vertical Rainwater Piping

Size of Drain Leader or Pipe, mm	Flow, L/s	Maximum Allowable Horizontal Projected Roof Areas Square Meters at Various Rainfall Rates					
		25mm/hr	50mm/hr	75mm/hr	100mm/hr	125mm/hr	150mm/hr
50	1.5	202	101	67	51	40	34
80	4.2	600	300	200	150	120	100
100	9.1	1286	643	429	321	257	214
125	16.5	2334	1117	778	583	467	389
150	26.8	3790	1895	1263	948	758	632
200	57.6	8175	4088	2725	2044	1635	1363

Notes:
1. The sizing data for vertical conductors, leaders, and drains is based on the pipes flowing 7/24 full.
2. For rainfall rates other than those listed, determine the allowable roof area by dividing the area given in the 1 inch/hour (25 mm/hour) column by the desired rainfall rate.
3. Vertical piping may be round, square, or rectangular. Square pipe shall be sized to enclose its equivalent round pipe. Rectangular pipe shall have at least the same cross-sectional area as its equivalent round pipe, except that the ratio of its side dimensions shall not exceed 3 to 1.

FIGURE 11.3 The chart shows the decreasing ability of a roof to hold rain at increasing levels. *(Reprinted from the 2000 Uniform Plumbing Code (UPC) with the permission of the International Association of Plumbing and Mechanical Officials (IAPMO).)*

If I divide 10,000 by 2.4, I get 4,167. All of a sudden, I have solved the mystery of computing storm water piping needs. With this simple conversion, I know that if my surface area was 4,167 square feet, I would need a 4-inch pipe. But, my surface area is 15,000 square feet, so, what size pipe do I need? Well, I know it will have to be larger than 4 inches. So, I look down my conversion chart and find the appropriate surface area. My 15,000 square feet of surface area will require a storm-water drain with a diameter of 8 inches. I found this by dividing the surface areas of the numbers in the table found in my code book by 2.4, until I reached a number equal to, or greater than, my surface area. I could almost get by with a 6-inch pipe, but not quite.

Now, let's recap this exercise. To size a horizontal storm drain or sewer, decide what pitch you will put on the pipe. Next, determine what your area's rainfall is for a one-hour storm, occurring each 100 years. If you live in a city, your city may be listed, with its rainfall amount, in your code book. Using a standardized chart, rated for one inch of rainfall per hour, divide the surface area by a factor equal to your rainfall index, in my case it was 2.4. This division process converts a generic table into a customized table for your area.

Sizing of Horizontal Rainwater Piping

Size of Pipe, mm	Flow at 10 mm/m Slope, L/s	Maximum Allowable Horizontal Projected Roof Areas Square Meters at Various Rainfall Rates					
		25mm/hr	50mm/hr	75mm/hr	100mm/hr	125mm/hr	150mm/hr
80	2.1	305	153	102	76	61	51
100	4.9	700	350	233	175	140	116
125	8.8	1241	621	414	310	248	207
150	14.0	1988	994	663	497	398	331
200	30.2	4273	2137	1424	1068	855	713
250	54.3	7692	3846	2564	1923	1540	1282
300	87.3	12,375	6187	4125	3094	2476	2062
375	156.0	22,110	11,055	7370	5528	4422	3683

Size of Pipe, mm	Flow at 20 mm/m Slope, L/s	Maximum Allowable Horizontal Projected Roof Areas Square Meters at Various Rainfall Rates					
		25mm/hr	50mm/hr	75mm/hr	100mm/hr	125mm/hr	150mm/hr
80	3.0	431	216	144	108	86	72
100	6.9	985	492	328	246	197	164
125	12.4	1754	877	585	438	351	292
150	19.8	2806	1403	935	701	561	468
200	42.7	6057	3029	2019	1514	1211	1009
250	76.6	10,851	5425	3618	2713	2169	1807
300	123.2	17,465	8733	5816	4366	3493	2912
375	220.2	31,214	15,607	10,405	7804	6248	5202

Size of Pipe, mm	Flow at 40 mm/m Slope, L/s	Maximum Allowable Horizontal Projected Roof Areas Square Meters at Various Rainfall Rates					
		25mm/hr	50mm/hr	75mm/hr	100mm/hr	125mm/hr	150mm/hr
80	4.3	611	305	204	153	122	102
100	9.8	1400	700	465	350	280	232
125	17.5	2482	1241	827	621	494	413
150	28.1	3976	1988	1325	994	797	663
200	60.3	8547	4273	2847	2137	1709	1423
250	108.6	15,390	7695	5128	3846	3080	2564
300	174.6	24,749	12,374	8250	6187	4942	4125
375	312.0	44,220	22,110	14,753	11,055	8853	7367

Notes:
1. The sizing data for horizontal piping is based on the pipes flowing full.
2. For rainfall rates other than those listed, determine the allowable roof area by dividing the area given in the 1 inch/hour (25 mm/hour) column by the desired rainfall rate.

FIGURE 11.4 The chart illustrates metric sizing of horizontal rainwater piping. (*Reprinted from the* 2000 *Uniform Plumbing Code (UPC) with the permission of the International Association of Plumbing and Mechanical Officials (IAPMO).*)

Once the math is done, look down the table for the surface area that most closely matches the area you have to drain.

To be safe, go with a number slightly higher than your projected number. It is better to have a pipe sized one size too large than one size too small. When you have found the appropriate surface area, look across the table to see what size pipe you need.

See how easy that was? Well, maybe it's not easy, but it is a chore you can handle.

Sizing of Horizontal Rainwater Piping

Size of Pipe, Inches	Flow at 1/8"/ft. Slope, gpm	Maximum Allowable Horizontal Projected Roof Areas Square Feet at Various Rainfall Rates					
		1"/hr	2"/hr	3"/hr	4"/hr	5"/hr	6"/hr
3	34	3288	1644	1096	822	657	548
4	78	7520	3760	2506	1880	1504	1253
5	139	13,360	6680	4453	3340	2672	2227
6	222	21,400	10,700	7133	5350	4280	3566
8	478	46,000	23,000	15,330	11,500	9200	7670
10	860	82,800	41,400	27,600	20,700	16,580	13,800
12	1384	133,200	66,600	44,400	33,300	26,650	22,200
15	2473	238,000	119,000	79,333	59,500	47,600	39,650

Size of Pipe, Inches	Flow at 1/4"/ft. Slope, gpm	Maximum Allowable Horizontal Projected Roof Areas Square Feet at Various Rainfall Rates					
		1"/hr	2"/hr	3"/hr	4"/hr	5"/hr	6"/hr
3	48	4640	2320	1546	1160	928	773
4	110	10,600	5300	3533	2650	2120	1766
5	196	18,880	9440	6293	4720	3776	3146
6	314	30,200	15,100	10,066	7550	6040	5033
8	677	65,200	32,600	21,733	16,300	13,040	10,866
10	1214	116,800	58,400	38,950	29,200	23,350	19,450
12	1953	188,000	94,000	62,600	47,000	37,600	31,350
15	3491	336,000	168,000	112,000	84,000	67,250	56,000

Size of Pipe, Inches	Flow at 1/2"/ft. Slope, gpm	Maximum Allowable Horizontal Projected Roof Areas Square Feet at Various Rainfall Rates					
		1"/hr	2"/hr	3"/hr	4"/hr	5"/hr	6"/hr
3	68	6576	3288	2192	1644	1310	1096
4	156	15,040	7520	5010	3760	3010	2500
5	278	26,720	13,360	8900	6680	5320	4450
6	445	42,800	21,400	14,267	10,700	8580	7140
8	956	92,000	46,000	30,650	23,000	18,400	15,320
10	1721	165,600	82,800	55,200	41,400	33,150	27,600
12	2768	266,400	133,200	88,800	66,600	53,200	44,400
15	4946	476,000	238,000	158,700	119,000	95,200	79,300

Notes:
1. The sizing data for horizontal piping is based on the pipes flowing full.
2. For rainfall rates other than those listed, determine the allowable roof area by dividing the area given in the 1 inch/hour (25 mm/hour) column by the desired rainfall rate.

FIGURE 11.5 The chart illustrates sizing of horizontal rainwater piping. *(Courtesy of International Code Council, Inc. and International Plumbing Code 2000.)*

SIZING RAIN LEADERS AND GUTTERS

When you are required to size rain leaders or downspouts, you use the same procedure described above, with one exception. You use a table, supplied in your code book, to size the vertical piping. Determine the amount of surface area your leader will drain and use the appropriate table to establish your pipe size. The conversion factors are the same. Sizing gutters is essentially the same as sizing horizontal storm

Size of Gutters

Diameter of Gutter in Inches	Maximum Rainfall in Inches per Hour				
1/16"/ft. Slope	**2**	**3**	**4**	**5**	**6**
3	340	226	170	136	113
4	720	480	360	288	240
5	1250	834	625	500	416
6	1920	1280	960	768	640
7	2760	1840	1380	1100	918
8	3980	2655	1990	1590	1325
10	7200	4800	3600	2880	2400

Diameter of Gutter in Inches	Maximum Rainfall in Inches per Hour				
1/8"/ft. Slope	**2**	**3**	**4**	**5**	**6**
3	480	320	240	192	160
4	1020	681	510	408	340
5	1760	1172	880	704	587
6	2720	1815	1360	1085	905
7	3900	2600	1950	1560	1300
8	5600	3740	2800	2240	1870
10	10,200	6800	5100	4080	3400

Diameter of Gutter in Inches	Maximum Rainfall in Inches per Hour				
1/4"/ft. Slope	**2**	**3**	**4**	**5**	**6**
3	680	454	340	272	226
4	1440	960	720	576	480
5	2500	1668	1250	1000	834
6	3840	2560	1920	1536	1280
7	5520	3680	2760	2205	1840
8	7960	5310	3980	3180	2655
10	14,400	9600	7200	5750	4800

Diameter of Gutter in Inches	Maximum Rainfall in Inches per Hour				
1/2"/ft. Slope	**2**	**3**	**4**	**5**	**6**
3	960	640	480	384	320
4	2040	1360	1020	816	680
5	3540	2360	1770	1415	1180
6	5540	3695	2770	2220	1850
7	7800	5200	3900	3120	2600
8	11,200	7460	5600	4480	3730
10	20,000	13,330	10,000	8000	6660

FIGURE 11.6 The chart illustrates the link between the amount of rain and the size of a gutter. *(Courtesy of International Code Council, Inc. and International Plumbing Code 2000.)*

SIZE OF SEMICIRCULAR ROOF GUTTERS

DIAMETER OF GUTTERS (inches)	HORIZONTALLY PROJECTED ROOF AREA (square feet)					
	RAINFALL RATE (inches per hour)					
	1	2	3	4	5	6
1/16 unit vertical in 12 units horizontal (0.5-percent slope)						
3	680	340	226	170	136	113
4	1,440	720	480	360	288	240
5	2,500	1,250	834	625	500	416
6	3,840	1,920	1,280	960	768	640
7	5,520	2,760	1,840	1,380	1,100	918
8	7,960	3,980	2,655	1,990	1,590	1,325
10	14,400	7,200	4,800	3,600	2,880	2,400
1/8 unit vertical in 12 units horizontal (1-percent slope)						
3	960	480	320	240	192	160
4	2,040	1,020	681	510	408	340
5	3,520	1,760	1,172	880	704	587
6	5,440	2,720	1,815	1,360	1,085	905
7	7,800	3,900	2,600	1,950	1,560	1,300
8	11,200	5,600	3,740	2,800	2,240	1,870
10	20,400	10,200	6,800	5,100	4,080	3,400
1/4 unit vertical in 12 units horizontal (2-percent slope)						
3	1,360	680	454	340	272	226
4	2,880	1,440	960	720	576	480
5	5,000	2,500	1,668	1,250	1,000	834
6	7,680	3,840	2,560	1,920	1,536	1,280
7	11,040	5,520	3,860	2,760	2,205	1,840
8	15,920	7,960	5,310	3,980	3,180	2,655
10	28,800	14,400	9,600	7,200	5,750	4,800
1/2 unit vertical in 12 units horizontal (4-percent slope)						
3	1,920	960	640	480	384	320
4	4,080	2,040	1,360	1,020	816	680
5	7,080	3,540	2,360	1,770	1,415	1,180
6	11,080	5,540	3,695	2,770	2,220	1,850
7	15,600	7,800	5,200	3,900	3,120	2,600
8	22,400	11,200	7,460	5,600	4,480	3,730
10	40,000	20,000	13,330	10,000	8,000	6,660

For SI: 1 inch = 25.4 mm, 1 square foot = 0.0929 m².

FIGURE 11.7 This givers information about diameter of gutters and a roof. (*Courtesy of International Code Council, Inc. and International Plumbing Code 2000.*)

Controlled Flow Maximum Roof Water Depth

Roof Rise,*		Max Water Depth at Drain,	
Inches	(mm)	Inches	(mm)
Flat	(Flat)	3	(76)
2	(51)	4	(102)
4	(102)	5	(127s)
6	(152)	6	(152)

*Vertical measurement from the roof surface at the drain to the highest point of the roof surface served by the drain, ignoring any local depression immediately adjacent to the drain.

FIGURE 11.8 This illustrates depth of water on a roof along with the rise in both inches and millimeters. *(Courtesy of International Code Council, Inc. and International Plumbing Code 2000.)*

drains. You will use a different table, provided in your code book, but the mechanics are the same.

ROOF DRAINS

Roof drains are often the starting point of a storm-water drainage system. As the name implies, roof drains are located on roofs. On most roofs, the drains are equipped with strainers that protrude upward, at least four inches, to catch leaves and other debris. Roof drains should be at least twice the size of the piping connected to them. All roofs that do not drain to hanging gutters are required to have roof drains. A minimum of two roof drains should be installed on roofs with a surface area of 10,000 square feet, or less. If the surface area exceeds 10,000 square feet, a minimum of four roof drains should be installed.

When a roof is used for purposes other than just shelter, the roof drains may have a strainer that is flush with the roof's surface. Roof drains obviously should be sealed to prevent water from leaking around them. The size of the roof drain can be instrumental in the flow rates designed into a storm-water system. When a controlled flow from roof drains is wanted, the roof structure must be designed to accommodate the controlled flow.

MORE SIZING INFORMATION

If a combined storm-drain and sewer arrangement is approved, it must be sized properly. This requires converting fixture-unit loads into drainage surface area. For example, 256 fixture units will be treated as 1,000 square feet of surface area. Each additional fixture unit, in excess of 256, will be assigned a value of 3.9 square feet. In the case of sizing for continuous flow, each gpm is rated as 96 square feet of drainage area.

SOME FACTS ABOUT STORM-WATER PIPING

Storm-water piping requires the same number of cleanouts, with the same frequency as a sanitary system. Just as regular plumbing pipes must be protected, so must storm-water piping. For example, if a downspout is in danger of being crushed by automobiles, you must install a guard to protect the downspout. Back-water valves installed in a storm drainage system must conform to local code requirements.

As I said earlier, storm-water systems and sanitary systems should not be combined. There may be some cities where the two are combined, but they are the exception, rather than the rule. Areaway drains or floor drains must be trapped. Rain leaders and storm drains connected to a sanitary sewer must be trapped. The trap must be equal in size to the drain it serves. Traps must be accessible for cleaning the drainage piping. Storm-water piping may not be used for conveying sanitary drainage.

SUMP PUMPS

Sump pumps are used to remove water collected in building sub-drains. These pumps must be placed in a sump, but the sump need not be covered with a gastight lid or vented.

Many people are not sure what to do with the water pumped out of their basement by a sump pump. Do you pump it into your sewer? No, the discharge from a sump pump should not be pumped into a sanitary sewer. The water from the pump should be pumped to a storm-water drain, or in some cases, to a point on the property where it will not cause a problem.

All sump-pump discharge pipes should be equipped with a check valve. The valve prevents previously pumped water from running down the discharge pipe and refilling the sump, forcing the pump to pull double duty. When I speak of sump pumps, I am talking about pumps removing groundwater, not waste or sewage.

VARIATIONS

There are some variations in local codes for storm water drainage. For example, approved materials can be different from one jurisdiction to another. This can be true of both above-ground and underground materials. After storm-water piping extends at least two feet from a building, any approved material may be used in most regions.

Another example of a variation could be that the inlet area of a roof drain generally is required only to be one-and-one-half times the size of the piping connected to the roof drain. When positioned on roofs used for purposes other than weather protection, however, roof drain openings must be twice as large as the drain connecting to them.

Some regions provide different tables for sizing purposes. When computing the drainage area, you must take into account the effect vertical walls will have on the drainage area. For example, a vertical wall that reflects water onto the drainage area must be considered in your surface-area computations. In the case of a single vertical wall, add one-half of the wall's total square footage to the surface area.

Two vertical walls that are adjacent to each other require you to add 35 percent of the combined wall square footage to your surface area.

If you have two walls of the same height that are opposite each other, no added space is needed. In this case, each wall protects the other and does not allow extra water to collect on the roof area.

When you have two opposing walls with different heights, you must make a surface-area adjustment. Take the square footage of the highest wall, as it extends above the other wall, and add half of the square footage to your surface area.

When you encounter three walls, you use a combination of the above instructions to reach your goal. Four walls of equal height do not require an adjustment. If the walls are not of equal height, use the procedures above to compute your surface area.

Additional code variations may occur with sump pits. Some sump pits are required to have a minimum diameter of 18 inches. In some regions, floor drains may not connect to drains intended solely for storm water. When computing surface area to be drained for vertical walls, such as walls enclosing a roof-top stairway, use one-half of the total square footage from the vertical wall surface that reflects water onto the drainage surface in certain regions.

Some roof designs require a backup drainage system, in case of emergencies. These roofs generally are roofs surrounded by vertical sections. If these vertical sections are capable of retaining water on the roof if the primary drainage system fails, a secondary drainage system is required. In these cases, the secondary system must have independent piping and discharge locations. These special systems are sized with the use of different rainfall rates. The ratings are based on a 15-minute rainfall. Otherwise, the hundred-year conditions still apply.

Some regions have requirements for sizing a continuous flow requiring a rating of 24 square feet of surface area to be given for every gpm generated. For regular sizing, based on 4 inches of rain per hour, 256 fixture units equal 1,000 square feet of surface area. Each additional fixture unit is rated at 3.9 inches. If the rainfall rate varies, a conversion must be done.

To convert the fixture-unit ratings to a higher or lower rainfall, you must do some math. Take the square-foot rating assigned to fixture units and multiply it by four. For example, 256 fixture units equal 1,000 square feet. Multiply 1,000 by four, and get 4,000. Now, divide 4,000 by the rate of rainfall for one hour. Say, for example, that the hourly rainfall was two inches, the converted surface area would be 2,000.

Distance of Scupper Bottoms Above Roof

Roof Rise,* Inches	(mm)	Maximum Distance of Scupper Bottom Above Roof Level at Drains, Inches	(mm)
Flat	(Flat)	3	(76.2)
2	(51)	4	(102)
4	(102)	5	(127.0)
6	(152)	6	(152)

*Vertical measurement from the roof surface at the drain to the highest point of the roof surface served by the drain, ignoring any local depression immediately adjacent to the drain.

FIGURE 11.9 Scupper bottoms and their height are examined here.. *(Reprinted from the* 2000 *Uniform Plumbing Code (UPC) with the permission of the International Association of Plumbing and Mechanical Officials (IAPMO).)*

Well, you have made it past a section of code regulations that gives professional plumbers the most trouble. Some plumbers strongly dislike storm-water drains, because they have little knowledge of how to compute them. With the aid of this chapter, you should be able to design a suitable system with minimal effort.

CHAPTER 12

SPECIAL PIPING
AND STORAGE SYSTEMS

Medical gas systems and nonmedical oxygen systems are covered in the plumbing code under the provisions for special piping and storage systems or under a category of health-care facilities and medical gas and vacuum systems, depending upon which local code you use. The distinction is between medical gases and oxygen systems that are not used for medical purposes.

The general provisions of special piping and storage systems governs the design and installation of piping and storage systems for nonflammable medical gas systems and non-medical oxygen systems. It is important that you note the limitations of this part of the code. Pay attention to the part about "nonflammable medical gas systems" and "nonmed-ical oxygen systems." These two elements are all that are covered under the special piping and storage systems as discussed in the plumbing code.

GENERAL REQUIREMENTS

Drinking fountains, valves, and other items that might protrude from a wall must be flush-mounted or full-recessed in corridors and other areas where patients might be transported on a gurney, hospital bed, or wheelchair. Piping and traps in psychiatric patients' rooms must be concealed. All fixtures and fittings in these rooms must be vandal-proof. All ice makers or ice storage chests must be installed in a nurses' station or other similarly super-vised area that is not subject to contamination.

STERILIZERS

Sterilizers and bedpan steamers must be connected indirectly to the sanitary drainage sys-tem. The indirect waste pipe must not be less than the size of the drain connection on the fixture being served. The length of the piping shall not exceed 15 feet. Receptors must be located in the same room as the equipment served. With the exception of bedpan steam-ers, such indirect waste pipes do not require traps. A trap with a minimum seal of three inches must be provided in the indirect waste pipe for a bedpan steamer. Sterilizers that have provisions for a vapor vent and the vent is required by the manufacturer must extend

STACK SIZES FOR BEDPAN STEAMERS AND BOILING-TYPE STERILIZERS
(Number of Connections of Various Sizes Permitted to Various-sized Sterilizer Vent Stacks)

STACK SIZE (inches)	CONNECTION SIZE		
	1 1/2"		2"
$1^1/_2{}^a$	1	or	0
2^a	2	or	1
2^b	1	and	1
3^a	4	or	2
3^b	2	and	2
4^a	8	or	4
4^b	4	and	4

For SI: 1 inch = 25.4 mm.
a. Total of each size.
b. Combination of sizes.

FIGURE 12.1 These are stack and connection sizes permitted to sterilizer vent stacks. *(Courtesy of International Code Council, Inc. and International Plumbing Code 2000.)*

STACK SIZES FOR PRESSURE STERILIZERS
(Number of Connections of Various Sizes Permitted to Various-sized Vent Stacks)

STACK SIZE (inches)	CONNECTION SIZE			
	3/4"	1"	1 1/4"	1 1/2"
$1^1/_2{}^a$	3 or	2 or	1	
$1^1/_2{}^b$	2 and	1		
2^a	6 or	3 or	2 or	1
2^b	3 and	2		
2^b	2 and	1 and	1	
2^b	1 and	1 and		1
3^a	15 or	7 or	5 or	3
3^b		1 and	2 and	2
	1 and	5 and		1

For SI: 1 inch = 25.4 mm.
a. Total of each size.
b. Combination of sizes.

FIGURE 12.2 These figures show sizes for stacks and connections for various vent stacks. *(Courtesy of International Code Council, Inc. and International Plumbing Code 2000.)*

Minimum Outlets/Inlets per Station

Location	Oxygen	Medical Vacuum	Medical Air	Nitrous Oxide	Nitrogen	Helium	Carbon Dioxide
Patient rooms for medical/surgical, obstetrics and pediatrics	1/bed	1/bed	1/bed	—	—	—	—
Examination/treatment for nursing units	1/bed	1/bed	—	—	—	—	—
Intensive care (all)	3/bed	3/bed	2/bed	—	—	—	—
Nursery[1]	2/bed	2/bed	1/bed	—	—	—	—
General operating rooms	2/room	3/room[4]	2/room	1/room	1/room	—	—
Cystoscopic and invasive special procedures	2/room	3/room[4]	2/room	—	—	—	—
Recovery Delivery and labor/delivery/ recovery rooms[2]	2/bed 2/room	2/bed 3/room[4]	1/bed 1/room	—	—	—	—
Labor rooms	1/bed	1/bed	1/bed	—	—	—	—
First aid and emergency treatment[3]	1/bed	1/bed[4]	1/bed	—	—	—	—
Autopsy	—	1/station	1/station	—	—	—	—
Anesthesia workroom	1/station	—	1/station	—	—	—	—

1 Includes pediatric nursery.
2 Includes obstetric recovery.
3 Emergency trauma rooms used for surgical procedures shall be classified as general operating rooms.
4 Vacuum inlets required are in addition to any inlets used as part of a scavenging system for removal of anesthetizing gases.

FIGURE 12.3 There are minimum requirements for numbers of outlets in various sections of a hospital. *(Reprinted from the 2000 Uniform Plumbing Code (UPC) with the permission of the International Association of Plumbing and Mechanical Officials (IAPMO).)*

System Sizing – Flow Factors for More than One Inlet/Outlet Terminal Unit System Minimum Flow Requirements[1]

Number of Inlet/Outlet Terminal Units per Facility	Diversity Percentage of Average Flow per Inlet/Outlet Terminal Units	Minimum Permissible System Flow[2] SCFM (liters/minute)	
		All Pressurized Medical Gas Systems	Vacuum Systems
1 – 10	100%	Actual Demand	See Table 13-5
11 – 25	75%	7.0 (200)	
26 – 50	50%	13.1 (375)	
51 – 100	50%	17.5 (500)	

1 Flow rates of individual inlet/outlet terminals per Table 13-2.
2 The minimum system flow is the average inlet/outlet flow times the number of inlet/outlet terminal units times the diversity percentage.

FIGURE 12.4 Above are minimum flow factors of system sizing. *(Reprinted from the 2000 Uniform Plumbing Code (UPC) with the permission of the International Association of Plumbing and Mechanical Officials (IAPMO).)*

Operating Pressures for Medical Gas and Medical Vacuum Systems

System	Symbol	Minimum Pressure	Maximum Pressure
Oxygen	O_2	50 psig (0.34 MPa)	50 (+5-0) psig
Medical Vacuum	Vac	12" Hg	
Nitrous Oxide	N_2O	50 psig (0.34 MPa)	50(+5-0) psig
Medical Compressed Air	Med Air	50 psig (0.34 MPa)	50(+5-0) psig
Nitrogen	N_2	160 psig (1.10 MPa)	199 psig
Helium	He	50 psig (0.34 MPa)	50(+5-0) psig
Carbon Dioxide	CO_2	50 psig (0.34 MPa)	50(+5-0) psig
Non-Standard Nitrogen	N_2	200 psig (1.36 MPa)	300 psig (2.06 MPa)

FIGURE 12.5 Medical gases require certain operating pressures. *(Reprinted from the 2000 Uniform Plumbing Code (UPC) with the permission of the International Association of Plumbing and Mechanical Officials (IAPMO).)*

Minimum Flow Rates

Oxygen	.71 CFM per outlet[1] (20 LPM)
Nitrous Oxide	.71 CFM per outlet[1] (20 LPM)
Medical Compressed Air	.71 CFM per outlet[1] (20 LPM)
Nitrogen	15 CFM (0.42 m^3/min.) free air per outlet
Vacuum	1 SCFM (0.03 Sm^3/min.) per inlet[2]
Carbon Dioxide	.71 CFM per outlet[1] (20 LPM)
Helium	.71 CFM per outlet (20 LPM)

[1] Any room designed for a permanently located respiratory ventilator or anesthesia machine shall have an outlet capable of a flow rate of 180 LPM (6.36 CFM) at the station outlet.

[2] For testing and certification purposes, individual station inlets shall be capable of a flow rate of 3 SCFM, while maintaining a system pressure of not less than 12 inches (305 mm) at the nearest adjacent vacuum inlet.

FIGURE 12.6 Minimum flow rates for medical gases are shown above. *(Reprinted from the 2000 Uniform Plumbing Code (UPC) with the permission of the International Association of Plumbing and Mechanical Officials (IAPMO).)*

Outlet Rating for Vacuum Piping Systems

Location of Medical-Surgical Vacuum Outlets	Free-Air Allowance, Expressed as CFM (LPM) at 1 Atmosphere		Zone Allowances Corridors-Risers Main Supply Line-Valves	
	Per Room	Per Outlet	Simultaneous Usage, Factor Percent	Air to Be Transported CFM (LPM)*
Operating Rooms				
Major "A" (Radical, Open Heart)	3.5 (99.1)	–	100	3.5 (99.1)
(Organ Transplant)	3.5 (99.1)	–	100	3.5 (99.1)
(Radical Thoracic)	3.5 (99.1)	–	100	3.5 (99.1)
Major "B" (All Other Major ORs)	2.0 (56.6)	–	100	2.0 (56.6)
Minor	1.0 (28.3)	–	100	1.0 (28.3)
Delivery Rooms	1.0 (28.3)	–	100	1.0 (28.3)
Recovery Rooms (Post-Anesthesia) and Intensive Care Units (a minimum of 2 outlets per bed in each such department)				
1st outlet at each bed	–	3.0 (85.0)	50	1.5 (42.5)
2nd outlet at each bed	–	1.0 (28.3)	50	0.5 (14.2)
3rd outlet at each bed	–	1.0 (28.3)	10	0.1 (2.8)
All others at each bed	–	1.0 (28.3)	10	0.1 (2.8)
Emergency Rooms	–	1.0 (28.3)	100	1.0 (28.3)
Patient Rooms				
Surgical	–	1.0 (28.3)	50	0.5 (14.2)
Medical	–	1.0 (28.3)	10	0.1 (2.8)
Nurseries	–	1.0 (28.3)	10	0.1 (2.8)
Treatment and Examining Rooms	–	0.5 (14.2)	10	0.05 (1.4)
Autopsy Area	–	2.0 (56.6)	20	0.4 (11.3)
Inhalation Therapy, Central Supply and Instructional Areas	–	1.0 (28.3)	10	0.1 (2.8)

*Free air at 1 atmosphere

FIGURE 12.7 Different sections of a hospital are required to have a certain number of vacuum outlets. *(Reprinted from the* 2000 Uniform Plumbing Code (UPC) *with the permission of the International Association of Plumbing and Mechanical Officials (IAPMO).)*

Size of Gas/Vacuum Piping

| ft x 304.8 = mm |
| in x 25.4 = mm |

Medical Gas System	Pipe Size Inch[2]	Maximum Delivery Capacity[3] in SCFM (LPM)				
		Length of Piping in Feet (m)[1]				
		100 (30)	250 (76)	500 (152)	750 (228)	1000 (304)
Oxygen	1/2	15.0 (425)	10.6 (300)	7.4 (209)	5.9 (167)	5.1 (144)
	3/4	40.0 (1133)	28.3 (801)	19.6 (555)	15.7 (445)	13.3 (377)
	1	50.0 (1416)	50.0 (1416)	40.2 (1138)	32.2 (912)	27.7 (784)
Nitrous Oxide	1/2	15.0 (425)	9.5 (269)	6.5 (184)	5.3 (150)	4.5 (127)
	3/4	30.0 (849)	24.7 (699)	17.1 (484)	13.7 (388)	11.7 (331)
	1	40.0 (1113)	40.0 (1133)	34.7 (983)	28.2 (798)	24.3 (688)
Medical Air	1/2	18.1 (512)	11.1 (314)	7.8 (221)	6.3 (177)	5.3 (151)
	3/4	40.0 (1133)	29.9 (847)	21.0 (595)	16.5 (467)	14.1 (399)
	1	50.0 (1416)	50.0 (1416)	42.1 (1192)	35.8 (1013)	29.2 (826)
Vacuum	1	22.8 (645)	13.7 (388)	9.5 (269)	7.6 (215)	6.5 (184)
	1-1/4	40.1 (1135)	24.5 (694)	16.7 (473)	13.3 (377)	11.2 (317)
	1-1/2	63.7 (1804)	38.9 (1101)	26.8 (759)	21.1 (600)	17.9 (507)
	2	132.7 (3758)	81.4 (2305)	56.0 (1586)	45.0 (1274)	38.3 (1084)
Nitrogen	1/2	25.0 (708)	25.0 (708)	25.0 (708)	23.8 (674)	20.6 (583)
	3/4	60.0 (1699)	60.0 (1699)	60.0 (1699)	60.0 (1699)	54.2 (1535)
	1	110.0 (3115)	110.0 (3115)	110.0 (3115)	110.0 (3115)	110.0 (3115)

[1] Length of piping includes a 30% allowance for fittings.
[2] One-half inch (12.7 mm) diameter pipe is the minimum size allowed in medical gas systems.
[3] Based on the following maximum pressure drops:
    Oxygen, nitrous oxide, and medical air – 5 psig (10" Hg)
    Vacuum – 1.96 psig (4" Hg)
    Nitrogen – 20 psig (41" Hg)

FIGURE 12.8 The size of gas and vacuum piping varies with its use in a hospital. *(Reprinted from the 2000 Uniform Plumbing Code (UPC) with the permission of the International Association of Plumbing and Mechanical Officials (IAPMO).)*

the vapor vent to the outdoors above the roof. These vents are not allowed to be connected to any drainage system vent.

ASPIRATORS

All aspirators or other water-supplied suction devices can be installed only in strict accordance to code requirements. Aspirators used for removing body fluids must be equipped with a collecting bottle or similar fluid trap. Aspirators must discharge indirectly to the sanitary drainage system through an air gap. The potable water supply to an aspirator must be protected by a vacuum breaker or equivalent backflow protection.

MEDICAL GASES

Medical gases covered under this portion of the code are "nonflammable." Key components of the code pertain to nonflammable medical gas systems, inhalation anesthetic systems, and vacuum-piping systems. These systems might exist in hospitals, dental offices, or other facilities. The code rulings are simple. These systems must be designed and installed in accordance with NFPA 99C. There are, however, two exceptions.

The special piping and storage systems section of the code does not apply to portable systems or cylinder storage. Vacuum system exhaust must comply with the local mechanical code. That's all there is to the medical gases part of this section for one code.

OXYGEN SYSTEMS

Oxygen systems that are not used for medical purposes must be designed and installed in accordance with NFPA 50 and NFPA 51. It might strike you as strange, but this is all that one code has to say about nonmedical oxygen systems.

MORE DETAILED REQUIREMENTS

Some codes have much more detailed code requirements for health care issues, and we are going to review those issues now. As you would expect, all medical gas and medical vacuum systems must be installed in accordance with all code requirements. The code requires all installers to be competent. Medical gas and vacuum systems must be supplied with at least two sources. For example, a system would be required to have at least two cylinder banks with at least two cylinders in each bank, a minimum of two air compressors, or a minimum of two vacuum pumps. Two supply pipelines, however, are not required. Operating pressures, minimum flow rates, and minimum station outlets and inlets are regulated by code requirements.

The sizing of medical gas and vacuum systems is recommended to be done by a mechanical engineer, rather than someone using the plumbing code. This recommendation appears in the code book. Since the code suggests that only mechanical engineers be responsible for the design of the systems, I see no reason to cover sizing methods in this book. The local code book does offer information about sizing requirements and practices, but again, the code clearly recommends that the sizing done by an engineer.

Plans and specs must be provided to the local code officer prior to installing any medical gas or medical vacuum system. The plans and specs must be approved before a permit is issued. An approval package usually will contain a plot plan of the site, drawn to scale, that indicates the location of existing or new cylinder storage areas, property lines, driveways, and existing or proposed buildings. There will be a piping layout of all proposed piping systems and alterations. Full specifications of the materials to be used are required. A record of as-built plans and valve identification records must remain on the site of the system at all times. Always check your local code requirements prior to performing any work. Codes vary and you must check your local code requirements to be sure of what your obligations are.

CHAPTER 13
FUEL-GAS PIPING

The installation of fuel-gas piping is a substantial responsibility. If a plumber has a leak in a water pipe, the result is something getting wet. A similar leak in a gas pipe can result in a serious explosion. All code requirements have importance and should be followed; this is especially true when working with gas piping and gas connections. The installation, modification, and maintenance of fuel-gas systems all are addressed in the plumbing code.

Starting with the point of delivery, the code covers all piping from that point to the connections with each utilization device. Code coverage includes design, materials, components, fabrication, assembly, installation, testing, inspection, operation, and maintenance of such piping systems. What is the point of delivery? It is the outlet of the service meter assembly or the outlet of the service regulator or service shutoff valve where a meter is not provided, or where the service meter assembly is located within a building, at the entrance of the supply pipe into the building. When working with undiluted liquefied petroleum (LP) gas systems, the point of delivery is the outlet of the first stage pressure regulator. Any LP gas storage system must be designed and installed in accordance with the fire prevention code and NFPA 58.

When a gas system is modified or added to, all pipe sizing must conform with the sizing requirements of the code. If an additional gas appliance is being added, the existing piping must be checked to determine if it has the required capacity for all appliances being served. When an existing gas system is not large enough for additional appliances, the system must be upgraded to meet demand capacities. This can be done by enlarging the existing system or by running a new system for the appliances to be served.

All exposed gas piping, except for black steel pipe, must be marked and identified by a yellow label that shows the word GAS in black letters. This identification marker must be spaced at intervals that do not exceed five feet. No marking is required on piping that is in the same room as the equipment being served. Any tubing that carries medium-pressure gas must be marked with a label at the beginning and end of each section of tubing used to convey gas.

It is a violation of code to make interconnections between multiple meters. When multiple meters are installed on the same premises to serve multiple users, each meter must be served by an independent gas system. It is not acceptable to connect gas piping from one user's system to the system of another user. To avoid mistaken interconnections, piping from multiple meter installations must be marked with an approved permanent identification by the installer. The markings must be clear and readily identifiable.

Size of Gas Piping

Maximum Delivery Capacity of Cubic Feet of Gas Per Hour of IPS Pipe Carrying
Natural Gas of 0.60 Specific Gravity Based on Pressure Drop of 0.5 Inch Water Column

Pipe Size, Inches	Length In Feet										
	10	20	30	40	50	60	70	80	90	100	125
1/2	174	119	96	82	73	66	61	56	53	50	44
3/4	363	249	200	171	152	138	127	118	111	104	93
1	684	470	377	323	286	259	239	222	208	197	174
1-1/4	1404	965	775	663	588	532	490	456	428	404	358
1-1/2	2103	1445	1161	993	880	798	734	683	641	605	536
2	4050	2784	2235	1913	1696	1536	1413	1315	1234	1165	1033
2-1/2	6455	4437	3563	3049	2703	2449	2253	2096	1966	1857	1646
3	11,412	7843	6299	5391	4778	4329	3983	3705	3476	3284	2910
3-1/2	16,709	11,484	9222	7893	6995	6338	5831	5425	5090	4808	4261
4	23,277	15,998	12,847	10,995	9745	8830	8123	7557	7091	6698	5936

Pipe Size, Inches	150	200	250	300	350	400	450	500	550	600
1/2	40	34	30	28	25	24	22	21	20	19
3/4	84	72	64	58	53	49	46	44	42	40
1	158	135	120	109	100	93	87	82	78	75
1-1/4	324	278	246	223	205	191	179	169	161	153
1-1/2	486	416	369	334	307	286	268	253	241	230
2	936	801	710	643	592	551	517	488	463	442
2-1/2	1492	1277	1131	1025	943	877	823	778	739	705
3	2637	2257	2000	1812	1667	1551	1455	1375	1306	1246
3-1/2	3861	3304	2929	2654	2441	2271	2131	2013	1912	1824
4	5378	4603	4080	3697	3401	3164	2968	2804	2663	2541

FIGURE 13.1 Here, the relationship between pipe size and natural gas carrying capacity is explained. *(Reprinted from the 2000 Uniform Plumbing Code (UPC) with the permission of the International Association of Plumbing and Mechanical Officials (IAPMO).)*

Size of Gas Piping

Maximum Delivery Capacity in Liters Per Second of IPS Pipe Carrying Natural Gas of 0.60 Specific Gravity Based on Pressure Drop of 12.7 mm Water Column

Pipe Size, mm	914	1829	2743	3719	4633	5547	6492	7315	8352	9266	11582
					Length in mm						
15	1.4	0.9	0.8	0.6	0.6	0.5	0.5	0.4	0.4	0.4	0.3
20	2.9	2.0	1.6	1.3	1.2	1.1	1.0	0.9	0.9	0.8	0.7
25	5.4	3.7	3.0	2.5	2.3	2.0	1.9	1.7	1.6	1.5	1.4
32	11.0	7.6	6.1	5.2	4.6	4.2	3.9	3.6	3.4	3.2	2.8
40	16.5	11.4	9.1	7.8	6.9	6.3	5.8	5.4	5.0	4.8	4.2
50	31.9	21.9	17.6	15.1	13.3	12.1	11.1	10.3	9.7	9.2	8.1
65	50.8	34.9	28.0	24.0	21.3	19.3	17.7	16.5	15.5	14.6	13.0
80	89.8	61.7	49.6	42.4	37.6	34.1	31.3	29.1	27.3	25.8	22.9
90	131.4	90.3	72.5	62.1	55.0	49.9	45.9	42.7	40.0	37.8	33.5
100	183.1	125.9	101.1	86.5	76.7	69.5	63.9	59.5	55.8	52.7	46.7

Pipe Size, mm	13899	18532	23165	27798	32431	37064	41697	46330	50963	55596
15	0.3	0.3	0.2	0.2	0.2	0.2	0.2	0.2	0.2	0.1
20	0.7	0.6	0.6	0.5	0.4	0.4	0.4	0.3	0.3	0.3
25	1.2	1.1	0.9	0.9	0.8	0.7	0.7	0.6	0.6	0.6
32	2.6	2.2	1.9	1.8	1.6	1.5	1.4	1.3	1.3	1.2
40	3.8	3.3	2.9	2.6	2.4	2.2	2.1	2.0	1.9	1.8
50	7.4	6.3	5.6	5.1	4.7	4.3	4.1	3.8	3.6	3.5
65	11.7	10.0	8.9	8.1	7.4	6.9	6.5	6.1	5.8	5.5
80	20.7	17.8	15.7	14.3	13.1	12.2	11.4	10.8	10.3	9.8
90	30.4	26.0	23.0	20.9	19.2	17.9	16.8	15.8	15.0	14.3
100	42.3	36.2	32.1	29.1	26.8	24.9	23.4	22.1	21.0	20.0

FIGURE 13.2 Here, the metric relationship between pipe size and natural gas carrying capacity is explained. (Reprinted from the 2000 Uniform Plumbing Code (UPC) with the permission of the International Association of Plumbing and Mechanical Officials (IAPMO).)

Medium Pressure Natural Gas Systems for Sizing Gas Piping Systems Carrying Gas of 0.60 Specific Gravity

Capacity of Pipes of Different Diameters and Lengths in Cubic Feet Per Hour for Gas Pressure for 2.0 psi with a Drop to 1.5 psi

Pipe Size, Inches	50	100	150	200	250	300	350	400	450	500	550	600
							Length in Feet					
1/2	466	320	257	220	195	177	163	151	142	134	127	121
3/4	974	669	537	460	408	369	340	316	297	280	266	254
1	1834	1261	1012	866	768	696	640	595	559	528	501	478
1-1/4	3766	2588	2078	1799	1577	1429	1314	1223	1147	1084	1029	982
1-1/2	5642	3878	3114	2665	2362	2140	1969	1832	1719	1624	1542	1471
2	10,867	7469	5998	5133	4549	4122	3792	3528	3310	3127	2970	2833
2-1/2	17,320	11,904	9559	8181	7251	6570	6044	5623	5276	4984	4733	4515
3	30,618	21,044	16,899	14,463	12,818	11,614	10,685	9940	9327	8810	8367	7983

	650	700	750	800	850	900	950	1000	1100	1200	1300	1400
1/2	116	112	108	104	101	97	95	92	87	83	80	77
3/4	243	234	225	217	210	204	198	193	183	174	167	161
1	458	440	424	409	396	384	373	363	345	329	315	302
1-1/4	940	903	870	840	813	788	766	745	707	675	646	621
1-1/2	1409	1353	1304	1259	1218	1181	1147	1116	1060	1011	968	930
2	2713	2606	2511	2425	2347	2275	2209	2149	2041	1947	1865	1791
2-1/2	4324	4154	4002	3865	3740	3626	3522	3425	3253	3103	2972	2855
3	7644	7344	7075	6832	6612	6410	6225	6055	5751	5486	5254	5047
4	15,592	14,979	14,430	13,935	13,486	13,075	12,698	12,351	11,730	11,190	10,716	10,295

	1500	1600	1700	1800	1900	2000	2100	2200	2300	2400	2500	2600
1/2	74	71	69	67	65	63	62	60	59	57	56	55
3/4	155	149	145	140	136	132	129	126	123	120	117	115
1	291	281	272	264	256	249	243	237	231	226	221	216
1-1/4	598	578	559	542	526	512	499	486	475	464	454	444
1-1/2	896	865	837	812	788	767	747	728	711	695	680	665
2	1726	1667	1613	1564	1519	1477	1439	1403	1369	1338	1309	1282
2-1/2	2751	2656	2570	2492	2420	2354	2293	2236	2183	2133	2086	2043
3	4862	4696	4544	4406	4279	4162	4053	3953	3859	3771	3688	3611
4	9918	9578	9269	8986	8727	8488	8267	8062	7870	7691	7523	7365
5	17,943	17,327	16,768	16,258	15,789	15,357	14,957	14,585	14,238	13,914	13,610	13,325
6	29,054	28,057	27,151	26,325	25,566	24,866	24,218	23,616	23,055	22,531	22,038	21,576

FIGURE 13.3 The delivery capacity per hour in cubic feet is examined. *(Reprinted from the* 2000 Uniform Plumbing Code (UPC) *with the permission of the International Association of Plumbing and Mechanical Officials (IAPMO).)*

PIPE SIZING

The sizing of gas pipe depends on many factors. Code books contain tables that are used to determine proper pipe sizing. All pipe used for an installation, extension, or alteration must be sized to supply the full number of outlets for the intended purposes. When a gas supply has a pressure of 0.5 psig or less, and the gas meter is located within 3 feet of the building exterior, all building piping from the meter outlet downstream, including the pipe outlets, must have a minimum diameter of one-half inch. The hourly volume of gas demand at each gas outlet must not be less than the maximum hourly demand, as specified by the manufacturer of the appliance being served.

Calculating gas demand can seem intimidating, but it's really not all that difficult to do. The goal is to determine the cubic feet per hour of gas required. In order to do this, you must divide the maximum Btu/h input of an appliance by the average Btu/h heating value per cubic foot of gas being served. This is a simple formula, as long as you know what the Btu/h rating is for the appliances being served. Many appliances will be labeled with a Btu/h rating, but some will not bear the labeling. If you are faced with appliances that are not labeled with a rating, you can refer to a table in your code book that will allow you to estimate the rating needed to perform the gas demand calculation. Whether you are

Medium Pressure Natural Gas Systems for Sizing Gas Piping Systems Carrying Gas of 0.60 Specific Gravity
Capacity of Pipes of Different Diameters and Lengths in Liters Per Second for Gas Pressure of 13.8 kPa with a Drop to 10.3 kPa

Pipe Size, mm	Length in Meters											
	15.2	30.4	45.6	60.8	76.0	91.2	106.4	121.6	136.8	152.0	167.2	182.4
15	3.7	2.5	2.0	1.7	1.5	1.4	1.3	1.2	1.1	1.1	1.0	1.0
20	7.7	5.3	4.2	3.6	3.2	2.9	2.7	2.5	2.3	2.2	2.1	2.0
25	14.4	9.9	8.0	6.8	6.0	5.5	5.0	4.7	4.4	4.2	3.9	3.8
32	29.6	20.4	16.4	14.0	12.4	11.2	10.3	9.6	9.0	8.5	8.1	7.7
40	44.4	30.5	24.5	21.0	18.6	16.8	15.5	14.4	13.5	12.8	12.1	11.6
50	85.5	58.8	47.2	40.4	35.8	32.4	29.8	27.8	26.0	24.6	23.4	22.3
65	136.3	93.6	75.2	64.4	57.0	51.7	47.6	44.2	41.5	39.2	37.2	35.5
80	240.9	165.6	132.9	113.8	100.8	91.4	84.1	78.2	73.4	69.3	65.8	62.8
	197.6	212.8	228.0	243.2	258.4	273.6	288.8	304.0	334.4	364.8	395.2	425.6
15	0.9	0.9	0.8	0.8	0.8	0.8	0.7	0.7	0.7	0.7	0.6	0.6
20	1.9	1.8	1.8	1.7	1.7	1.6	1.6	1.5	1.4	1.4	1.3	1.3
25	3.6	3.5	3.3	3.2	3.1	3.0	2.9	2.9	2.7	2.6	2.5	2.4
32	7.4	7.1	6.8	6.6	6.4	6.2	6.0	5.9	5.6	5.3	5.1	4.9
40	11.1	10.6	10.3	9.9	9.6	9.3	9.0	8.8	8.3	8.0	7.6	7.3
50	21.3	20.5	19.8	19.1	18.5	17.9	17.4	16.9	16.1	15.3	14.7	14.1
65	34.0	32.7	31.5	30.4	29.4	28.5	27.7	26.9	25.6	24.4	23.4	22.5
80	60.1	57.8	55.7	53.7	52.0	50.4	49.0	47.6	45.2	43.2	41.3	39.7
100	122.7	117.8	113.5	109.6	106.1	102.9	99.9	97.2	92.3	88.0	84.3	81.0
	456.0	486.4	516.8	547.2	577.6	608.0	638.4	668.8	669.2	729.6	760.0	790.4
15	0.6	0.6	0.5	0.5	0.5	0.5	0.5	0.5	0.5	0.5	0.4	0.4
20	1.2	1.2	1.1	1.1	1.1	1.0	1.0	1.0	1.0	0.9	0.9	0.9
25	2.3	2.2	2.1	2.1	2.0	2.0	1.9	1.9	1.9	1.8	1.7	1.7
32	4.7	4.5	4.4	4.3	4.1	4.0	3.9	3.8	3.7	3.6	3.6	3.5
40	7.0	6.8	6.6	6.4	6.2	6.0	5.9	5.7	5.6	5.5	5.3	5.2
50	13.6	13.1	12.7	12.3	11.9	11.6	11.3	11.0	10.8	10.5	10.3	10.1
65	21.6	20.9	20.2	19.6	19.0	18.5	18.0	17.6	17.2	16.8	16.4	16.1
80	38.3	36.9	35.7	34.7	33.7	32.7	31.9	31.1	30.4	29.7	29.0	28.4
100	78.0	75.3	72.9	70.7	68.7	66.8	65.0	63.4	61.9	60.5	59.2	57.9
125	141.2	136.3	131.9	127.9	124.2	120.8	117.7	114.7	112.0	109.5	107.1	104.8
150	228.6	220.7	213.6	207.1	201.1	195.6	190.5	185.8	181.4	177.2	173.4	169.7

FIGURE 13.4 This is a metric version of the previous figure. *(Reprinted from the* 2000 Uniform Plumbing Code (UPC) *with the permission of the International Association of Plumbing and Mechanical Officials (IAPMO).)*

working with known ratings or estimates from a table, you must size the gas piping to maintain and supply gas in a capacity that is not less than the actual demand of the installed appliances. The sizing of all gas piping must be in keeping with code requirements.

MATERIALS FOR GAS PIPING

Materials for gas piping are available in various types. Some of the piping options available include:

- Aluminum-alloy pipe
- Aluminum-alloy tubing
- Brass pipe
- Copper pipe
- Copper-alloy pipe
- Copper tubing (Type K or L)

**Medium Pressure Natural Gas Systems for Sizing Gas Piping Systems
Carrying Gas of 0.60 Specific Gravity**

Capacity of Pipes of Different Diameters and Lengths in Cubic Feet Per Hour for
Gas Pressure of 3.0 psi with a Drop to 1.5 psi

Pipe Size, Inches						Length in Feet						
	50	100	150	200	250	300	350	400	450	500	550	600
1/2	857	589	473	405	359	325	299	278	261	247	234	224
3/4	1793	1232	990	847	751	680	626	582	546	516	490	467
1	3377	2321	1864	1595	1414	1281	1179	1096	1029	972	923	881
1-1/4	6934	4766	3827	3275	2903	2630	2420	2251	2112	1995	1895	1808
1-1/2	10,389	7140	5734	4908	4349	3941	3626	3373	3165	2989	2839	2709
2	20,008	13,752	11,043	9451	8377	7590	6983	6496	6095	5757	5468	5216
2-1/2	31,890	21,918	17,601	15,064	13,351	12,097	11,129	10,353	9714	9176	8715	8314
3	56,376	38,747	31,115	26,631	23,602	21,385	19,674	18,303	17173	16,222	15,406	14,698
	650	700	750	800	850	900	950	1000	1100	1200	1300	1400
1/2	214	206	198	191	185	180	174	170	161	154	147	141
3/4	448	430	414	400	387	375	365	355	337	321	308	296
1	843	810	780	754	729	707	687	668	634	605	580	557
1-1/4	1731	1663	1602	1547	1497	1452	1410	1371	1302	1242	1190	1143
1-1/2	2594	2492	2401	2318	2243	2175	2112	2055	1951	1862	1783	1713
2	4995	4799	4623	4465	4321	4189	4068	3957	3758	3585	3433	3298
2-1/2	7962	7649	7369	7116	6886	6677	6484	6307	5990	5714	5472	5257
3	14,075	13,522	13,027	12,580	12,174	11,803	11,463	11,149	10,589	10,102	9674	9294
4	28,709	27,581	26,570	25,658	24,831	24,074	23,380	22,741	21,598	20,605	19,731	18,956
	1500	1600	1700	1800	1900	2000	2100	2200	2300	2400	2500	2600
1/2	136	131	127	123	120	117	114	111	108	106	103	101
3/4	285	275	266	258	251	244	237	231	226	221	216	211
1	536	518	501	486	472	459	447	436	426	416	407	398
1-1/4	1101	1063	1029	998	969	942	918	895	874	854	835	818
1-1/2	1650	1593	1542	1495	1452	1412	1375	1341	1309	1279	1252	1225
2	3178	3069	2970	2879	2796	2720	2649	2583	2522	2464	2410	2360
2-1/2	5064	4891	4733	4589	4457	4335	4222	4117	4019	3927	3842	3761
3	8953	8646	8367	8112	7878	7663	7463	7278	7105	6943	6791	6649
4	18,262	17,635	17,066	16,546	16,069	15,629	15,222	14,844	14,491	14,161	13,852	13,561
5	33,038	31,904	30,875	29,935	29,072	28,276	27,539	26,855	26,217	25,620	25,060	24,534
6	53,496	51,660	49,993	48,471	47,074	45,785	44,593	43,484	42,451	41,485	40,579	39,727

FIGURE 13.5 This chart shows the figures for a higher gas pressure of 3.0 psi. *(Reprinted from the* 2000 *Uniform Plumbing Code (UPC) with the permission of the International Association of Plumbing and Mechanical Officials (IAPMO).)*

- Copper-alloy tubing (Type K or L)
- Copper tube (Type ACR)
- Corrugated stainless steel tubing
- Ductile iron pipe
- Plastic pipe
- Plastic tubing
- Steel pipe
- Steel tubing

All corrugated stainless steel tubing used for gas piping must be tested, listed, and installed in accordance with ANSI LC-1. Plastic pipe or tubing and compatible fittings must be installed only underground outside of buildings. Only polyethylene pipe shall be used with LP gases, and such applications must comply with NFPA 58. Plastic pipe is allowed to terminate above ground outside of buildings where installed in premanufac-

Medium Pressure Natural Gas Systems for Sizing Gas Piping Systems Carrying Gas of 0.60 Specific Gravity

Capacity of Pipes of Different Diameters and Lengths in Liters Per Second for Gas Pressure of 20.7 kPa with a Drop to 10.3 kPa

Pipe Size, mm					Length in Meters							
	15.2	30.4	45.6	60.8	76.0	91.2	106.4	121.6	136.8	152.0	167.2	182.4
15	6.7	4.6	3.7	3.2	2.8	2.6	2.4	2.2	2.1	1.9	1.8	1.8
20	14.1	9.7	7.8	6.7	5.9	5.4	4.9	4.6	4.3	4.1	3.9	3.7
25	26.6	18.3	14.7	12.6	11.1	10.1	9.3	8.6	8.1	7.6	7.3	6.9
32	54.5	37.5	30.1	25.8	22.8	20.7	19.0	17.7	16.6	15.7	14.9	14.2
40	81.7	56.2	45.1	38.6	34.2	31.0	28.5	26.5	24.9	23.5	22.3	21.3
50	157.4	108.2	86.9	74.4	65.9	59.7	54.9	51.1	47.9	45.3	43.0	41.0
65	250.9	172.4	138.5	118.5	105.0	95.2	87.6	81.5	76.4	72.2	68.6	65.4
80	443.5	304.8	244.8	209.5	185.7	168.2	154.8	144.0	135.1	127.6	121.2	115.6
	197.6	212.8	228.0	243.2	258.4	273.6	288.8	304.0	334.4	364.8	395.2	425.6
15	1.7	1.6	1.6	1.5	1.5	1.4	1.4	1.3	1.3	1.2	1.2	1.1
20	3.5	3.4	3.3	3.1	3.0	3.0	2.9	2.8	2.6	2.5	2.4	2.3
25	6.6	6.4	6.1	5.9	5.7	5.6	5.4	5.3	5.0	4.8	4.6	4.4
32	13.6	13.1	12.6	12.2	11.8	11.4	11.1	10.8	10.2	9.8	9.4	9.0
40	20.4	19.6	18.9	18.2	17.6	17.1	16.6	16.2	15.4	14.6	14.0	13.5
50	39.3	37.8	36.4	35.1	34.0	33.0	32.0	31.1	29.6	28.2	27.0	25.9
65	62.6	60.2	58.0	56.0	54.2	52.5	51.0	49.6	47.1	45.0	43.0	41.4
80	110.7	106.4	102.5	99.0	95.8	92.9	90.2	87.7	83.3	79.5	76.1	73.1
100	225.9	217.0	209.0	201.9	195.3	189.4	183.9	178.9	169.9	162.1	155.2	149.1
	456.0	486.4	516.8	547.2	577.6	608.0	638.4	668.8	699.2	729.6	760.0	790.4
15	1.1	1.0	1.0	1.0	0.9	0.9	0.9	0.9	0.9	0.8	0.8	0.8
20	2.2	2.2	2.1	2.0	2.0	1.9	1.9	1.8	1.8	1.7	1.7	1.7
25	4.2	4.1	3.9	3.8	3.7	3.6	3.5	3.4	3.3	3.3	3.2	3.1
32	8.7	8.4	8.1	7.8	7.6	7.4	7.2	7.0	6.9	6.7	6.6	6.4
40	13.0	12.5	12.1	11.8	11.4	11.1	10.8	10.6	10.3	10.1	9.8	9.6
50	25.0	24.1	23.4	22.6	22.0	21.4	20.8	20.3	19.8	19.4	19.0	18.6
65	39.8	38.5	37.2	36.1	35.1	34.1	33.2	32.4	31.6	30.9	30.2	29.6
80	70.4	68.0	65.8	63.8	62.0	60.3	58.7	57.3	55.9	54.6	53.4	52.3
100	143.7	138.7	134.3	130.2	126.4	123.0	119.8	116.8	114.0	111.4	109.0	106.7
125	259.9	251.0	242.9	235.5	228.7	222.4	216.7	211.3	206.2	201.6	197.2	193.0
150	420.8	406.4	393.3	381.3	370.3	360.2	350.8	342.1	334.0	326.4	319.2	312.5

FIGURE 13.6 This figure shows the metric figures for a gas pressure of 3.0 psi. *(Reprinted from the 2000 Uniform Plumbing Code (UPC) with the permission of the International Association of Plumbing and Mechanical Officials (IAPMO).)*

tured anodeless risers or service head adapter risers that are installed in accordance with the manufacturer's installation instructions.

Copper tubing has some restrictions on its use. For example, copper tubing cannot be used to convey gas that contains more than an average of 0.3 grain of hydrogen sulfide for 100 standard cubic feet of gas. Copper tubing systems must be identified with an appropriate label, with black letters on a yellow field, to indicate the piping system conveys fuel gas. The labels must be attached permanently to the tubing within 1 foot of the penetration of a wall, floor, or partition and at maximum intervals throughout the length of the tubing run. The labels must be visible to facilitate inspections.

Aluminum-alloy pipe and tubing is not approved for underground use or for use outside of a structure. Fittings used in gas applications must be approved for use with fuel-gas systems. All fittings must be compatible with, or shall be of the same material as, the pipe or tubing being used. At no time are bushings allowed to be used. All flange fittings must conform to ASME B16.1 or ASME B15.5. Any gasket material used for flanged fittings must be approved and compatible with the fuel gas being conveyed.

If used pipe is to be reused for gas piping, the used pipe must not have been used previously for any purpose other than conveying gas. Reused pipe must be clean, in good

**Medium Pressure Natural Gas Systems for Sizing Gas Piping Systems
Carrying Gas of 0.60 Specific Gravity**
Capacity of Pipes of Different Diameters and Lengths in Cubic Feet Per Hour for
Gas Pressure of 5.0 psi with a Drop to 1.5 psi

Pipe Size, Inches					Length in Feet							
	50	100	150	200	250	300	350	400	450	500	550	600
1/2	1399	961	772	661	586	531	488	454	426	402	382	365
3/4	2925	2010	1614	1381	1224	1109	1021	949	891	842	799	762
1	5509	3786	3041	2602	2306	2090	1923	1789	1678	1585	1506	1436
1-1/4	11,311	7774	6243	5343	4735	4291	3947	3672	3445	3255	3091	2949
1-1/2	16,947	11,648	9353	8005	7095	6429	5914	5502	5162	4876	4631	4418
2	32,638	22,432	18,014	15,417	13,664	12,381	11,390	10,596	9942	9391	8919	8509
2-1/2	52,020	35,753	28,711	24,573	21,779	19,733	18,154	16,889	15,846	14,968	14,216	13,562
3	91,962	63,205	50,756	43,441	38,501	34,884	32,093	29,865	28,013	26,461	25,131	23,976

	650	700	750	800	850	900	950	1000	1100	1200	1300	1400
1/2	349	335	323	312	302	293	284	277	263	251	240	231
3/4	730	701	676	653	632	612	595	578	549	524	502	482
1	1375	1321	1273	1229	1190	1153	1120	1089	1035	987	945	908
1-1/4	2824	2713	2614	2524	2442	2368	2300	2237	2124	2027	1941	1865
1-1/2	4231	4065	3916	3781	3659	3548	3446	3351	3183	3037	2908	2794
2	8149	7828	7542	7283	7048	6833	6636	6455	6130	5848	5600	5380
2-1/2	12,988	12,477	12,020	11,608	11,233	10,891	10,577	10,288	9771	9321	8926	8575
3	22,960	22,057	21,249	20,520	19,858	19,253	18,698	18,187	17,273	16,478	15,780	15,160
4	46,830	44,990	43,342	41,855	40,504	39,271	38,139	37,095	35,231	33,611	32,186	30,921

	1500	1600	1700	1800	1900	2000	2100	2200	2300	2400	2500	2600
1/2	222	214	208	201	195	190	185	181	176	172	168	165
3/4	464	449	434	421	409	398	387	378	369	360	352	345
1	875	845	818	798	770	749	729	711	694	678	664	650
1-1/4	1796	1735	1679	1628	1581	1537	1497	1460	1425	1393	1363	1334
1-1/2	2691	2599	2515	2439	2368	2303	2243	2188	2136	2087	2042	1999
2	5183	5005	4844	4696	4561	4436	4321	4213	4113	4020	3932	3849
2-1/2	8261	7978	7720	7485	7270	7071	6886	6715	6556	6406	6267	6135
3	14,605	14,103	13,648	13,233	12,851	12,500	12,174	11,871	11,589	11,326	11,078	10,846
4	29,789	28,766	27,838	26,991	26,213	25,495	24,831	24,214	23,639	23,100	22,596	22,121
5	53,892	52,043	50,363	48,830	47,422	46,124	44,923	43,806	42,765	41,792	40,879	40,021
6	87,263	84,269	81,550	79,067	76,787	74,686	72,740	70,932	69,247	67,671	66,193	64,803

FIGURE 13.7 This chart shows capacity of different-sized pipes at a pressure of 5.0 psi. (*Reprinted from the* 2000 Uniform Plumbing Code (UPC) *with the permission of the International Association of Plumbing and Mechanical Officials (IAPMO).*)

condition, free from internal obstructions, and burred ends must be reamed to the full bore of the pipe. Before any pipe, tubing, fittings, valves, or other devices can be reused, they must be cleaned thoroughly, inspected, and determined to be equivalent to new materials.

All joints and connections must be of an approved type and tested to be gastight at required test pressures. When connections are made between different types of materials, the joints must be made with approved adapter fittings. Any connections made between different metallic piping materials must be made with dielectric fittings. This is to isolate electrically above-ground piping from underground piping or to isolate electrically different metallic piping materials joined underground.

PREPARATION AND INSTALLATION

Preparation and installation of gas-pipe materials varies, depending upon the type of material being used. All pipe has to be prepared in a similar fashion. Ends of piping must b cut squarely. The pipe must be reamed and chamfered. Any burrs or obstructions in or

**Medium Pressure Natural Gas Systems for Sizing Gas Piping Systems
Carrying Gas of 0.60 Specific Gravity**
Capacity of Pipes of Different Diameters and Lengths in Liters Per Second for
Gas Pressure of 34.5 kPa with a Drop to 10.3 kPa

Pipe Size, mm	Length in Meters											
	15.2	30.4	45.6	60.8	76.0	91.2	106.4	121.6	136.8	152.0	167.2	182.4
15	11.0	7.6	6.1	5.2	4.6	4.2	3.8	3.6	3.4	3.2	3.0	2.9
20	23.0	15.8	12.7	10.9	9.6	8.7	8.0	7.5	7.0	6.6	6.3	6.0
25	46.3	29.8	23.9	20.5	18.1	16.4	15.1	14.1	13.0	12.5	11.8	11.3
32	89.0	61.2	49.1	42.0	37.3	33.8	31.1	28.9	27.1	25.6	24.3	23.2
40	133.3	91.6	73.6	63.0	55.8	50.6	46.5	43.3	40.6	38.4	36.4	34.8
50	256.8	176.5	141.7	121.3	107.5	97.4	89.6	83.4	78.2	73.9	70.2	66.9
65	409.2	281.3	225.9	193.3	171.3	155.2	142.8	132.9	124.7	117.8	111.8	106.7
80	723.5	497.2	399.3	341.7	302.9	274.4	252.5	234.9	220.4	208.2	197.7	188.6
	197.6	212.8	228.0	243.2	258.4	273.6	288.0	304.0	334.4	364.8	395.2	425.6
15	2.7	2.6	2.5	2.5	2.4	2.3	2.2	2.2	2.1	2.0	1.9	1.8
20	5.7	5.5	5.3	5.1	5.0	4.8	4.7	4.6	4.3	4.1	3.9	3.8
1	10.8	10.4	10.0	9.7	9.4	9.1	8.8	8.6	8.1	7.8	7.4	7.1
32	22.2	21.3	20.6	19.9	19.2	18.6	18.1	17.6	16.7	15.9	15.3	14.7
40	33.3	32.0	30.8	39.7	28.8	27.9	27.1	26.4	25.0	23.9	22.9	22.0
50	64.1	61.6	59.3	57.3	55.4	53.8	52.2	50.8	48.2	46.0	44.1	42.3
65	102.2	98.2	94.6	91.3	88.4	85.7	83.2	80.9	76.9	73.3	70.2	67.5
80	180.6	173.5	167.2	161.4	156.2	151.5	147.1	143.1	135.9	129.6	124.1	119.3
100	368.4	353.9	341.0	329.3	318.6	308.9	300.0	291.8	277.2	264.4	253.2	243.3
	456.0	486.4	516.8	547.2	577.6	608.0	638.4	668.8	699.2	729.6	760.0	790.4
15	1.7	1.7	1.6	1.6	1.5	1.5	1.5	1.4	1.4	1.4	1.3	1.3
20	3.7	3.5	3.4	3.3	3.2	3.1	3.0	3.0	2.9	2.8	2.8	2.7
25	6.9	6.6	6.4	6.2	6.1	5.9	5.7	5.6	5.5	5.3	5.2	5.1
32	14.1	13.6	13.2	12.8	12.4	12.1	11.8	11.5	11.2	11.0	10.7	10.5
40	21.2	20.4	19.8	19.2	18.6	18.1	17.6	17.2	16.8	16.4	16.1	15.7
50	40.8	39.4	38.1	36.9	35.9	34.9	34.0	33.1	32.4	31.6	30.9	30.3
65	65.0	62.8	60.7	58.9	57.2	55.6	54.2	52.8	51.6	50.4	49.3	48.3
80	114.9	111.0	107.4	104.1	101.1	98.3	95.8	93.4	91.2	89.1	87.2	85.3
100	234.3	226.3	219.0	212.3	206.2	200.6	195.3	190.5	186.0	181.7	177.8	174.0
125	424.0	409.4	396.2	384.1	373.1	362.9	353.4	344.6	336.4	328.8	321.6	314.8
150	686.5	662.9	641.6	622.0	604.1	587.6	572.2	558.0	544.8	532.4	520.7	509.8

FIGURE 13.8 This is the metric version of the previous chart. (*Reprinted from the* 2000 Uniform Plumbing Code (UPC) *with the permission of the International Association of Plumbing and Mechanical Officials (IAPMO).*)

on the pipe ends must be removed. Every pipe end must have a full-bore opening and must not be undercut. In addition to pipe preparation, all joints must be prepared properly.

When working with brazed joints, you must ensure that all joint surfaces are cleaned. An approved flux must be applied where it is required. Filler used to make a brazed joint must conform to AWS A5.8. Brazing materials must have a melting point that is in excess of 1000°F, and alloys used for brazing must not contain more than 0.05 percent phosphorous. All brazed joints must be made by certified braziers.

Flared joints must be created with tools designed for the purpose of flaring joints. When mechanical joints are needed, they must be made with fittings that are specified by the manufacturer for the gas service and must be installed according to the manufacturer's recommendations.

Pipe-joint compound or sealant tape must be used on threaded joints. All threads used with threaded joints must conform to ASME B1.20.1. Joint surfaces must be cleaned by an approved procedure when joints are to be welded. Any filler material used to weld a joint must be of an approved type.

Joints made between aluminum-alloy pipe and tubing or fittings must be flared or made as mechanical joints. Brass piping allows for connections to be made by brazing,

Maximum Capacity of Pipe in Thousands of BTU per Hour of Undiluted Liquified Petroleum Gases

(Based on a Pressure Drop of 0.5 Inch Water Column)
Low Pressure 11" Water Column

Length in Feet

Pipe Size, Inches	10	20	30	40	50	60	70	80	90	100	125	150	200
1/2	275	189	152	129	114	103	96	89	83	78	69	63	55
3/4	567	393	315	267	237	217	196	185	173	162	146	132	112
1	1071	732	590	504	448	409	378	346	322	307	275	252	213
1-1/4	2205	1496	1212	1039	913	834	771	724	677	630	567	511	440
1-1/2	3307	2299	1858	1559	1417	1275	1181	1086	1023	976	866	787	675
2	6221	4331	3465	2992	2646	2394	2205	2047	1921	1811	1606	1496	1260

FIGURE 13.9 The maximum capacity of pipes in thousands of BTU per hour for petroleum gas is examined here. *(Reprinted from the 2000 Uniform Plumbing Code (UPC) with the permission of the International Association of Plumbing and Mechanical Officials (IAPMO).)*

Maximum Capacity of Pipe in Thousands of Watts of Undiluted Liquified Petroleum Gases

(Based on a Pressure Drop of 12.7 mm Water Column)
Low Pressure 279 mm Water Column

Length in Millimeters

Pipe Size, mm	3048	6096	9144	12192	15240	18288	21336	24384	27432	30480	38100	45720	60960
15	80.6	55.4	44.5	37.8	33.4	30.2	28.1	26.1	24.3	22.9	20.2	18.5	16.1
20	166.1	115.2	92.3	78.2	69.4	63.6	57.4	54.2	50.7	47.5	42.8	38.7	32.8
25	313.8	214.5	172.9	147.7	131.3	119.8	110.8	101.4	94.4	90.0	80.6	73.8	62.4
32	646.1	438.3	355.1	304.4	267.5	244.4	225.9	212.1	198.4	184.6	166.1	149.7	128.9
40	969.0	673.6	544.4	456.8	415.2	373.6	346.0	318.2	299.7	286.0	253.7	230.6	197.8
50	1822.8	1269.0	1015.2	876.7	775.3	701.4	646.1	600.0	562.9	530.6	470.6	438.3	369.2

FIGURE 13.10 This is the metric version of the previous chart. (Reprinted from the 2000 Uniform Plumbing Code (UPC) with the permission of the International Association of Plumbing and Mechanical Officials (IAPMO).)

For Undiluted Liquified Petroleum Gas Pressure of 10.0 psi with Maximum Pressure Drop of 3.0 psi

Maximum Delivery Capacity in Cubic Feet of Gas per Hour of IPS Pipe of Different Diameters and Lengths Carrying Undiluted Liquified Petroleum Gas of 1.52 Specific Gravity

Pipe Size, Inches	Length in Feet											
	50	100	150	200	250	300	350	400	450	500	550	600
1/2	1000	690	550	470	420	390	350	325	300	285	272	260
3/4	2070	1423	1142	978	867	785	722	672	631	596	566	540
1	3899	2680	2152	1842	1632	1479	1361	1266	1188	1122	1066	1017
1-1/4	8005	5502	4418	3782	3351	3037	2794	2599	2439	2303	2188	2087
1-1/2	11,994	8244	6620	5666	5022	4550	4186	3894	3654	3451	3278	3127
2	23,100	15,877	12,750	10,912	9671	8763	8062	7500	7037	6647	6313	6023
2-1/2	36,818	25,305	20,321	17,392	15,414	13,966	12,849	11,953	11,215	10,594	10,062	9599
3	65,088	44,734	35,923	30,746	27,249	24,690	22,714	21,131	19,827	18,728	17,787	16,969

	650	700	750	800	850	900	950	1000	1100	1200	1300	1400
1/2	250	240	230	222	215	208	202	198	188	180	171	164
3/4	517	496	478	462	447	433	421	409	389	371	355	341
1	973	935	901	870	842	816	793	771	732	699	669	643
1-1/4	1999	1920	1850	1786	1729	1676	1628	1583	1504	1434	1374	1320
1-1/2	2995	2877	2772	2676	2590	2511	2439	2372	2253	2149	2058	1977
2	5767	5541	5338	5155	4988	4836	4697	4568	4339	4139	3964	3808
2-1/2	9192	8831	8507	8215	7950	7708	7486	7281	6915	6597	6318	6069
3	16,250	15,611	15,040	14,523	14,055	13,627	13,234	12,872	12,225	11,663	11,169	10,730
4	33,145	31,842	30,676	29,623	28,667	27,795	26,993	26,255	24,935	23,789	22,780	21,885

	1500	1600	1700	1800	1900	2000	2100	2200	2300	2400	2500	2600
1/2	158	152	148	143	139	136	133	130	127	124	121	118
3/4	329	317	307	298	289	281	274	267	261	255	249	244
1	619	598	579	561	545	530	516	503	491	480	470	460
1-1/4	1271	1228	1138	1152	1119	1088	1060	1033	1009	986	964	944
1-1/2	1905	1839	1780	1726	1676	1630	1588	1548	1512	1477	1445	1415
2	3669	3543	3428	3324	3228	3140	3058	2982	2911	2845	2783	2724
2-1/2	5847	5646	5464	5298	5145	5004	4874	4753	4640	4534	4435	4342
3	10,337	9982	9660	9366	9096	8847	8616	8402	8203	8016	7841	7676
4	21,083	20,360	19,705	19,103	18,552	18,045	17,575	17,138	16,731	16,350	15,993	15,657
5	38,143	36,834	35,645	34,560	33,564	32,645	31,795	31,005	30,268	29,579	28,933	28,325
6	61,762	59,643	57,718	55,961	54,348	52,860	51,483	50,204	49,011	47,895	46,849	45,865

FIGURE 13.11 Pipe size and length for carrying undiluted liquified petroleum gas of 1.52 specific gravity is examined. *(Reprinted from the* 2000 Uniform Plumbing Code (UPC) *with the permission of the International Association of Plumbing and Mechanical Officials (IAPMO).)*

mechanical joints, threaded connections, or welded joints. When copper or copper-alloy pipe is used, joints must be brazed, mechanical, threaded, or welded. Joints made between copper or copper-alloy tubing or fittings must be brazed, flared, or mechanical. Flanged joints are required when ductile iron pipe is used for gas piping. All connections must conform to the manufacturer's instructions.

JOINING PLASTIC PIPE AND FITTINGS

The installation of plastic pipe and fittings requires joints as described in the general plumbing code. When plastic pipe and fittings are used, joints between them may be made with a solvent-cement method, a heat-fusion method, or with compression couplings or flanges. The method of joining the pipe and fittings must be compatible with the materials and the manufacturer's recommendations. Threading plastic pipe or tubing is prohibited. When

For Undiluted Liquified Petroleum Gas Pressure of 68.9 kPa with Maximum Pressure Drop of 20.7 kPa

Maximum Delivery Capacity in Liters per Second of IPS Pipe of Different Diameters
and Lengths Carrying Undiluted Liquified Petroleum Gas of 1.52 Gravity

Pipe Size, mm	Length in Meters											
	15.2	30.4	45.6	60.8	76.0	91.2	106.4	121.6	136.8	152.0	167.2	182.4
15	8.0	5.5	4.4	3.8	3.4	3.0	2.8	2.6	2.4	2.3	2.2	2.1
20	16.6	11.4	9.1	7.8	6.9	6.3	5.8	5.4	5.1	4.8	4.5	4.3
25	31.2	21.4	17.2	14.7	13.1	11.8	10.9	10.1	9.5	9.0	8.5	8.1
32	64.0	44.0	35.3	30.3	26.8	24.3	22.4	20.8	19.5	18.4	17.5	16.7
40	96.0	66.0	53.0	45.3	40.2	36.4	33.5	31.2	29.2	27.6	26.2	25.0
50	184.8	127.0	102.0	87.3	77.4	70.1	64.5	60.0	56.3	53.2	50.5	48.2
65	294.5	202.4	162.6	139.1	123.3	111.7	102.8	95.6	89.7	84.8	80.5	76.8
80	520.7	357.9	287.4	246.0	218.0	197.5	181.7	169.1	158.6	149.8	142.3	135.8

	197.6	212.8	228.0	243.2	258.4	273.6	288.8	304.0	334.4	364.8	395.2	425.6
15	2.0	1.9	1.8	1.8	1.7	1.7	1.6	1.6	1.5	1.4	1.4	1.3
20	4.1	4.0	3.8	3.7	3.6	3.5	3.4	3.3	3.1	3.0	2.8	2.7
25	7.8	7.5	7.2	7.0	6.7	6.5	6.3	6.2	5.9	5.6	5.4	5.1
32	16.0	15.4	14.8	14.3	13.8	13.4	13.0	12.7	12.0	11.5	11.0	10.6
40	24.0	23.0	22.2	21.4	20.7	20.1	19.5	19.0	18.0	17.2	16.5	15.8
50	46.1	44.3	42.7	41.2	39.9	38.7	37.6	36.5	34.7	33.1	31.7	30.5
65	73.5	70.7	68.1	65.7	63.6	61.7	60.0	58.2	55.3	52.8	50.5	48.6
80	130.0	124.9	120.3	116.2	112.4	109.0	105.9	103.0	97.8	93.3	89.4	85.8
100	265.2	254.7	245.4	237.0	229.3	222.4	215.9	210.0	199.5	190.3	182.2	175.1

	456.0	486.4	516.8	547.2	577.6	608.0	638.4	668.8	699.2	729.6	760.0	790.4
15	1.3	1.2	1.2	1.1	1.1	1.1	1.1	1.0	1.0	1.0	1.0	0.9
20	2.6	2.5	2.5	2.4	2.3	2.2	2.2	2.1	2.1	2.0	2.0	2.0
25	5.0	4.8	4.6	4.5	4.4	4.2	4.1	4.0	3.9	3.8	3.8	3.7
32	10.2	9.8	9.1	9.2	9.0	8.7	8.5	8.3	8.1	7.9	7.7	7.6
40	15.2	14.7	14.2	13.8	13.4	13.0	12.7	12.4	12.1	11.8	11.6	11.3
50	29.4	28.3	27.4	26.6	25.8	25.1	24.5	23.9	23.3	22.8	22.3	21.6
65	46.8	45.2	43.7	42.4	41.2	40.0	39.0	38.0	37.1	36.3	35.5	34.7
80	82.7	79.9	77.3	74.9	72.8	70.8	68.9	67.2	65.6	64.1	62.7	61.4
100	168.7	162.9	157.6	152.8	148.4	144.4	140.6	137.1	133.8	130.8	127.9	125.3
125	305.1	294.7	285.2	276.5	268.5	261.2	254.4	248.0	242.1	236.6	231.5	226.6
150	494.1	477.1	461.7	447.7	434.8	422.9	411.9	401.6	392.1	383.2	374.8	366.9

FIGURE 13.12 This is the metric version of the previous chart. *(Reprinted from the 2000 Uniform Plumbing Code (UPC) with the permission of the International Association of Plumbing and Mechanical Officials (IAPMO).)*

polyethylene pipe, tubing, or fittings are used, joints shall be made by means of either heat-fusion or mechanical joints. Joints of this type must be made to sustain effectively the longitudinal pull-out forces caused by contraction of the piping or by external loading.

Heat-fusion joints are allowed only on polyolefin pipe or tubing. This includes polyethylene and polybutylene pipe and tubing. All joint surfaces must be clean and dry. The joint surfaces have to be heated to melt temperature and joined. Joints made this way must not be disturbed until they cool to a tolerance allowed by the manufacturer.

Plastic gas piping can be joined with the use of mechanical compression fittings. These fittings must be designed and approved for use with plastic pipe that carries natural gas or LP-gas vapor. Mechanical fittings used with polyethylene pipe must conform with ASTM D 2513 category 1, full-restraint, full-seal joints and must be so marked.

The gasket material in a compression-type mechanical fitting must be compatible with the plastic piping and the gas being distributed. An internal tubular rigid stiffener must be used in conjunction with the fitting, and the stiffener must be flush with the end of the pipe or tubing. The stiffener also must extend at least to the outside end of the compression fitting when it is installed. No stiffener that has rough or sharp edges may be used. Split tubular stiffeners are prohibited, and no stiffener shall be force fitted into the plastic.

Capacities of Listed Metal Appliance Connectors for Use with Gas Pressures Not Less Than an Eight Inch Water Column

Capacities for Various Lengths, in Thousands Btu/h
(Based on Pressure Drop of 0.2" Water Column Natural Gas of 1100 Btu/cu. ft.)

Semi-Rigid Connector O.D., Inches	Flexible Connector Nominal I.D., Inches	1 foot	1-1/2 feet	2 feet	2-1/2 feet	3 feet	4 feet	5 feet	6 feet
		All Gas Appliances					Ranges and Dryers Only		
3/8	1/4	40	33	29	27	25			
1/2	3/8	93	76	66	62	58			
5/8	1/2	189	155	134	125	116	101	90	80
	3/4	404	330	287	266	244			
	1	803	661	573	534	500			

Notes:

1. Flexible connector listings are based on the nominal internal diameter.
2. Semi-rigid connector listings are based on the outside diameter.
3. Gas connectors are certified by the testing agency as complete assemblies, including the fittings and valves. Capacities shown are based on the use of fittings and valves supplied with the connector.
4. Capacities for LPG are 1.6 times the natural gas capacities shown.

Example: Capacity of a 1/4" flexible connector one (1) foot long is 40,000 x 1.6 = 64,000 Btu/h

FIGURE 13.13 The capacities of various metal appliance connectors are shown. (*Reprinted from the 2000 Uniform Plumbing Code (UPC) with the permission of the International Association of Plumbing and Mechanical Officials (IAPMO).*)

Capacities of Listed Metal Appliance Connectors for Use with Gas Pressures Not Less Than a 203 mm Water Column

Capacities for Various Lengths, in Thousands Watts
(Based on Pressure Drop of 50 Pa mm Water Column Natural Gas of 11.4 W/L)

Semi-Rigid Connector O.D., mm	Flexible Connector Nominal I.D., mm	All Gas Appliances					Ranges and Dryers Only		
		305 mm	457 mm	610 mm	762 mm	914 mm	1219 mm	1524 mm	1829 mm
10	6.4	11.7	9.7	8.5	7.9	7.3			
15	9.5	27.3	22.3	19.3	18.2	17.0			
18	12.7	55.4	45.4	39.3	36.6	34.0	29.6	26.4	23.4
	19.1	118.4	96.7	84.1	77.9	71.5			
	25.4	235.3	193.7	167.9	156.5	146.5			

Note: See Table 12-9 (English units).

Example: Capacity of a 6.4 mm flexible connector 0.3 m long is 11720W x 1.6 = 18752W

FIGURE 13.141 This is the metric version of the previous chart. (*Reprinted from the 2000 Uniform Plumbing Code (UPC) with the permission of the International Association of Plumbing and Mechanical Officials (IAPMO).*)

Copper Tube – Low Pressure
Maximum Delivery Capacity* of Cubic Feet of Gas Per Hour of Copper Tube Carrying Natural Gas of 0.60** Specific Gravity at Low Pressure (Less than 14 inches Water Column) Based on Pressure Drop of 0.50 Inch Water Column:

Length of Tube, feet	Outside Diameter of Tube, inches						
	3/8 (10 mm)	1/2 (15 mm)	5/8 (18 mm)	3/4 (20 mm)	7/8 (22 mm)	1-1/8 (28 mm)	1-3/8 (34 mm)
10	24	50	101	176	250	535	963
20	17	34	69	121	172	368	662
30	13	27	56	97	138	295	531
40	11	23	48	83	118	253	455
50	10	21	42	74	105	224	403
60	9.1	19	38	67	95	203	365
70	8.4	17	35	62	84	197	336
80	7.8	16	33	57	81	174	313
90	7.3	15	31	54	76	163	293
100	6.9	14	29	51	72	154	277
125	6.1	13	26	45	64	136	245
150	5.6	11	23	41	58	124	222
175	5.1	11	21	38	53	114	205
200	4.8	10	20	35	50	106	190
250	4.2	8.7	18	31	44	94	169

*Includes 20% factor for fittings.
**For other pressure drops values see Table 12-12.

FIGURE 13.15 This illustrates the use of copper tube for carrying gas at low pressure. *(Reprinted from the 2000 Uniform Plumbing Code (UPC) with the permission of the International Association of Plumbing and Mechanical Officials (IAPMO).)*

PVC pipe, tubing, and fittings used for gas piping can be joined with a solvent-cement joint. The procedure for making this type of joint is essentially the same as those used for making joints in drain, waste, and vent piping where PVC pipe is used. All services to be joined must be clean and dry. An application of an approved primer to both a pipe end and a fitting hub is required prior to cementing a joint. Any solvent cement used to make a connection must be of an approved type. All solvent-cement joints must be made while the cement is wet.

Polyolefin plastic pipe (polybutylene and polybutylene), tubing, and fittings are not allowed to be joined with solvent cement. Heat fusion and mechanical joints are acceptable ways of joining polyolefin plastic pipe, tubing, and fittings. It is not acceptable to use solvent cement or heat fusion to join different types of plastic.

Flanges and special joints must be qualified by the manufacturer for the intended use with plastic pipe and tubing. Any connection made with a flange or other special joint must be approved by a code official. Polyethylene pipe and tubing must

Specific Gravity
Multipliers to be Used with Copper Tube when Specific Gravity of Gas is other than 0.60.

Specific Gravity	Multiplier	Specific Gravity	Multiplier
.35	1.31	1.00	.78
.40	1.23	1.10	.74
.45	1.16	1.20	.71
.50	1.10	1.30	.68
.55	1.04	1.40	.66
.60	1.00	1.50	.63
.65	.96	1.60	.59
.70	.93	1.70	.58
.75	.90	1.80	.56
.80	.87	1.90	.56
.85	.84	2.00	.55
.90	.82	2.10	.54

Adjustment for a gas with an average specific gravity (relative density) other than 0.60 is achieved by multiplying the CFH values of Tables 12-11, 12-13, or 12-14 by the appropriate multiplier.

FIGURE 13.16 Certain figures need to be used when specific gravity of gas is other than 060. *(Reprinted from the 2000 Uniform Plumbing Code (UPC) with the permission of the International Association of Plumbing and Mechanical Officials (IAPMO).)*

not be flared at any time. PVC pipe and tubing usually is not allowed to be flared. However, there can be times when a manufacturer specifies that PVC pipe can be flared for use in underground installations.

JOINING STEEL PIPE AND TUBING

Steel pipe and fittings can be joined with threaded joints, welded joints, or mechanical joints. Mechanical joints must be made with an approved elastomeric seal. All mechanical joints must be installed within the guidelines of the manufacturer's recommendations.

Copper Tube – Medium Pressure
Maximum Delivery Capacity* of Cubic Feet of Gas Per Hour of Copper Tube Carrying Natural Gas of 0.60** Specific Gravity at Medium Pressure (2.0 psig) Based on Pressure Drop of 1.0 psig:

Length of Tube, feet	Outside Diameter of Tube, inches						
	3/8 (10 mm)	1/2 (15 mm)	5/8 (18 mm)	3/4 (20 mm)	7/8 (22 mm)	1-1/8 (28 mm)	1-3/8 (34 mm)
10	222	458	932	1629	2311	4937	8889
20	153	315	641	1120	1589	3393	6109
30	123	253	515	899	1276	2725	4906
40	105	216	440	770	1092	2332	4199
50	93	192	390	682	968	2067	3721
60	84	174	354	618	877	1873	3372
70	78	160	325	569	807	1723	3102
80	72	149	303	529	750	1603	2886
90	68	140	254	496	704	1504	2708
100	64	132	268	469	665	1421	2558
125	57	117	238	415	589	1259	2267
150	51	106	215	376	534	1141	2054
175	44	91	184	322	457	976	1758
200	39	80	163	286	405	865	1558
250							

*Includes 20% factor for fittings.
**For other pressure drops values see Table 12-12.

FIGURE 13.17 This illustrates the use of copper tube at medium pressure. *(Reprinted from the* 2000 *Uniform Plumbing Code (UPC) with the permission of the International Association of Plumbing and Mechanical Officials (IAPMO).)*

Mechanical joints for steel gas piping usually are allowed only outside and underground. Steel tubing can be joined with welded joints or mechanical joints, but not with threaded joints.

Connecting metal pipe to plastic pipe is allowed only when the connection is made outside and underground. There is one exception to this rule. Plastic pipe is permitted to terminate above ground outside of buildings where installed in premanufactured anodeless risers or service head adapter risers that are installed in accordance with the manufacturer's installation instructions. Underground connections between metallic piping and plastic piping are to be made with mechanical fittings or factory-assembled, plastic-to-steel transition fittings when they are specified by the manufacturer for use in gas piping applications.

When installing a compression-end riser transition, you must be sure that there is at least twelve inches of horizontal length of metallic piping underground at the end of any plastic piping installed. Metallic pipe installed below ground for the purpose of conveying gas must be protected from corrosion. The length and size of metallic pipe joined with plastic pipe must be adequate to prevent stress or strain on the plastic piping.

Copper Tube – High Pressure
Maximum Delivery Capacity* of Cubic Feet of Gas Per Hour of Copper Tube Carrying Natural Gas of 0.60** Specific Gravity at High Pressure (5.0 psig) Based on Pressure Drop of 3.50 psig:

Length of Tube, feet	Outside Diameter of Tube, inches						
	3/8 (10 mm)	1/2 (15 mm)	5/8 (18 mm)	3/4 (20 mm)	7/8 (22 mm)	1-1/8 (28 mm)	1-3/8 (34 mm)
10	462	954	1941	3392	4812	10279	18504
20	318	656	1334	2331	3307	7064	12718
30	255	527	1071	1872	2656	5673	10213
40	218	451	917	1602	2273	4855	8741
50	194	399	812	1420	2015	4303	8741
60	175	362	736	1287	1825	3899	7019
70	161	333	677	1184	1679	3587	6458
80	150	310	630	1101	1562	3337	6008
90	141	291	591	1033	1466	3131	5637
100	133	274	558	976	1385	2958	5324
125	118	243	495	865	1227	2621	4719
150	107	220	448	784	1112	2375	4276
175	98	203	413	721	1023	2185	3934
200	91	189	384	671	952	2033	3659
250	81	167	340	594	843	1802	3243

*Includes 20% factor for fittings.
**For other pressure drops values see Table 12-12.

FIGURE 13.18 This shows the use of copper tube at high pressure. *(Reprinted from the* 2000 Uniform Plumbing Code (UPC) *with the permission of the International Association of Plumbing and Mechanical Officials (IAPMO).)*

Anodeless risers and service head adapter risers are the only types of risers allowable when polyethylene gas pipe is terminated above ground. Any above-ground portion of polyethylene must be centered in the metallic casing to ensure that the temperature of the polyethylene pipe does not exceed 150°F.

All metallic pipe and tubing exposed to corrosive action must be protected from corrosion. An example of such a situation would be steel pipe installed below ground. Zinc coatings, the type which are used to galvanize pipe, are not acceptable protection for underground gas piping. Consult a local code official for advice in choosing a suitable means of protection.

Protective coatings and wrapping for gas piping must be of an approved type and installed, for the most part, with a machine. When wrapping is done in the field, the wrapping is to be limited to fittings, short sections of piping, and piping where the factory wrap has been damaged or stripped for threading or welding.

When joint compound is used to seal threaded joints, the compound must be applied only to male threads. The joint compound must be resistant to the action of

Capacities of Listed Metal Appliance Connectors for Use with Gas Pressures Less Than an Eight Inch Water Column

Capacities for Various Lengths, in Thousands Btu/h
(Based on Pressure Drop of 0.2" Water Column Natural Gas of 1100 Btu/cu. ft.)

Semi-Rigid Connector O.D., Inches	Flexible Connector Nominal I.D., Inches	1 foot	1-1/2 feet	2 feet	2-1/2 feet	3 feet	4 feet	5 feet	6 feet
			All Gas Appliances				Ranges and Dryers Only		
3/8	1/4	28	23	20	19	17			
1/2	3/8	66	54	47	44	41			
5/8	1/2	134	110	95	88	82	72	63	57
—	3/4	285	233	202	188	174			
—	1	567	467	405	378	353			

Notes:

1. Flexible connector listings are based on the nominal internal diameter.
2. Semi-rigid connector listings are based on the outside diameter.
3. Gas connectors are certified by the testing agency as complete assemblies, including the fittings and valves.
4. Capacities shown are based on the use of fittings and valves supplied with the connector.
5. Capacities for LPG are 1.6 times the natural gas capacities shown.

Example: Capacity of a 1/4" flexible connector 1 foot long is 28,000 x 1.6 = 44,800 Btu/h

FIGURE 13.19 The capacities of different metal appliance connectors is listed. *(Reprinted from the 2000 Uniform Plumbing Code (UPC) with the permission of the International Association of Plumbing and Mechanical Officials (IAPMO).)*

Capacities of Listed Metal Appliance Connectors for Use with Gas Pressures Less Than a 203 mm Water Column

Semi-Rigid Connector O.D., mm	Flexible Connector Nominal I.D., mm	Capacities for Various Lengths, in Thousands Watts (Based on Pressure Drop of 50 Pa Water Column Natural Gas of 11.4 W/L)							
		305 mm	457 mm	610 mm	762 mm	914 mm	1219 mm	1524 mm	1829 mm
		All Gas Appliances					Ranges and Dryers Only		
10	6.4	8.2	6.7	5.9	5.6	5			
15	9.5	19.3	15.8	13.8	12.9	12			
18	12.7	39.3	32.2	27.8	25.8	24	21.1	18.5	16.7
	19.1	83.5	68.3	59.2	55.1	51			
	25.9	166.1	136.8	118.7	110.8	103.4			

Note: See Table 12-10 (English units).

Example: Capacity of a 6.4 mm flexible connector 305 mm long is 8204 x 1.6 = 13126.4W

FIGURE 13.20 This is the metric version of the previous chart. (*Reprinted from the 2000 Uniform Plumbing Code (UPC) with the permission of the International Association of Plumbing and Mechanical Officials (IAPMO).*)

Support of Piping				
Size of Pipe				
	Inches	mm	Feet	mm
	1/2	15	6	1829
	3/4 or 1	20 or 25	8	2438
Horizontal	1-1/4 or larger	32 or larger	10	3048
Vertical	1-1/4 or larger	32 or larger	Every floor level	

FIGURE 13.21 Different size pipes are needed for either horizontal or vertical use each floor level. *(Reprinted from the 2000 Uniform Plumbing Code (UPC) with the permission of the International Association of Plumbing and Mechanical Officials (IAPMO).)*

liquefied petroleum gases. All threads on pipe and fittings must comply with ASME B1.20.1 and all code requirements. Any threads that are damaged must not be used. Welds that open or are defective must be cut out and replaced by a new section of piping and connection.

INSTALLING GAS PIPING SYSTEMS

There are a number of rules to follow when installing a gas piping system. There are interior locations where gas piping may not be installed. It is a code violation to install gas piping inside any of the following:

• Air ducts
• Clothes chutes
• Chimneys
• Vents
• Ventilating ducts
• Dumbwaiters
• Elevator shafts
• Concealed plenums

A gas pipe installed in a concealed location must not be fitted with unions, tubing fittings, or running threads. Concealed piping is not allowed in solid partitions and walls, unless the gas piping is installed in a chase or casing. It is never acceptable to allow a gas pipe to be used as a grounding electrode.

Gas pipe that penetrates masonry work must be protected by being encased in a sleeve. The sleeve opening on the outside of a building must be sealed so that water will not run through the sleeve and into a building. Penetrating a building foundation below grade is not allowed when installing gas piping.

Gas piping that is going to be concealed should be run with black or galvanized steel pipe. When some other type of pipe or tubing is used and the pipe or tubing is not protected by at least 1.25 inches of wood or framing member, the installation must be protected from damage with the use of shield plates. The plates must be made of steel and have a minimum thickness of $1/16$ inch. Shield plates must cover the area of the gas pipe or tubing where it is not protected adequately by the framing member. Shield plates must extend a minimum of four inches above sole plates, below top plates, and to each side of a stud, joist, or rafter.

Minimum Demand of Typical Gas Appliances in Btus (Watts) Per Hour

Appliance	Demand in	
	Btu/h	Watts/h
Barbecue (residential)	50,000	14,650
Bunsen Burner	3,000	879
Domestic Clothesdryer	35,000	10,255
Domestic Gas Range	65,000	19,045
Domestic Recessed Oven Section	25,000	7,325
Domestic Recessed Top Burner Section	40,000	11,720
Fireplace Log Lighter (commercial)	50,000	14,650
Fireplace Log Lighter (residential)	25,000	7,325
Gas Engines (per horsepower)	10,000	2,930
Gas Refrigerator	3,000	879
Mobile Homes (see Appendix E)	—	—
Steam Boilers (per horsepower)	50,000	14,650
Storage Water Heater		
up to 30 gallon (114 L) tank	30,000	8,790
Storage Water Heater		
40 (151 L) to 50 gallon (189 L) tank	50,000	14,650

FIGURE 13.22 Appliances place demands in Btu per hour. Here are some of the numbers. (*Reprinted from the 2000 Uniform Plumbing Code (UPC) with the permission of the International Association of Plumbing and Mechanical Officials (IAPMO).*)

Unless a code official determines that no other option exists, you must not install gas piping in solid floor slabs, such as concrete. In the case of solid floors, house piping is to be installed above the floor. The piping can be installed in open or furred spaces, hollow partitions, hollow walls, attic space, or pipe chases.

When a code official determines that gas piping in a solid floor cannot be avoided, special rules must be followed. There are two basic options to consider. One way to install such piping is to sleeve it in Schedule 40 steel pipe that has tightly sealed ends and joints. Both ends of the casing must extend at least two inches beyond the point where the pipe emerges from the solid floor. The other option is to install piping in a channel in the floor. The channel in the floor must be covered so that access to the gas piping is available. The covering must prevent the entrance of corrosive materials, or the channel must be filled with a noncorrosive material that will cause a minimum of damage to a floor when the material is removed.

Gas pipe that penetrates solid floors or solid walls must be protected, unless otherwise approved by a code official. The proper type of protection is a sleeve. Gas pipe installed through a solid wall or solid floor must either be installed in a casing or through an opening of adequate size. Piping must be encased in a 1:3 mixture of cement and sand. The coating must not be less than $3/4$-inch thick.

Except where otherwise approved, gas piping is not allowed to be installed in a common trench with water, sewer, or drainage piping. All gas piping that is installed in contact with earth or other corrosive material must be protected from corrosion. Any dissimilar metals that are joined together underground require the use of an insulated coupling. It is prohibited to place metallic piping in contact with cinders. The minimum allowable depth for burying gas pipe is eighteen inches, unless otherwise specified by a code official. When individual gas lines are installed for outside lights, grills, or other appliances the gas piping must be buried at least eight inches deep. Additional depth may be required if it is believed that the piping will be subject to physical damage at an eight-inch depth. Buried gas piping must not be installed so as to be in contact with the ground or fill under a building or floor slab.

Pipe hangers and supports shall be in compliance with standard code requirements. Copper tubing conveying gas and running parallel to joists must be secured to the center of the joist at a maximum interval of six feet. If the copper tubing is running at an angle to joists, it must be installed either through holes in the joists that are at least one-and-a-half times the outside diameter of the tubing or secured to and supported at maximum intervals of six feet. If the tubing is installed closer than $1^{3}/4$ inches to the face of a joist, a steel nail plate must be installed on the face of the joist to prevent the tubing from being damaged by nails or staples. Copper tubing installed vertically in walls must be supported at the floor and ceiling levels and protected by nail plates. The tubing must not be supported at any other point in the wall. Nail plates must have a minimum thickness of 0.0508 inch (16 gauge) and a minimum length of four inches beyond concealed penetrations.

Changes in direction for gas piping must be made by the use of fittings or bends that are approved by the standard code. Bends used with metallic pipe must be smooth and free from buckling, cracks, or other evidence of mechanical damage. The bends must be made only with bending equipment and procedures that are intended for the purpose of making bends. Longitudinal welds of a pipe are required to be near the neutral axis of the bend. No bend shall exceed a 90° angle. An inside radius of a bend must not be less than six times the outside diameter of the pipe. Plastic piping must be bent in accordance with the manufacturer's recommendations.

Gas outlets may not be installed behind doors. Any unthreaded portion of gas piping outlets must extend at least one inch through finished ceilings and walls. When outlets are extending through floors, slabs, or outdoor patios, the outlets must extend at least two inches above the floor, slab, or patio. Outlet fittings or piping must be fastened securely.

All outlet locations must allow for the use of proper wrenches without any strain, bend, or damage to the piping. Flush-mounted-type-quick-disconnect devices are an exception. These devices are to be installed in compliance with the manufacturer's installation instructions. Any gas outlet that is not connected to an appliance must be capped to a gastight condition.

Plastic piping used as a gas pipe must not be used within or under any building or slab. The pipe is not allowed to operate at pressures greater than 100 psig for natural gas or 30 psig for LP gas. Plastic pipe buried as a gas pipe must be accompanied by a yellow insulated copper tracer wire or other approved conductor. The tracer wire must be installed adjacent to the underground plastic pipe and access must be provided to the tracer wire or the wire must terminate above ground at each end of the plastic gas piping. Sizing for the tracer wire shall not be smaller than 18 AWG and the insulation type must be suitable for direct burial.

TESTING INSTALLATIONS

The testing of installations is required before any system of gas piping is put into use. All components of the gas system must be tested to ensure that the system is gas tight. System components that will be concealed must be tested prior to concealment. The testing of a gas system is to be done with either air or an inert gas. No other type of gas or liquid is allowed for testing purposes. The test pressure must be measured with a device that is acceptable to local code officials.

SHUTOFF VALVES

Shutoff valves must not be installed in concealed locations or any space that is used as a plenum. All gas outlets are required to have individual shutoff valves. The valves must be in the same room and within six feet of any appliance being served by a gas outlet. Access to the shutoff valve is required. With the exception of buildings meant to be used as single-family or two-family dwellings, cutoff valves are required for every gas system that serves a separate tenant. All cutoff valves must be accessible to the tenant who is served by the gas system. When a common gas system is used to service a number of individual buildings, shutoff valves are required outside each building being served. Every gas meter must be equipped with a shut-off valve on the supply side of the meter.

Any shutoff valve used in a gas system must be of an approved type. The valves must be constructed of materials that are compatible with gas piping. All shutoff valves must be accessible. Shutoff valves that control separate piping systems must be placed an adequate distance from each other so they will be readily accessible for operation. They also must be installed in a way that protects them from damage. All shutoff valves must be marked plainly with an identification tag attached by the installer of the valves. The tagging is required to ensure that each valve is readily identifiable.

REGULATIONS PERTAINING TO 2-PSI AND HIGHER GAS PIPING

What is a 2-psi gas piping system? It is a gas piping system that is equipped with a service regulator that is set to deliver gas at two psi. Sizing gas pipe that runs from the point of delivery to a medium pressure regulator requires consideration of many factors. The first step is to determine the distance between the point of delivery and the most remote medium pressure regulator. This distance will be used in conjunction with a sizing table provided in your local code book.You will look at the table and find the length of piping needed to reach the medium pressure regulator. If your required distance falls between two numbers, choose the larger number to indicate your length.

You will have to decide what the outside diameter of your piping will be. The gas demand figures in the table will help you to establish the pipe size. If the demand that you need is not listed, choose the next higher rating. By determining pipe length and gas demand, you can see easily what the minimum pipe diameter must be. This type of sizing procedure works for pipe and tubing. You simply use the proper table and your sizing work is simple.

SPECIAL CONDITIONS

Special conditions sometimes apply to pipe sizing for gas systems. There can be times when systems call for a more sophisticated sizing solution. The maximum design operating pressure for piping systems located within buildings is five psi, with the possible exception of a variance from a code official. There are also some conditions that must be met to allow such a high pressure. Generally, one or more of the following conditions must be met:

• Piping system is located in an industrial processing or heating structure

• Piping system is located in a research structure

• Piping system is in an area that contains boilers or mechanical equipment only

• Piping system is welded steel pipe

• Piping system is temporary for structures under construction

GAS FLOW CONTROLS

The gas flow controls on a piping system are very important. Gas pressure regulators, or gas equipment pressure regulators, are required when a gas appliance is designed to operate at a lower pressure than that of the main gas system. Access to the regulators is required. All regulators must be protected from physical damage. Any regulator installed on the exterior of a building must be rated for such an installation. Second-stage regulators for undiluted LP gas systems must be listed and labeled in accordance with UL 144.

Medium pressure regulators installed in two-psi portions of systems must meet certain specifications. All of these regulators must be approved and suitable for the inlet and outlet gas pressure required by the application. The device must have a reduced outlet pressure under lockup conditions. Any medium pressure regulator used must have a rated capacity capable of serving the appliances installed. Access must be provided for all regulators. If a medium pressure regulator is installed indoors, it must be vented to the outdoors or equipped with a leak-limiting device. Medium pressure regulators must not be installed in locations where they will be concealed. Any piping system served by one gas meter and containing multiple medium pressure regulators must have listed shutoff valves installed immediately ahead of each medium pressure regulator.

All regulators that require venting must be equipped with an independent vent to the outside air. This vent must be designed to prevent the entry of water, foreign objects, or any other form of obstruction. One exception to this rule is that second-stage regulators equipped with and labeled for utilization with approved vent-limiting devices do not have to be vented to outside air.

UNDILUTED LIQUEFIED PETROLEUM GASES

Undiluted liquefied petroleum gases are butane and propane. The sizing procedure for gas systems conveying undiluted liquefied petroleum gases are about the same as those already discussed. You simply will use the tables in your local code book to determine the proper pipe or tubing size. Regulators used for these systems must be listed for the use and must be installed in accordance with their listing.

There are various options in the ways that appliance connections may be made. Rigid metallic pipe and fittings are suitable for such connections. Unless there is an official code variance, semirigid metallic tubing and metallic fittings can be used as long as they don't exceed 6 feet. These materials, when used, must be located entirely in the same room as the appliance. It is not acceptable to have semirigid metallic tubing enter a motor-operated appliance through an unprotected knockout opening.

Listed and labeled gas appliance connectors can be used to make connections. They must be installed in accordance with the manufacturer's instructions and located entirely in the same room as the appliance to which they are connected. Quick-disconnect devices that are listed and labeled used with listed and labeled gas appliance connectors can be used for making connections. Any listed and labeled gas convenience outlet used in conjunction with a listed and labeled gas appliance connector also can be used to make connections to appliances. Another option is a listed and labeled gas appliance connector that complies with ANSI Z21.69 and is designed for use with food-service equipment that has casters or is otherwise subject to movement for cleaning. This same type of device can be used for connections to large movable gas utilization equipment. All connectors, tubing, and piping must be protected from physical damage.

The diameter of fuel connectors must not be smaller than the inlet connection to the appliance. Most connectors cannot be more than three feet in length. Connectors for ranges and domestic clothes dryers can be up to six feet long. At no time can a connector be concealed in a wall, nor can it be extended through a wall. This same ruling applies to floors, partitions, ceilings or appliance housings. An approved shutoff valve, not less than the nominal size of the connector, must be accessible at the gas piping outlet immediately ahead of the connector. Any connector used must be sized to provide the total demand of the connected appliance, as described in a table in your local code book.

Gas appliances that are intended to be moved for routine cleaning, maintenance, or other reasons must be connected to a gas system with a flexible connector. The connector must be labeled for the application and protected from physical damage.

GAS-DISPENSING SYSTEMS

All gas dispensers are required to have an emergency switch to shut off all power to the dispenser. An approved backflow device that prevents the reverse flow of gas must be installed on the gas supply pipe or in the gas dispenser. Any gas-dispensing system installed inside a building must be ventilated by mechanical means in accordance with the International Mechanical Code. Compressed natural gas fuel-dispensing systems for vehicles fueled by the gas must be installed in accordance with NFPA 52 and the fire prevention code.

SUPPLEMENTAL GAS

Any time air, oxygen, or other special supplementary gas is introduced into a gas system an approved backflow preventer must be installed. This device must be on the gas line to the equipment supplied by the special gas and located between the source of the special gas and the gas meter.

Sometimes there is a need for standby gas. When LP gas or some other standby gas is interconnected with primary gas piping systems, an approved three-way, two-port valve or approved backcheck safeguard shall be installed to prevent backflow into any system.

Gas piping should not be intimidating. But working with gas is serious business. Learn and understand the gas code completely. Follow it. Don't cut corners. A gas leak has the potential to be much more devastating than a water leak.

CHAPTER 14
REFERENCED STANDARDS

ANSI

American National Standards Institute
11 West 42nd Street
New York, NY 10036

Standard reference number	Title	Referenced in code section number
ICC/ANSI A117.1—98	Accessible and Usable Building and Facilities	202, 401.1, 404.1, 404.2.1, 404.2.6
Z4.3—95	Minimum Requirements for Nonsewered Waste-Disposal Systems	311.1
Z21.22—86	Relief Valves and Automatic Gas Shutoff Devices for Hot Water Supply Systems—with 1990 Addendum	504.2, 504.5
Z124.1—95	Plastic Bathtub Units	407.1
Z124.2—95	Plastic Shower Receptors and Shower Stalls	417.1
Z124.3—95	Plastic Lavatories	416.1, 416.2
Z124.4—96	Plastic Water Closet Bowls and Tanks	420.1
Z124.6—97	Plastic Sinks	415.1, 418.1

ARI

Air-Conditioning & Refrigeration Institute
4301 N. Fairfax Drive, Suite 425
Arlington, VA 22203

Standard reference number	Title	Referenced in code section number
1010—94	Self-Contained, Mechanically-Refrigerated Drinking-Water Coolers	410.1

ASME

American Society of Mechanical Engineers
Three Park Avenue
New York, NY 10016-5990

Standard reference number	Title	Referenced in code section number
A112.1.2—98	Air Gaps in Plumbing Systems	Table 608.1
A112.3.1—93	Performance Standard and Installation Procedures for Stainless Steel Drainage Systems for Sanitary, Storm and Chemical Applications, Above and Below Ground	412.1, Table 702.1, Table 702.2, Table 702.3, Table 702.4, 708.2, Table 1102.4, Table 1102.5, 1102.6, Table 1102.7
A112.6.1—88	Supports for Off-the-Floor Plumbing Fixtures for Public Use	405.4.3
A112.14.1—98	Backwater Valves	715.3
A112.18.1M—96	Plumbing Fixture Fittings—with 1995 Errata	424.1
A112.18.3—96	Performance Requirements for Backflow Protection Devices and Systems in Plumbing Fixture Fittings	424.5, 608.15.4.3
A112.19.1M—94	Enameled Cast Iron Plumbing Fixtures	407.1, 410.1, 415.1, 416.1, 418.1
A112.19.2M—98	Vitreous China Plumbing Fixtures—with 1996 Errata	401.2, 408.1, 410.1, 416.1, 418.1, 419.1, 420.1
A112.19.3M—(R1996)	Stainless Steel Plumbing Fixtures (Designed for Residential Use)	415.1, 416.1, 418.1
A112.19.4M—94	Porcelain Enameled Formed Steel Plumbing Fixtures	407.1, 416.1, 418.1
A112.19.5—98	Trim for Water-Closet Bowls, Tanks, and Urinals	425.4
A112.19.6—95	Hydraulic Performance Requirements for Water Closets and Urinals	419.1, 420.1
A112.19.7M—95	Whirlpool Bathtub Appliances	421.1
A112.19.8—96	Suction Fittings for Use in Swimming Pools, Wading Pools, Spas, Hot Tubs, and Whirlpool Bathtub Appliances	421.4
A112.19.9—91	Non-Vitreous Ceramic Plumbing Fixtures	407.1, 408.1, 410.1, 415.1, 416.1, 417.1, 418.1, 420.1
A112.21.1M—98	Floor Drains	412.1
A112.21.2M—91	Roof Drains	1102.6

(Note: A112.26.1—84 has been discontinued and has not been replaced by ASME.)

FIGURE 14.1 The information in this chapter lists standards that are referred to in various sections of this book. Included in the list are the agency issuing them, and date and title. *(Courtesy of International Code Council, Inc. and International Plumbing Code 2000.)*

ASME—continued

A112.36.2—91 (R1998)	Cleanouts	708.2
B1.20.1—83	Pipe Threads, General Purpose (inch)	605.10.3, 605.12.3, 605.14.4, 605.16.3, 605.17.1, 605.20.3, 705.2.3, 705.4.3, 705.7.4, 705.10.1, 705.12.3
B16.3—92	Malleable Iron Threaded Fittings	Table 605.6, Table 702.4, Table 1102.7
B16.4—92	Cast Iron Threaded Fittings	Table 605.6, Table 702.4, Table 1102.7
B16.9—93	Factory-Made Wrought Steel Buttwelding Fittings	Table 605.6, Table 702.4, Table 1102.7
B16.11—97	Forged Steel Fittings, Socket-Welding and Threaded	Table 605.6, Table 702.4, Table 1102.7
B16.12—91	Cast-Iron Threaded Drainage Fittings	Table 605.6, Table 702.4, Table 1102.7
B16.15—(R1994)	Cast Bronze Threaded Fittings	Table 605.6, Table 702.4, Table 1102.7
B16.18—R94	Cast Copper Alloy Solder Joint Pressure Fittings	Table 605.6, Table 702.4, Table 1102.7
B16.22—95	Wrought Copper and Copper Alloy Solder Joint Pressure Fittings	Table 605.6, Table 702.4, Table 1102.7
B16.23—92	Cast Copper Alloy Solder Joint Drainage Fittings (DWV)—with 1994 Errata	Table 605.6, Table 702.4, Table 1102.7
B16.26—88	Cast Copper Alloy Fittings for Flared Copper Tubes	Table 605.6, Table 702.4, Table 1102.7
B16.28—94	Wrought Steel Buttwelding Short Radius Elbows and Returns	Table 605.6, Table 702.4, Table 1102.7
B16.29—94	Wrought Copper and Wrought Copper Alloy Solder Joint Drainage Fittings—DWV	Table 605.6, Table 702.4, Table 1102.7

ASSE

American Society of Sanitary Engineering
28901 Clemens Road
Westlake, OH 44145

Standard reference number	Title	Referenced in code section number
1001—90	Performance Requirements for Pipe Applied Atmospheric Type Vacuum Breakers	425.2, Table 608.1, 608.13.6
1002.79—99	Performance Requirements for Water Closet Flush Tank Ball Cocks	425.3.1, Table 608.1
1003—95	Performance Requirements for Water Pressure Reducing Valves for Domestic Water Supply Systems	604.8
1004—90	Performance Requirements for Commercial Dishwashing Machines	409.1
1005—86	Performance Requirements for Water Heater Drain Valves—with 1986 Revision	501.3
1006—89	Performance Requirements for Residential Use (Household) Dishwashers	409.1
1007.73—92	Performance Requirements for Home Laundry Equipment	406.1, 406.2
1008.73—89	Performance Requirements for Household Food Waste Disposer Units	413.1
1009—90	Performance Requirements for Commercial Food Waste Grinder Units	413.1
1010—98	Performance Requirements for Water Hammer Arresters	604.9
1011—95	Performance Requirements for Hose Connection Vacuum Breakers	Table 608.1, 608.13.6
1012—95	Performance Requirements for Backflow Preventers with Intermediate Atmospheric Vent	Table 608.1, 608.13.3, 608.16.1, 608.16.2
1013—93	Performance Requirements for Reduced Pressure Principle Backflow Preventers	Table 608.1, 608.13.2, 608.16.2
1014—90	Performance Requirements for Handheld Showers	424.3
1015—93	Performance Requirements for Double Check Backflow Prevention Assembly—with 1993 Revision	Table 608.1, 608.13.7
1016—96	Performance Requirements for Individual Thermostatic, Pressure Balancing and Combination Control Valves for Bathing Facilities	424.4
1018—86	Performance Requirements for Trap Seal Primer Valves; Water Supply Fed—with 1986 Revision	1002.4
1019—97	Performance Requirements for Wall Hydrants, Freezeless, Automatic Draining, Anti-Backflow Types	Table 608.1, 608.13.6
1020—97	Performance Requirements for Pressure Vacuum Breaker Assembly	Table 608.1, 608.13.5
1024—98	Performance Requirements for Dual Check Valve Type Backflow Preventers (for Residential Supply Service or Individual Outlets)	605.9, Table 608.1
1025—78	Performance Requirements for Diverters for Plumbing Faucets with Hose Spray, Anti-Siphon Type, Residential Applications	424.2
1032—80	Performance Requirements for Dual Check Valve Type Backflow Preventers; for Carbonated Beverage Dispensers—Post Mix-Types	Table 608.1, 608.16.1
1035—95	Performance Requirements for Laboratory Faucet Backflow Preventers	Table 608.1, 608.13.6
1037	Performance Requirements for Pressurized Flushing Devices for Plumbing Fixtures	425.2
1044—86	Performance Requirements for Trap Seal Primer Valves; Drainage Type	1002
1047—95	Performance Requirements for Reduced Pressure Detector Assembly Backflow Preventers	Table 608.1, 608.13.2
1048—95	Performance Requirements for Double Check Detector Assembly Backflow Preventer	Table 608.1, 608.13.7
1050—98	Performance Requirements for Outdoor Enclosures for Backflow Devices	608.14.1
1051—98	Performance Requirements for Air Admittance Valves for Plumbing Drainage Systems, Fixture and Branch Devices	917.1
1052—94	Performance Requirements for Hose Connection Backflow Preventers	Table 608.1, 608.13.6
1055—98	Performance Requirements for Backflow Devices for Chemical Dispersors	608.13.9
1056—95	Back Siphonage Vacuum Breaker	Table 608.1, 608.13.5, 608.13.8
5010-1013-1—91	Field Test Procedure for a Reduced Pressure Principle Assembly Using a Differential Pressure Gauge—with August 1992 Revisions	312.9.2

FIGURE 14.1 The information in this chapter lists standards that are referred to in various sections of this book. Included in the list are the agency issuing them, and date and title. *(Courtesy of International Code Council, Inc. and International Plumbing Code 2000.) (Continued)*

ASTM—continued

FIGURE 14.1 The information in this chapter lists standards that are referred to in various sections of this book. Included in the list are the agency issuing them, and date and title. (*Courtesy of International Code Council, Inc. and International Plumbing Code 2000.*) (*Continued*)

5010-1015-1—91	Field Test Procedure for a Double Check Valve Assembly Using a Duplex Gauge—with August 1992 Revisions 312.9.2
5010-1015-2—91	Field Test Procedure for a Double Check Valve Assembly Using a Differential Pressure Gauge—High- and Low-Pressure Hose Method—with August 1992 Revisions 312.9.2
5010-1015-3—91	Field Test Procedure for a Double Check Valve Assembly Using a Differential Pressure Gauge—High Pressure Hose Method—with August 1992 Revisions 312.9.2
5010-1015-4—91	Field Test Procedure for a Double Check Valve Assembly Using a Sight Tube—with August 1992 Revisions 312.9.2
5010-1020-1—91	Field Test Procedure for a Pressure Vacuum Breaker Assembly 312.9.2
5010-1047-1—91	Field Test Procedure for a Reduced Pressure Detector Assembly Using a Differential Pressure Gauge—with August 1992 Revisions 312.9.2
5010-1048-1—91	Field Test Procedure for a Double Check Detector Assembly Using a Duplex Gauge—with August 1992 Revisions 312.9.2
5010-1048-2—91	Field Test Procedure for a Double Check Detector Assembly Using a Differential Pressure Gauge—High- and Low-Pressure Hose Method—with August 1992 Revisions 312.9.2
5010-1048-3—91	Field Test Procedure for a Double Check Detector Assembly Using a Differential Pressure Gauge—High-Pressure Hose Method—with August 1992 Revisions 312.9.2
5010-1048-4—91	Field Test Procedure for a Double Check Detector Assembly Using a Sight Tube—with August 1992 Revisions 312.9.2

ASTM

American Society for Testing and Materials
100 Barr Harbor Drive
West Conshohocken, PA 19428-2959

Standard reference number	Title	Referenced in code section number
A 53—98	Specification for Pipe, Steel, Black and Hot-Dipped, Zinc-Coated Welded and Seamless .... Table 605.4, Table 605.5, Table 702.1	
A 74—98	Specification for Cast Iron Soil Pipe and Fittings .... Table 702.1, Table 702.2, Table 702.3, Table 702.4 708.2, Table 1102.4, Table 1102.5	
A 733—93	Specification for Welded and Seamless Carbon Steel and Austenitic Stainless Steel Pipe Nipples .... Table 605.8	
A 888—98e1	Specification for Hubless Cast Iron Soil Pipe and Fittings for Sanitary and Storm Drain, Waste, and Vent Piping Application .... Table 702.1, Table 702.2, Table 702.3, Table 702.4, Table 1102.4, Table 1102.5, Table 1102.7	
B 32—96	Specification for Solder Metal 605.14.3, 605.15.4, 705.7.3, 705.8.3	
B 42—96	Specification for Seamless Copper Pipe, Standard Sizes .... Table 605.4, Table 605.5, Table 702.1	
B 43—98	Specification for Seamless Red Brass Pipe, Standard Sizes .... Table 605.4, Table 605.5, Table 702.1	
B 75—97	Specification for Seamless Copper Tube .... Table 605.4, Table 605.5, Table 702.1, Table 702.2, Table 702.3, Table 1102.4	
B 88—96	Specification for Seamless Copper Water Tube .... Table 605.4, Table 605.5, Table 702.1, Table 702.2, Table 702.3, Table 1102.4	
B 152—97A	Specification for Copper Sheet, Strip Plate and Rolled Bar 402.3, 425.3.3, 417.5.2.4, 902.2	
B 251—97	Specification for General Requirements for Wrought Seamless Copper and Copper-Alloy Tube .... Table 605.4, Table 605.5 Table 702.1, Table 702.2, Table 702.3, Table 1102.4	
B 302—98	Specification for Threadless Copper Pipe .... Table 605.4, Table 605.5, Table 702.1	
B 306—96	Specification for Copper Drainage Tube (DWV) .... Table 702.1, Table 702.2, Table 1102.4	
B 447—97	Specification for Welded Copper Tube .... Table 605.4, Table 605.5	
B 687—96	Specification for Brass, Copper, and Chromium-Plated Pipe Nipples .... Table 605.8	
B 813—93	Standard Specification for Liquid and Paste Fluxes for Soldering Applications of Copper and Copper Alloy Tube 605.14.3, 605.15.4, 705.7.3, 705.8.3	
B 828—92-E01	Practice for Making Capillary Joints by Soldering of Copper and Copper Alloy Tube and Fittings 605.14.3, 605.15.4, 705.7.3, 705.8.3	
C 4—98	Specification for Clay Drain Tile .... Table 702.3, Table 1102.4	
C 14—95	Specification for Concrete Sewer, Storm Drain, and Culvert Pipe .... Table 702.3, Table 1102.4	
C 76—98	Specification for Reinforced Concrete Culvert, Storm Drain, and Sewer Pipe .... Table 702.3, Table 1102.4	
C 296—93	Specification for Asbestos-Cement Pressure Pipe .... Table 605.4	
C 425—98	Specification for Compression Joints for Vitrified Clay Pipe and Fittings 705.13, 705.14	
C 428—97	Specification for Asbestos-Cement Nonpressure Sewer Pipe .... Table 702.2, Table 702.3, Table 1102.4	
C 443—98	Specification for Joints for Circular Concrete Sewer and Culvert Pipe, Using Rubber Gaskets 705.6, 705.14	
C 508—97	Specification for Asbestos-Cement Underdrain Pipe .... Table 1102.5	
C 564—97	Specification for Rubber Gaskets for Cast Iron Soil Pipe and Fittings 705.5.2, 705.5.3, 705.14	
C 700—97	Specification for Vitrified Clay Pipe, Extra Strength, Standard Strength, and Perforated .. Table 702.3, Table 1102.4, Table 1102.5	
C 1053—90 (1995)	Specification for Borosilicate Glass Pipe and Fittings for Drain, Waste, and Vent (DWV) Applications .. Table 702.1, Table 702.4	
C 1173—97	Specification for Flexible Transition Couplings for Underground Piping System 705.14	
C 1277—97	Specification for Shielded Coupling Joining Hubless Cast Iron Soil Pipe and Fittings 705.5.3	
D 1527—96a	Specification for Acrylonitrile-Butadiene-Styrene (ABS) Plastic Pipe, Schedules 40 and 80 .... Table 605.4	
D 1785—96b	Specification for Poly (Vinyl Chloride) (PVC) Plastic Pipe, Schedules 40, 80 and 120 .... Table 605.4	

FIGURE 14.1 The information in this chapter lists standards that are referred to in various sections of this book. Included in the list are the agency issuing them, and date and title. *(Courtesy of International Code Council, Inc. and International Plumbing Code 2000.) (Continued)*

F 1282—98	Specification for Polyethylene/Aluminum/Polyethylene (PE-AL-PE) Composite Pressure Pipe . . . . Table 605.4
F 1488—98	Standard Specification for Coextruded Composite Pipe . . . . Table 702.1, Table 702.2, Table 702.3
F 1807—98	Standard Specifications for Metal Insert Fittings Utilizing a Copper Crimp Ring SDR9 Cross-linked Polyethylene (PEX) Tubing . . . . Table 605.6, 605.17.2

AWS

American Welding Society
550 N.W. LeJeune Road
P. O. Box 351040
Miami, FL 33135

Standard reference number	Title	Referenced in code section number
A5.8—92	Specifications for Filler Metals for Brazing 605.12.1, 605.14.1, 605.15.1, 705.4.1, 705.7.1, 705.8.1

AWWA

American Water Works Association
6666 West Quincy Avenue
Denver, CO 80235

Standard reference number	Title	Referenced in code section number
C110—98	Standard for Ductile-Iron and Gray-Iron Fittings, 3 Inch through 48 Inches, for Water and Other Liquids . . . . Table 605.6, Table 702.4, Table 1102.7
C151—96	Standard for Ductile-Iron Pipe, Centrifugally Cast for Water or Other Liquids . . . . Table 605.4
C153—88	Ductile-Iron Compact Fittings . . . . Table 605.6
C510—92	Double Check Valve Backflow-Prevention Assembly . . . . Table 608.1, 608.13.7
C511—92	Reduced-Pressure Principle Backflow-Prevention Assembly . . . . Table 608.1, 608.13.2, 608.16.2
C651—92	Standard for Disinfecting Water Mains 610.1
C652—92	Standard for Disinfection of Water-Storage Facilities 610.1

CSA

Canadian Standards Association
178 Rexdale Blvd.
Rexdale (Toronto), Ontario, Canada M9W 1R3

Standard reference number	Title	Referenced in code section number
B45.1—94	Ceramic Plumbing Fixtures 408.1, 416.1, 418.1, 419.1, 420.1
B45.2—94	Enameled Cast-Iron Plumbing Fixtures 407.1, 415.1, 416.1, 418.1
B45.3—94	Porcelain Enameled Steel Plumbing Fixtures 407.1, 416.1, 418.1
B45.4—94	Stainless-Steel Plumbing Fixtures 415.1, 416.1, 418.1, 420.1
B45.5—94	Plastic Plumbing Fixtures 407.1, 416.2, 417.1, 419.1, 420.1, 421.1
B64.7—94	Vacuum Breakers, Laboratory Faucet Type (LFVB) . . . . Table 608.1, 608.13.6
B79—94	Floor, Area and Shower Drains, and Cleanouts for Residential Construction 412.1
B125—94	Plumbing Fittings . . . . 424.1, 424.2, 424.3, 424.4, 425.3.1, 425.4, Table 608.1
B137.1—95	Polyethylene Pipe, Tubing and Fittings for Cold Water Pressure Services . . . . Table 605.4
B137.2—93	PVC Injection-Moulded Gasketed Fittings for Pressure Applications . . . . Table 605.6, Table 1102.7
B137.3—93	Rigid Poly (Vinyl Chloride) (PVC) Pipe for Pressure Applications . . . . Table 605.4, 605.20.2, 705.7.2, 705.12.2
B137.5—97	Cross-Linked Polyethylene (PBX) Tubing Systems for Pressure Applications— with Revisions through September 1992 . . . . Table 605.4, Table 605.5
B137.6—96	CPVC Pipe, Tubing and Fittings for Hot and Cold Water Distribution Systems— with Revisions through May 1986 . . . . Table 605.4, Table 605.5
B181.1—96	ABS Drain, Waste, and Vent Pipe and Pipe Fittings . . . . Table 702.1, Table 702.2, Table 702.4, 705.2.2, 705.6.2, 715.3
B181.2—96	PVC Drain, Waste, and Vent Pipe and Pipe Fittings— with Revisions through December 1993 . . . . Table 702.1, Table 702.2, 705.7.2, 705.12.2, 715.3
B182.1—96	Plastic Drain and Sewer Pipe and Pipe Fittings 705.7.2, 705.12.2
B182.2—95	PVC Sewer Pipe and Fittings (PSM Type) . . . . Table 702.2, Table 702.3, Table 1102.4, Table 1102.5
CAN3-B137.8M—92	Polybutylene (PB) Piping for Pressure Applications—with Revisions through July 1992 . . . . Table 605.4, Table 605.5, 605.18.2, 605.18.3

‹FIGURE 14.1 The information in this chapter lists standards that are referred to in various sections of this book. Included in the list are the agency issuing them, and date and title. *(Courtesy of International Code Council, Inc. and International Plumbing Code 2000.) (Continued)*

CAN/CSA- A257.2M—92	Reinforced L Circular Concrete Culvert, Storm Drain, Sewer Pipe and Fittings	Table 702.3, Table 1102.4
CAN/CSA A257.3M—92	Joints for Circular Concrete Sewer and Culvert Pipe, Manhole Sections, and Fittings Using Rubber Gaskets	705.6, 705.14
CAN/CSA-B64.1.1—94	Vacuum Breakers, Atmospheric Type (AVB)	425.2, Table 608.1, 608.13.6
CAN/CSA-B64.2—94	Vacuum Breakers, Hose Connection Type (HCVB)	Table 608.1, 608.13.6
CAN/CSA-B64.2.2—94	Vacuum Breakers, Hose Connection Type (HCVB) with Automatic Draining Feature	Table 608.1, 608.13.6
CAN/CSA-B64.3—94	Backflow Preventers, Dual Check Valve Type with Atmospheric Port (DCAP)	Table 608.1, 608.13.2, 608.13.3, 608.15.2
CAN/CSA-B64.4—94	Backflow Preventers, Reduced Pressure Principle Type (RP)	Table 608.1, 608.15.2
CAN/CSA-B64.10—94	Manual for the Selection, Installation, Maintenance and Field Testing of Backflow Prevention Devices	312.9.2
CAN/CSA- B137.9M—98	Polyethylene/Aluminum/Polyethylene Composite Pressure Pipe Systems	Table 605.4
CAN/CSA- B137.10M—98	Crosslinked Polyethylene/Aluminum/Polyethylene Composite Pressure Pipe Systems	Table 605.4, Table 605.5
CAN/CSA-B181.2—96 (R92)	Polyolefin Laboratory Drainage Systems—with Revisions through October 1990	Table 702.1, Table 702.2
CAN/CSA-B182.4—97	Profile PVC Sewer Pipe and Fittings	Table 702.3, Table 1102.4, Table 1102.5
CAN/CSA-B602M—90	Mechanical Couplings for Drain, Waste, and Vent Pipe and Sewer Pipe	705.2.1, 705.5.3, 705.6, 705.12.1, 705.13, 705.14
CSA-B45 (Supplement 1)—88	Hydromassage Bathtubs	421.1

CISPI

Cast Iron Soil Pipe Institute
Suite 419
5959 Shallowford Road
Chattanooga, TN 37421

Standard reference number	Title	Referenced in code section number
301—99	Specification for Hubless Cast Iron Soil Pipe and Fittings for Sanitary and Storm Drain, Waste and Vent Piping Applications	Table 702.1, Table 702.2, Table 702.3, Table 702.4, Table 1102.4, Table 1102.5
310-97	Specification for Coupling for Use in Connection with Cast-Iron Soil Pipe and Fittings for Sanitary and Storm Drain Waste and Vent Piping Applications	705.5.3

FS

Federal Specification*
General Service Administration
7th & D Streets
407 E. Lenfant Plaza, SW, Suite 8100
Washington, DC 20024-2124

Standard reference number	Title	Referenced in code section number
TT-P-1536a—75	Federal Specification for Plumbing Fixture Setting Compound	405.4

* Standards are available from the Supt. of Documents, U.S. Government Printing Office, Washington, DC 20402-9325.

ICC

International Code Council
5203 Leesburg Pike, Suite 201
Falls Church, VA 22041

Standard reference number	Title	Referenced in code section number
ICC/ANSI A117.1—98	Accessible and Usable Building and Facilities	202, 401.1, 404.1, 404.2.1, 404.2.6
IBC—2000	International Building Code®	201.3, 305.4, 307.1, 307.2, 307.3, 308.2, 309.1:1, 309.1.2, 310.1, 310.3, 403.1, Table 403.2, 404.1, 404.3, 407.3, 417.6, 502.6, 606.5.2
IFGC—2000	International Fuel Gas Code®	315, 502.1, 1201.2
IMC—2000	International Mechanical Code®	201.3, 310.1, 422.9, 502.1, 1201.2, 1302.1

FIGURE 14.1 The information in this chapter lists standards that are referred to in various sections of this book. Included in the list are the agency issuing them, and date and title. *(Courtesy of International Code Council, Inc. and International Plumbing Code 2000.) (Continued)*

NFPA

National Fire Protection Association
Batterymarch Park
Quincy, MA 02269

Standard reference number	Title	Referenced in code section number
50—96	Bulk Oxygen Systems at Consumer Sites	1303.1
51—97	Oxygen-Fuel Gas Systems for Welding, Cutting, and Allied Processes	1303.1
70—93	National Electrical Code	502.1, 504.3, 1113.1.3
99C—99	Gas and Vacuum Systems	1302.1

NSF

National Sanitation Foundation
3475 Plymouth Road
P. O. Box 130140
Ann Arbor, MI 48113-0140

Standard reference number	Title	Referenced in code section number
3—96	Commercial Spray-Type Dishwashing Machines	409.1
18—96	Manual Food and Beverage Dispensing Equipment	426.1
14—98	Plastic Piping Components and Related Materials	303.3
42—98	Drinking Water Treatment Units—Aesthetic Effects	611.1
44—98	Cation Exchange Water Softeners	611.1
53—98	Drinking Water Treatment Units—Health Effects	611.1
58—97	Reverse Osmosis Drinking Water Treatment Systems	611.2
61—99	Drinking Water System Components—Health Effects	424.1, 605.4, 605.5, 605.6
62—97	Drinking Water Distillation Systems	611.1

PDI

Plumbing and Drainage Institute
1106 West 77th Street, South Drive
Indianapolis, IN 46260-3318

Standard reference number	Title	Referenced in code section number
G101—96	Testing and Rating Procedure for Grease Interceptors with Appendix of Sizing and Installation Data	1003.3.4

FIGURE 14.1 The information in this chapter lists standards that are referred to in various sections of this book. Included in the list are the agency issuing them, and date and title. *(Courtesy of International Code Council, Inc. and International Plumbing Code 2000.) (Continued)*

CHAPTER 15
RAINFALL RATE

RATES OF RAINFALL FOR VARIOUS CITIES

Rainfall rates, in inches per hour, are based on a storm of one-hour duration and a 100-year return period. The rainfall rates shown in the appendix are derived from Figure 1106.1.

Alabama:
Birmingham 3.8
Huntsville 3.6
Mobile 4.6
Montgomery 4.2

Alaska:
Fairbanks 1.0
Juneau 0.6

Arizona:
Flagstaff 2.4
Nogales 3.1
Phoenix 2.5
Yuma 1.6

Arkansas:
Fort Smith 3.6
Little Rock 3.7
Texarkana 3.8

California:
Barstow 1.4
Crescent City 1.5
Fresno 1.1
Los Angeles 2.1
Needles 1.6
Placerville 1.5
San Fernando 2.3
San Francisco ... 1.5
Yreka 1.4

Colorado:
Craig 1.5
Denver 2.4
Durango 1.8
Grand Junction 1.7
Lamar 3.0
Pueblo 2.5

Connecticut:
Hartford 2.7
New Haven 2.8
Putnam 2.6

Delaware:
Georgetown 3.0
Wilmington 3.1

District of Columbia:
Washington 3.2

Florida:
Jacksonville 4.3
Key West 4.3
Miami 4.7
Pensacola 4.6
Tampa 4.5

Georgia:
Atlanta 3.7
Dalton 3.4
Macon 3.9
Savannah 4.3
Thomasville 4.3

Hawaii:
Hilo 6.2
Honolulu 3.0
Wailuku 3.0

Idaho:
Boise 0.9
Lewiston 1.1
Pocatello 1.2

Illinois:
Cairo 3.3
Chicago 3.0
Peoria 3.3
Rockford 3.2
Springfield 3.3

Indiana:
Evansville 3.2
Fort Wayne 2.9
Indianapolis 3.1

Iowa:
Davenport 3.3
Des Moines 3.4
Dubuque 3.3
Sioux City 3.6

Kansas:
Atwood 3.3
Dodge City 3.3
Topeka 3.7
Wichita 3.7

Kentucky:
Ashland 3.0
Lexington 3.1
Louisville 3.2
Middlesboro 3.2
Paducah 3.3

Louisiana:
Alexandria 4.2
Lake Providence ... 4.0
New Orleans 4.8
Shreveport 3.9

Maine:
Bangor 2.2
Houlton 2.1
Portland 2.4

Maryland:
Baltimore 3.2
Hagerstown 2.8
Oakland 2.7
Salisbury 3.1

Massachusetts:
Boston 2.5
Pittsfield 2.8
Worcester 2.7

Michigan:
Alpena 2.5
Detroit 2.7
Grand Rapids ... 2.6
Lansing 2.8
Marquette 2.4
Sault Ste. Marie ... 2.2

Minnesota:
Duluth 2.8
Grand Marais 2.3
Minneapolis 3.1
Moorhead 3.2
Worthington 3.5

Mississippi:
Biloxi 4.7
Columbus 3.9
Corinth 3.6
Natchez 4.4
Vicksburg 4.1

Missouri:
Columbia 3.2
Kansas City 3.6
Springfield 3.4
St. Louis 3.2

Montana:
Ekalaka 2.5
Havre 1.6
Helena 1.5
Kalispell 1.2
Missoula 1.3

Nebraska:
North Platte 3.3
Omaha 3.8
Scottsbluff 3.1
Valentine 3.2

Nevada:
Elko 1.0
Ely 1.1
Las Vegas 1.4
Reno 1.1

New Hampshire:
Berlin 2.5
Concord 2.5
Keene 2.4

New Jersey:
Atlantic City 2.9
Newark 3.1
Trenton 3.1

New Mexico:
Albuquerque 2.0
Hobbs 3.0
Raton 2.5
Roswell 2.6
Silver City 1.9

New York:
Albany 2.5
Binghamton 2.3
Buffalo 2.3
Kingston 2.7
New York 3.0
Rochester 2.2

North Carolina:
Asheville 4.1
Charlotte 3.7
Greensboro 3.4
Wilmington 4.2

North Dakota:
Bismarck 2.8
Devils Lake 2.9
Fargo 3.1
Williston 2.6

Ohio:
Cincinnati 2.9
Cleveland 2.6
Columbus 2.8
Toledo 2.8

Oklahoma:
Altus 3.7
Boise City 3.3
Durant 3.8
Oklahoma City ... 3.8

Oregon:
Baker 0.9
Coos Bay 1.5
Eugene 1.3
Portland 1.2

Pennsylvania:
Erie 2.6
Harrisburg 2.8
Philadelphia 3.1
Pittsburgh 2.6
Scranton 2.7

FIGURE 15.1 A range of cities and rainfall rates is shown here. (*Courtesy of International Code Council, Inc. and International Plumbing Code 2000.*)

Rhode Island:
Block Island 2.75
Providence 2.6

South Carolina:
Charleston 4.3
Columbia 4.0
Greenville 4.1

South Dakota:
Buffalo 2.8
Huron 3.3
Pierre 3.1
Rapid City 2.9
Yankton 3.6

Tennessee:
Chattanooga 3.5
Knoxville 3.2
Memphis 3.7
Nashville 3.3

Texas:
Abilene 3.6
Amarillo 3.5
Brownsville 4.5
Dallas 4.0
Del Rio 4.0
El Paso 2.3
Houston 4.6
Lubbock 3.3
Odessa 3.2
Pecos 3.0
San Antonio 4.2

Utah:
Brigham City 1.2
Roosevelt 1.3
Salt Lake City 1.3
St. George 1.7

Vermont:
Barre 2.3
Bratteboro 2.7
Burlington 2.1
Rutland 2.5

Virginia:
Bristol 2.7
Charlottesville 2.8
Lynchburg 3.2
Norfolk 3.4
Richmond 3.3

Washington:
Omak 1.1
Port Angeles 1.1
Seattle 1.4
Spokane 1.0
Yakima 1.1

West Virginia:
Charleston 2.8
Morgantown 2.7

Wisconsin:
Ashland 2.5
Eau Claire 2.9
Green Bay 2.6
La Crosse 3.1
Madison 3.0
Milwaukee 3.0

Wyoming:
Cheyenne 2.2
Fort Bridger 1.3
Lander 1.5
New Castle 2.5
Sheridan 1.7
Yellowstone Park .. 1.4

For SI: 1 inch =25.4 mm.

Source: National Weather Service, National Oceanic and Atmospheric Administration, Washington, D.C.

FIGURE 15.1   This concludes the listing of cities and rainfall rates. *(Courtesy of International Code Council, Inc. and International Plumbing Code 2000.) (Continued)*

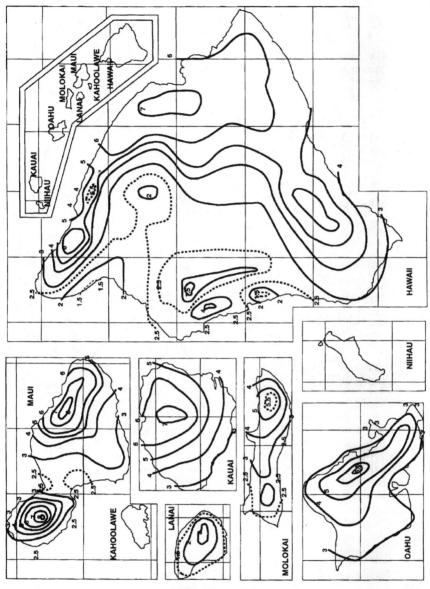

FIGURE 1106.1—continued
100-YEAR, 1-HOUR RAINFALL (INCHES)
HAWAII

For SI: 1 inch = 25.4 mm.

Source: National Weather Service, National Oceanic and Atmospheric Administration, Washington, DC.

FIGURE 15.2 Hawaii figures show a 100-year, one-hour rainfall rate. *(Courtesy of International Code Council, Inc. and International Plumbing Code 2000.)*

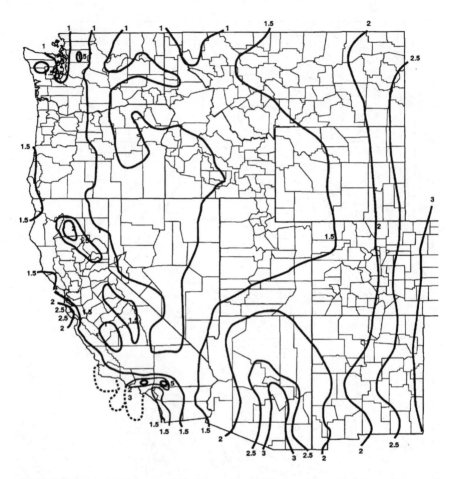

FIGURE 15.3 A chart of the western United States shows a 100-year, one-hour rainfall rate. *(Courtesy of International Code Council, Inc. and International Plumbing Code 2000.)*

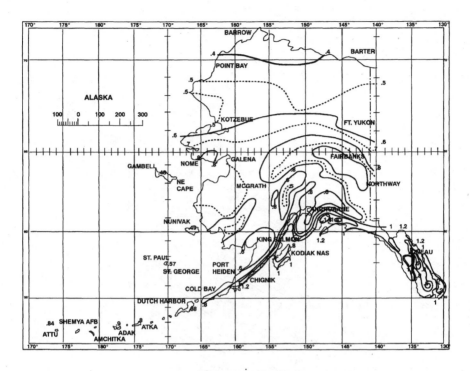

100-YEAR, 1-HOUR RAINFALL (INCHES)
ALASKA

For SI: 1 inch = 25.4 mm.

Source: National Weather Service, National Oceanic and Atmospheric Administration, Washington, DC.

FIGURE 15.4 This charts a 100-year, one-hour rainfall rate in Alaska. *(Courtesy of International Code Council, Inc. and International Plumbing Code 2000.)*

100-YEAR, 1-HOUR RAINFALL (INCHES)
EASTERN UNITED STATES

For SI: 1 inch = 25.4 mm.
Source: National Weather Service, National Oceanic and Atmospheric Administration, Washington, DC.

FIGURE 15.5   The eastern United States chart shows a 100-year, one-hour rainfall rate. *(Courtesy of International Code Council, Inc. and International Plumbing Code 2000.)*

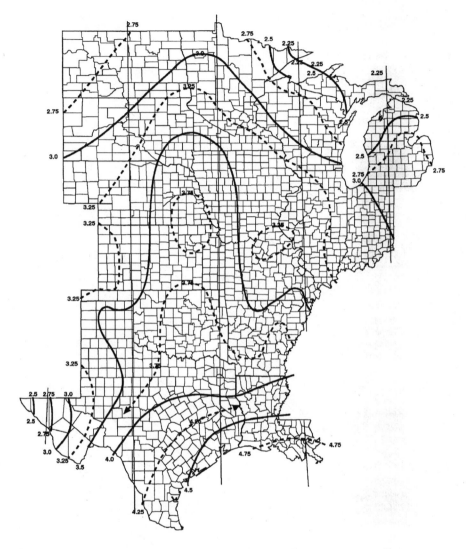

FIGURE 15.6 This is a chart showing the 100-year, one-hour rainfall rate in the central United States. *(Courtesy of International Code Council, Inc. and International Plumbing Code 2000.)*

CHAPTER 16
RECYCLING GRAY WATER

The recycling of gray water has become more popular and the plumbing code has respond-ed to this demand. What is gray water? It is defined as waste water that is discharged from lavatories, bathtubs, showers, clothes washers, and laundry sinks. Gray water is not a new issue. Dry wells have been used to collect gray water for years. However, the modern plumbing code is much more restrictive in the recycling of gray water. This is done for pub-lic health and safety and should not be considered a bad position.

The plumbing code generally states that all plumbing fixtures must discharge into a sanitary drainage system. There is the exception that allows for the recycling of gray water. Fixtures that are eligible for gray water recycling are as follows:

- Bathtubs
- Lavatories
- Laundry sinks

- Showers
- Clothes washers

What can gray water be used for when it is recycled? The water can be used to flush water closets and urinals that are located in the same building as the gray water recycling system. All of these systems must comply with the overall regulations of the plumbing code. In addition to being used for flushing toilets and urinals, gray water can be used for irrigation purposes when specific approval is given by the administrating authority.

Any installation involving the use of gray water, including drains, wastes, and vent pipes, must be installed in full compliance with the plumbing code. Gray water is to be collected in a reservoir, which must be constructed of durable, nonabsorbent and corro-sion-resistant materials. It is required that the reservoir be a closed and gastight vessel.

Access openings are required to allow inspection and cleaning of the reservoir interior. Each reservoir needs a holding capacity that is at a minimum twice the size of the volume of water required to meet the daily flushing requirements of the fixtures supplied with gray water, but not less than 50 gallons. Each reservoir must be sized to limit retention time of gray water to a maximum of 72 hours.

All gray water that enters a reservoir must pass through an approved filter such as a media, sand, or diatomaceous earth. A disinfection process is required for all gray water that is recycled. The disinfection process must be of an approved type. Three approved types of disinfectants include chlorine, iodine, and ozone.

Location of Gray-Water System

Minimum Horizontal Distance in Clear Required From:	Holding Tank		Irrigation/ Disposal Field	
	Feet	(mm)	Feet	(mm)
Building Structures[1]	5[2]	(1524 mm)	2[3]	(610 mm)
Property line adjoining private property	5	(1524 mm)	5	(1524 mm)
Water supply wells[4]	50	(15240 mm)	100	(30480 mm)
Streams and lakes[4]	50	(15240 mm)	50[5]	(15240 mm)
Sewage pits or cesspools	5	(1524 mm)	5	(1524 mm)
Disposal field and 100% expansion area	5	(1524 mm)	4[6]	(1219 mm)
Septic tank	0	(0)	5	(1524 mm)
On-site domestic water service line	5	(1524 mm)	5	(1524 mm)
Pressurized public water main	10	(3048 mm)	10[7]	(3048 mm)

Notes: When irrigation/disposal fields are installed in sloping ground, the minimum horizontal distance between any part of the distribution system and the ground surface shall be fifteen (15) feet (4572 mm).

[1] Including porches and steps, whether covered or uncovered, breezeways, roofed porte-cocheres, roofed patios, carports, covered walks, covered driveways and similar structures or appurtenances.

[2] The distance may be reduced to zero feet for above ground tanks when first approved by the Administrative Authority.

[3] Assumes a 45 degree (0.79 rad) angle from foundation.

[4] Where special hazards are involved, the distance required shall be increased as may be directed by the Administrative Authority.

[5] These minimum clear horizontal distances shall also apply between the irrigation/disposal field and the ocean mean higher high tide line.

[6] Plus two (2) feet (610 mm) for each additional foot of depth in excess of one (1) foot (305 mm) below the bottom of the drain line.

[7] For parallel construction/for crossings, approval by the Administrative Authority shall be required.

FIGURE 16.1 A gray-water system must be a certain distance from buildings, property lines, and disposal and water-service lines. (*Reprinted from the 2000 Uniform Plumbing Code (UPC) with the permission of the International Association of Plumbing and Mechanical Officials (IAPMO).*)

Design Criteria of Six Typical Soils

Type of Soil	Minimum square feet of irrigation/leaching area per 100 gallons of estimated graywater discharge per day	Maximum absorption capacity in gallons per square foot of irrigation/leaching area for a 24-hour period
Coarse sand or gravel	20	5.0
Fine sand	25	4.0
Sandy loam	40	2.5
Sandy clay	60	1.7
Clay with considerable sand or gravel	90	1.1
Clay with small amounts of sand or gravel	120	0.8

FIGURE 16.2 A minimum square footage figure for different soils, along with their absorption capacity, is listed. *(Reprinted from the* 2000 Uniform Plumbing Code (UPC) *with the permission of the International Association of Plumbing and Mechanical Officers (IAPMO).)*

Design Criteria of Six Typical Soils

Type of Soil	Minimum square meters of irrigation/leaching area per liter of estimated graywater discharge per day	Maximum absorption capacity in liters per square meter of irrigation/leaching area for a 24-hour period
Coarse sand or gravel	0.005	203.7
Fine sand	0.006	162.9
Sandy loam	0.010	101.8
Sandy clay	0.015	69.2
Clay with considerable sand or gravel	0.022	44.8
Clay with small amounts of sand or gravel	0.030	32.6

FIGURE 16.3 This is the metric version of the table above. *(Reprinted from the* 2000 Uniform Plumbing Code (UPC) *with the permission of the International Association of Plumbing and Mechanical Officers (IAPMO).)*

It's not always possible to collect and recycle enough gray water for the intended purposes. This requires the use of what is known as makeup water. Potable water is added to gray water in order to create the makeup factor. When this is the case, the potable water supply must be protected against backflow, in accordance with code requirements. Any potable water supply that feeds a gray water reservoir must be fitted with a full-open valve.

An overflow pipe must be installed in conjunction with a reservoir collection point. The overflow pipe must be of the same diameter as the influent pipe for the gray water. Any overflow pipe must be connected directly to a sanitary drainage system.

All collection reservoirs must be fitted with a drain at the lowest point of the reservoir. The drain must be connected directly to a sanitary drainage system. It is required that the drain be the same size as the overflow pipe required, and the drain must be equipped with a full-open valve.

A vent is required for a collection reservoir. The sizing of the vent is determined by normal code requirements, based on the size of the reservoir influent pipe. All gray water used in a recycling situation is required to be dyed either blue or green. The dying agent must be a food grade vegetable dye. Gray water must be dyed prior to the water being supplied to plumbing fixtures.

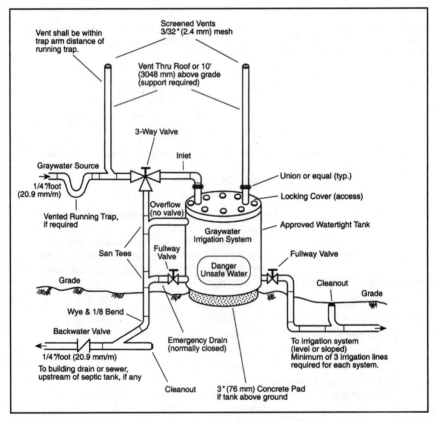

FIGURE 16.4 This is an illustration of a gravity-fed system. (*Reprinted from the* 2000 *Uniform Plumbing Code (UPC) with the permission of the International Association of Plumbing and Mechanical Officers (IAPMO).*)

Piping that is used for the distribution of gray water and reservoirs must be identified as conduits of nonpotable water. Requirements for marking and identifying said piping are covered in the general plumbing code.

Most plumbers don't encounter many gray water recycling systems. You might never deal with one, but if you do, you must be able to understand the code requirements associated with such a system. Remember that you cannot treat a gray water system exactly as you would a traditional system. The rules pertaining to gray water are not complex, so don't be intimidated by them. It's all a matter of understanding code requirements, and the regulations for gray water are fairly easy to comprehend.

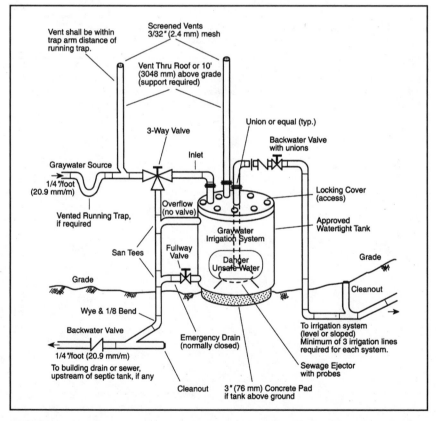

FIGURE 16.5 This illustrates a pumped system. *(Reprinted from the 2000 Uniform Plumbing Code (UPC) with the permission of the International Association of Plumbing and Mechanical Officers (IAPMO).)*

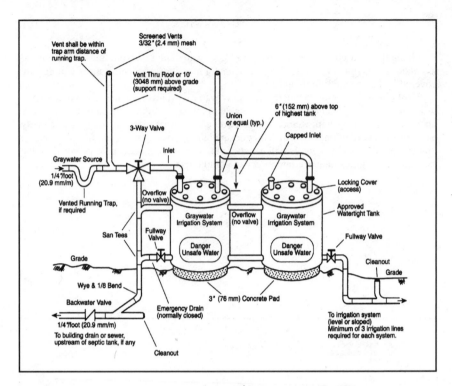

FIGURE 16.6 This is a gray-water multiple tank installation. *(Reprinted from the 2000 Uniform Plumbing Code (UPC) with the permission of the International Association of Plumbing and Mechanical Officers (IAPMO).)*

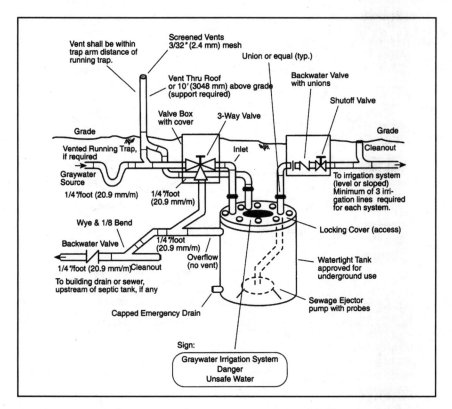

FIGURE 16.7 This illustrates an underground gray-water tank. *(Reprinted from the* 2000 Uniform Plumbing Code (UPC) *with the permission of the International Association of Plumbing and Mechanical Officers (IAPMO).)*

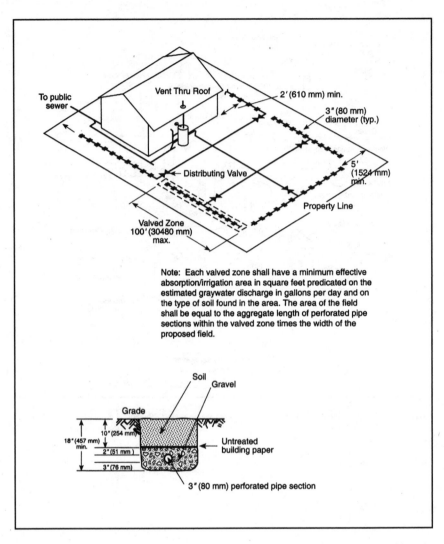

FIGURE 16.8 Gray water also may be used for irrigation, under certain conditions. *(Reprinted from the 2000 Uniform Plumbing Code (UPC) with the permission of the International Association of Plumbing and Mechanical Officers (IAPMO).)*

CHAPTER 17
DEGREE DAYS AND DESIGN TEMPERATURES

DEGREE DAY AND DESIGN TEMPERATURES[a] FOR CITIES IN THE UNITED STATES

STATE	STATION[b]	HEATING DEGREE DAYS (yearly total)	DESIGN TEMPERATURES			DEGREES NORTH LATITUDE[c]
			Winter 97¹/₂%	Summer Dry bulb 2¹/₂%	Summer Wet bulb 2¹/₂%	
AL	Birmingham	2,551	21	94	77	33°3'
	Huntsville	3,070	16	93	77	34°4'
	Mobile	1,560	29	93	79	30°4'
	Montgomery	2,291	25	95	79	32°2'
AK	Anchorage	10,864	-18	68	59	61°1'
	Fairbanks	14,279	-47	78	62	64°5'
	Juneau	9,075	1	70	59	58°2'
	Nome	14,171	-27	62	56	64°3'
AZ	Flagstaff	7,152	4	82	60	35°1'
	Phoenix	1,765	34	107	75	33°3'
	Tuscon	1,800	32	102	71	33°1'
	Yuma	974	39	109	78	32°4'
AR	Fort Smith	3,292	17	98	79	35°2'
	Little Rock	3,219	20	96	79	34°4'
	Texarkana	2,533	23	96	79	33°3'
CA	Fresno	2,611	30	100	71	36°5'
	Long Beach	1,803	43	80	69	33°5'
	Los Angeles	2,061	43	80	69	34°0'
	Los Angeles[d]	1,349	40	89	71	34°0'
	Oakland	2,870	36	80	64	37°4'
	Sacramento	2,502	32	98	71	38°3'
	San Diego	1,458	44	80	70	32°4'
	San Francisco	3,015	38	77	64	37°4'
	San Francisco[d]	3,001	40	71	62	37°5'
CO	Alamosa	8,529	-16	82	61	37°3'
	Colorado Springs	6,423	2	88	62	38°5'
	Denver	6,283	1	91	63	39°5'
	Grand Junction	5,641	7	94	63	39°1'
	Pueblo	5,462	0	95	66	38°2'
CT	Bridgeport	5,617	9	84	74	41°1'
	Hartford	6,235	7	88	75	41°5'
	New Haven	5,897	7	84	75	41°2'
DE	Wilmington	4,930	14	89	76	39°4'
DC	Washington	4,224	17	91	77	38°5'
FL	Daytona	879	35	90	79	29°1'
	Fort Myers	442	44	92	79	26°4'
	Jacksonville	1,239	32	94	79	30°3'
	Key West	108	57	90	79	24°3'
	Miami	214	47	90	79	25°5'
	Orlando	766	38	93	78	28°3'
	Pensacola	1,463	29	93	79	30°3'
	Tallahassee	1,485	30	92	78	30°2'
	Tampa	683	40	91	79	28°0'
	West Palm Beach	253	45	91	79	26°4'
GA	Athens	2,929	22	92	77	34°0'
	Atlanta	2,961	22	92	76	33°4'
	Augusta	2,397	23	95	79	33°2'
	Columbus	2,383	24	93	78	32°3'
	Macon	2,136	25	93	78	32°4'
	Rome	3,326	22	93	78	34°2'
	Savannah	1,819	27	93	79	32°1'
HI	Hilo	0	62	83	74	19°4'
	Honolulu	0	63	86	75	21°2'

FIGURE 17.1 The degree days and design temperatures for for selected cities in each state. *(Courtesy of International Code Council, Inc. and International Plumbing Code 2000.)*

17.1

STATE	STATION[b]	HEATING DEGREE DAYS (yearly total)	DESIGN TEMPERATURES			DEGREES NORTH LATITUDE[c]
			Winter	Summer		
			97$^1/_2$%	Dry bulb 2$^1/_2$%	Wet bulb 2$^1/_2$%	
ID	Boise	5,809	10	94	66	43°3'
	Lewiston	5,542	6	93	66	46°2'
	Pocatello	7,033	-1	91	63	43°0'
IL	Chicago (Midway)	6,155	0	91	75	41°5'
	Chicago (O'Hare)	6,639	-4	89	76	42°0'
	Chicago[d]	5,882	2	91	77	41°5'
	Moline	6,408	-4	91	77	41°3'
	Peoria	6,025	-4	89	76	40°4'
	Rockford	6,830	-4	89	76	42°1'
	Springfield	5,429	2	92	77	39°5'
IN	Evansville	4,435	9	93	78	38°0'
	Fort Wayne	6,205	1	89	75	41°0'
	Indianapolis	5,699	2	90	76	39°4'
	South Bend	6,439	1	89	75	41°4'
IA	Burlington	6,114	-3	91	77	40°5'
	Des Moines	6,588	-5	91	77	41°3'
	Dubuque	7,376	-7	88	75	42°2'
	Sioux City	6,951	-7	92	77	42°2'
	Waterloo	7,320	-10	89	77	42°3'
KS	Dodge City	4,986	5	97	73	37°5'
	Goodland	6,141	0	96	70	39°2'
	Topeka	5,182	4	96	78	39°0'
	Wichita	4,620	7	98	76	37°4'
KY	Covington	5,265	6	90	75	39°0'
	Lexington	4,683	8	91	76	38°0'
	Louisville	4,660	10	93	77	38°1'
LA	Alexandria	1,921	27	94	79	31°2'
	Baton Rouge	1,560	29	93	80	30°3'
	Lake Charles	1,459	31	93	79	30°1'
	New Orleans	1,385	33	92	80	30°0'
	Shreveport	2,184	25	96	79	32°3'
ME	Caribou	9,767	-13	81	69	46°5'
	Portland	7,511	-1	84	72	43°4'
MD	Baltimore	4,654	13	91	77	39°1'
	Baltimore[d]	4,111	17	89	78	39°2'
	Frederick	5,087	12	91	77	39°2'
MA	Boston	5,634	9	88	74	42°2'
	Pittsfield	7,578	-3	84	72	42°3'
	Worcester	6,969	4	84	72	42°2'
MI	Alpena	8,506	-6	85	72	45°0'
	Detroit (City)	6,232	6	88	74	42°2'
	Escanaba[d]	8,481	-7	83	71	45°4'
	Flint	7,377	1	87	74	43°0'
	Grand Rapids	6,894	5	88	74	42°5'
	Lansing	6,909	1	87	74	42°5'
	Marquette[d]	8,393	-8	81	70	46°3'
	Muskegon	6,696	6	84	73	43°1'
	Sault Ste. Marie	9,048	-8	81	70	46°3'
MN	Duluth	10,000	-16	82	70	46°5'
	Minneapolis	8,382	-12	89	75	44°5'
	Rochester	8,295	-12	87	75	44°0'
MS	Jackson	2,239	25	95	78	32°2'
	Meridian	2,289	23	95	79	32°2'
	Vicksburg[d]	2,041	26	95	80	32°2'

FIGURE 17.2 The degree days and design temperatures for for selected cities in each state. (*Courtesy of International Code Council, Inc. and International Plumbing Code 2000.*)

DEGREE DAY AND DESIGN TEMPERATURES[a] FOR CITIES IN THE UNITED STATES—continued

STATE	STATION[b]	HEATING DEGREE DAYS (yearly total)	DESIGN TEMPERATURES Winter 97¹/₂%	DESIGN TEMPERATURES Summer Dry bulb 2¹/₂%	DESIGN TEMPERATURES Summer Wet bulb 2¹/₂%	DEGREES NORTH LATITUDE[c]
MO	Columbia	5,046	4	94	77	39°0'
	Kansas City	4,711	6	96	77	39°1'
	St. Joseph	5,484	2	93	79	39°5'
	St. Louis	4,900	6	94	77	38°5'
	St. Louis[d]	4,484	8	94	77	38°4'
	Springfield	4,900	9	93	77	37°1'
MT	Billings	7,049	-10	91	66	45°5'
	Great Falls	7,750	-15	88	62	47°3'
	Helena	8,129	-16	88	62	46°4'
	Missoula	8,125	-6	88	63	46°5'
NE	Grand Island	6,530	-3	94	74	41°0'
	Lincoln[d]	5,864	-2	95	77	40°5'
	Norfolk	6,979	-4	93	77	42°0'
	North Platte	6,684	-4	94	72	41°1'
	Omaha	6,612	-3	91	77	41°2'
	Scottsbluff	6,673	-3	92	68	41°5'
NV	Elko	7,433	-2	92	62	40°5'
	Ely	7,733	-4	87	59	39°1'
	Las Vegas	2,709	28	106	70	36°1'
	Reno	6,332	10	92	62	39°3'
	Winnemucca	6,761	3	94	62	40°5'
NH	Concord	7,383	-3	87	73	43°1'
NJ	Atlantic City	4,812	13	89	77	39°3'
	Newark	4,589	14	91	76	40°4'
	Trenton[d]	4,980	14	88	76	40°1'
NM	Albuquerque	4,348	16	94	65	35°0'
	Raton	6,228	1	89	64	36°5'
	Roswell	3,793	18	98	70	33°2'
	Silver City	3,705	10	94	64	32°4'
NY	Albany	6,875	-1	88	74	42°5'
	Albany[d]	6,201	1	88	74	42°5'
	Binghamton	7,286	1	83	72	42°1'
	Buffalo	7,062	6	85	73	43°0'
	NY (Cent. Park)[d]	4,871	15	89	75	40°5'
	NY (Kennedy)	5,219	15	87	75	40°4'
	NY(LaGuardia)	4,811	15	89	75	40°5'
	Rochester	6,748	5	88	73	43°1'
	Schenectady[d]	6,650	1	87	74	42°5'
	Syracuse	6,756	2	87	73	43°1'
NC	Charlotte	3,181	22	93	76	35°1'
	Greensboro	3,805	18	91	76	36°1'
	Raleigh	3,393	20	92	77	35°5'
	Winston-Salem	3,595	20	91	75	36°1'
ND	Bismarck	8,851	-19	91	71	46°5'
	Devils Lake[d]	9,901	-21	88	71	48°1'
	Fargo	9,226	-18	89	74	46°5'
	Williston	9,243	-21	88	70	48°1'
OH	Akron-Canton	6,037	6	86	73	41°0'
	Cincinnati[d]	4,410	6	90	75	39°1'
	Cleveland	6,351	5	88	74	41°2'
	Columbus	5,660	5	90	75	40°0'
	Dayton	5,622	4	89	75	39°5'
	Mansfield	6,403	5	87	74	40°5'
	Sandusky[d]	5,796	6	91	74	41°3'
	Toledo	6,494	1	88	75	41°4'
	Youngstown	6,417	4	86	73	41°2'

FIGURE 17.3 The degree days and design temperatures for for selected cities in each state. (*Courtesy of International Code Council, Inc. and International Plumbing Code 2000.*)

17.3

STATE	STATION[b]	HEATING DEGREE DAYS (yearly total)	DESIGN TEMPERATURES			DEGREES NORTH LATITUDE[c]
			Winter	Summer		
			97 1/2%	Dry bulb 2 1/2%	Wet bulb 2 1/2%	
OK	Oklahoma City	3,725	13	97	77	35°2'
	Tulsa	3,860	13	98	78	36°1'
OR	Eugene	4,726	22	89	67	44°1'
	Medford	5,008	23	94	68	42°2'
	Portland	4,635	23	85	67	45°4'
	Portland[d]	4,109	24	86	67	45°3'
	Salem	4,754	23	88	68	45°0'
PA	Allentown	5,810	9	88	75	40°4'
	Erie	6,451	9	85	74	42°1'
	Harrisburg	5,251	11	91	76	40°1'
	Philadelphia	5,144	14	90	76	39°5'
	Pittsburgh	5,987	5	86	73	40°3'
	Pittsburgh[d]	5,053	7	88	73	40°3'
	Reading[d]	4,945	13	89	75	40°2'
	Scranton	6,254	5	87	73	41°2'
	Williamsport	5,934	7	89	74	41°1'
RI	Providence	5,954	9	86	74	41°4'
SC	Charleston	2,033	27	91	80	32°5'
	Charleston[d]	1,794	28	92	80	32°5'
	Columbia	2,484	24	95	78	34°0'
SD	Huron	8,223	-14	93	75	44°3'
	Rapid City	7,345	-7	92	69	44°0'
	Sioux Falls	7,839	-11	91	75	43°4'
TN	Bristol	4,143	14	89	75	36°3'
	Chattanooga	3,254	18	93	77	35°0'
	Knoxville	3,494	19	92	76	35°5'
	Memphis	3,232	18	95	79	35°0'
	Nashville	3,578	14	94	77	36°1'
TX	Abilene	2,624	20	99	74	32°3'
	Austin	1,711	28	98	77	30°2'
	Dallas	2,363	22	100	78	32°5'
	El Paso	2,700	24	98	68	31°5'
	Houston	1,396	32	94	79	29°4'
	Midland	2,591	21	98	72	32°0'
	San Angelo	2,255	22	99	74	31°2'
	San Antonio	1,546	30	97	76	29°3'
	Waco	2,030	26	99	78	31°4'
	Wichita Falls	2,832	18	101	76	34°0'
UT	Salt Lake City	6,052	8	95	65	40°5'
VT	Burlington	8,269	-7	85	72	44°3'
VA	Lynchburg	4,166	16	90	76	37°2'
	Norfolk	3,421	22	91	78	36°5'
	Richmond	3,865	17	92	78	37°3'
	Roanoke	4,150	16	91	74	37°2'
WA	Olympia	5,236	22	83	66	47°0'
	Seattle-Tacoma	5,145	26	80	64	47°3'
	Seattle[d]	4,424	27	82	67	47°4'
	Spokane	6,655	2	90	64	47°4'
WV	Charleston	4,476	11	90	75	38°2'
	Elkins	5,675	6	84	72	38°5'
	Huntington	4,446	10	91	77	38°2'
	Parkersburg[d]	4,754	11	90	76	39°2'

FIGURE 17.4 The degree days and design temperatures for for selected cities in each state. (*Courtesy of International Code Council, Inc. and International Plumbing Code 2000.*)

DEGREE DAY AND DESIGN TEMPERATURES[a] FOR CITIES IN THE UNITED STATES—continued

STATE	STATION[b]	HEATING DEGREE DAYS (yearly total)	Winter 97¹/₂%	Summer Dry bulb 2¹/₂%	Summer Wet bulb 2¹/₂%	DEGREES NORTH LATITUDE[c]
WI	Green Bay	8,029	-9	85	74	44°3'
	La Crosse	7,589	-9	88	75	43°5'
	Madison	7,863	-7	88	75	43°1'
	Milwaukee	7,635	-4	87	74	43°0'
WY	Casper	7,410	-5	90	61	42°5'
	Cheyenne	7,381	-1	86	62	41°1'
	Lander	7,870	-11	88	63	42°5'
	Sheridan	7,680	-8	91	65	44°5'

a. All data was extracted from the 1985 ASHRAE Handbook, Fundamentals Volume.

b. Design data developed from airport temperature observations unless noted.

c. Latitude is given to the nearest 10 minutes. For example, the latitude for Miami, Florida, is given as 25°5', or 25 degrees 50 minutes.

d. Design data developed from office locations within an urban area, not from airport temperature observations.

FIGURE 17.5 The degree days and design temperatures for for selected cities in each state. *(Courtesy of International Code Council, Inc. and International Plumbing Code 2000.)*

CHAPTER 18
MOBILE HOME
AND RV PARKS

Mobile home and RV parks may require special needs, which are addressed by the plumbing code. Requirements for parks intended to house mobile housing vary. The plumbing code, however, does provide plenty of direction for plumbers who are working with the needs of mobile home and RV parks. The first step in understanding the plumbing of mobile home and RV parks is to establish what constitutes a mobile home or RV. This matter is explained well in the code.

By code standards, a mobile home is determined to be a structure that is transportable in one, or more sections. In the travel mode, such a structure is required to be a minimum of 8 feet in width and no more than 40 feet in length, or when erected on a site, the structure must contain a minimum of 320 square feet that is built on a permanent chassis. The structure must be designed for use as a dwelling that does not require a permanent foundation when connected to required utilities, such as plumbing, heating, air-conditioning, and electrical systems. Federal regulation 24 CFR offers more descriptive details about what constitutes a mobile home.

In addition to mobile homes, the plumbing code deals with what it recognizes as mobile home accessory buildings and structures. Such structures are deemed to be an addition to a mobile home or a means of supplementing the mobile home. An accessory building shall not be self-contained, separate, or habitable. Examples of accessory buildings and structures can include awnings, ramadas, porches, carports, and storage areas.

What is considered to be a lot for a mobile home? Any portion of a mobile home park that is designed to accommodate a mobile home and its accessory buildings for the exclusive use of the mobile home's occupants is a mobile home lot. Bear in mind that private land that is not part of a park but that houses a mobile home is not considered to be a mobile home lot under the code requirements for mobile home parks.

How large does a development need to be in order to fall into the classification of a mobile home park? Not very big. Any piece of land or contiguous parcels that are designated and improved to contain two or more mobile home lots that are available to the general public for the placement of mobile homes for occupancy is said to be a mobile home park. Keep in mind the ruling that the lots must be available to the general public.

It is possible to place two mobile homes on a private parcel of land for personal use without triggering the code requirements of a mobile home park.

GENERAL REGULATIONS

General regulations state that mobile home parks must have plumbing and drainage facilities designed and installed in accordance with all of the requirements of the plumbing code. Just as with new construction, plans must be provided to the code enforcement office for review and approval before plumbing permits can be issued to install or alter plumbing systems in a mobile home park. All plans must be submitted in duplicate form. The same goes for specifications.

Plans and specifications submitted for permit approval are required to meet certain code standards. A plot plan of the mobile home park must be provided for code approval. The plot plan must be drawn to scale and indicate elevations, property lines, driveways, existing buildings, proposed buildings, and the sizes of any mobile home lots.

Specifications must be complete, and a piping layout of all proposed plumbing systems or alterations is required for an approval submission. Specs and layouts also are required for any proposed sewage disposal systems or alterations. All submissions are required to show clearly the nature and extent of all proposed work. As long as all proposed work and specifications are within the parameters of the plumbing code, approvals should be forthcoming.

PARK DRAINAGE SYSTEMS

All mobile home parks must be equipped with approved means of conveying and disposing of sewage. Ideally, such systems should be connected to public sewers, but when public utilities are not available, private sewage disposal systems may be used. All connections and installations must be performed in an approved manner.

Drainage Pipe Diameter and Number of Fixture Units on Drainage System		
Size of Drainage Pipe		Maximum Number
Inches	(mm)	of Fixture Units
2*	51	8
3	76	35
4	102	256
5	127	428
6	152	720
8	203	2640
10	254	4680
12	305	8200
*Except 6 unit fixtures		

FIGURE 18.1   The table covers the connection between drainage pipe diameter and fixture units. *(Reprinted from the* 2000 Uniform Plumbing Code (UPC) *with the permission of the International Association of Plumbing and Mechanical Officials (IAPMO).)*

Minimum Grade and Slope of Drainage Pipe							
Pipe Size		Slope per 100 ft. (30.5 m)		Pipe Size		Slope per 100 ft.(30.5 m)	
Inches	(mm)	Inches	(mm)	Inches	(mm)	Inches	(mm)
2	51	25	635	6	152	8	203
3	76	25	635	8	203	4	102
4	102	15	381	10	254	3 1/2	89
5	127	11	279	12	305	3	76

FIGURE 18.2 Grade and Drainage Pipe slope are covered above. *(Reprinted from the* 2000 Uniform Plumbing Code (UPC) *with the permission of the International Association of Plumbing and Mechanical Officials (IAPMO).)*

Underground installations in mobile home parks must be made with materials that are code-approved for their intended purposes. Drainage inlets and extensions to grade also must be made of materials approved for such use. Drainage systems installed in a mobile home park must be buried at a depth that protects them from damage from traffic or other movements that might stress the pipe and fittings. General code requirements for pipe sizing and slope apply to the drainage systems in mobile home parks.

Individual park lots must be provided with drainage inlets that are not less than three inches in diameter. Individual sewers must be installed with a minimum slope of one-quarter inch per foot. As you would expect, all joints must be watertight. Piping used between a mobile home and the lot drainage inlet must be made with semirigid, corrosion-resistant, nonabsorbent, durable material, and have a smooth inner surface.

Since mobile homes can be moved, and sometimes are moved often, drainage inlets must be fitted to accept suitable plugging or capping when a mobile home is not connected to the inlet. It is not allowable for surface water to run in close proximity to drainage inlets. Rims for drainage inlets extending above grade must not exceed a height of 4 inches above the finished grade. The location of a drainage inlet is required to be in the rear one-third of the park lot and not more than 44 feet from the proposed location of mobile homes that may be situated on the lot.

Pipe sizing for the drainage inlets of mobile home parks is based on a fixture-unit rating of 12 fixture units. Minimum pipe diameter for a drainage inlet is 3 inches. Refer to your local code book for available sizing tables. If a park system exceeds the recommended fixture-unit loading or cannot be installed with the minimum required slope, the system must be designed by a registered professional engineer.

DRAINAGE CONNECTORS

Pipe and fittings used for drainage connectors must have a minimum rating of Schedule 40. Pipe sizing for drainage connections must not be smaller than that of the drainage inlet to which it is connected. An approved cleanout must be installed between a mobile home and the drainage inlet. Any fitting used for the connection to a drainage inlet must be a directional fitting that will discharge flow into the drainage inlet. A minimum grade for a drainage connector is one-quarter inch per foot. All connectors must be gastight and sized so that the connector is no longer than what is needed to complete a satisfactory connection. It is acceptable to use a flexible connector at the lot drainage inlet area only. All drainage inlets must be capped in a gastight manner when the inlet is not connected to a mobile home.

PARK WATER SUPPLIES

Naturally, mobile home parks require a suitable water supply. It is preferable for this supply to come from a public water utility, but some parks function well with private water supplies. Every lot in a mobile home park must be provided with an approved water service. The minimum diameter of a water service for a park lot is three-quarters of an inch. It must be capable of delivering a minimum of 12 fixture units in its water rating. A minimum pressure rating for each lot's water service is 20 pounds per square inch (psi) at maximum operating conditions. As with a sewage connection, the water service must be in the rear third of the park lot and not more than 4 feet from the proposed location of a mobile home. Water services and sewers may not be placed in a common trench, except when the two piping systems are separated in height by a shelf for the water service, as is the case in the general plumbing code.

Water distribution systems in mobile home parks are required to provide for a minimum of 12 fixture units. The systems must be constructed of approved materials. Independent cutoff valves are required at each water service outlet on every park lot. The cutoff valve must be installed on the supply side of a backflow preventer when such a device is used. Connection between mobile homes and park water services are to be made with flexible materials, such as copper tubing. The minimum interior diameter of the connection material is three-quarters of an inch.

Backflow preventers are required on water services for mobile home parks when there may be a risk of cross connection. The installation of backflow devices shall be made at or near the water service outlet. Any hose bibb or outlet installed on the supply outlet riser in addition to a service connector must be fitted with an approved backflow preventer.

The installation of a backflow preventer on a park lot service outlet triggers the need for a pressure-relief valve. The relief valve is to be installed on the discharge side of the backflow preventer. The pressure-reducing valve must be set to release at a pressure that doesn't exceed 150 psi. All pressure relief valves must discharge toward the ground. Code requires all backflow preventers and pressure-relief valves to be installed at a point at least 12 inches above the ground. Water service outlets, backflow preventers, and pressure-relief valves in mobile home parks must be protected from damage that might occur from vehicles or other sources. It is permissible to use posts, fencing, or other permanent barriers to create suitable protection.

WATER CONDITIONING EQUIPMENT

Permits are required and must be approved when water conditioning equipment is installed on park lots. In cases where the water treatment equipment is of a regenerating type and the park drainage system discharges into a public sewer, the approval of the sanitary district or another agency holding jurisdiction is needed. All water conditioning equipment of a regenerating type must be labeled by an approved listing agency.

Demand Factors for Calculating Master Meter Gas Piping Systems in M/H Parks		
Number of M/H Lots	BTU/Hr. per M/H Lot	Watts/Hr. per M/H Lot
1	250,000	73,275
2	234,000	68,585
3	208,000	60,965
4	192,000	56,295
5	184,000	53,930
6	174,000	50,999
7	166,000	48,655
8	162,000	47,482
9	158,000	46,310
10	154,000	45,137
11-20	132,000	38,689
21-30	124,000	36,344
31-40	118,000	34,586
41-60	112,000	32,827
Over 60	102,000	29,896

FIGURE 18.3 This figure examines BTU and watts per hour with the number of mobile home lots in a park. *(Reprinted from the* 2000 Uniform Plumbing Code *(UPC) with the permission of the International Association of Plumbing and Mechanical Officials (IAPMO).)*

Regenerating units must discharge effluent into a trap that has a diameter of not less than 1½ inches in diameter. All effluent traps must be connected to the park drainage system. An approved air gap is required on the discharge line at a minimum height of 12 inches above the ground level. All installations must be tested, inspected, and approved by code officials, or other recognized inspectors.

FUEL-GAS EQUIPMENT

Standard code requirements for fuel-gas equipment apply to mobile home parks. There are, however, some differences in the gas codes when mobile home parks are involved. For starters, a permit is required before any gas equipment or installation is constructed or altered. Two sets of plans and specifications are needed when permit application is made. In addition to the plans and specs, a load calculation of the gas piping system must be provided for code approval.

Sizing tables in your local code book are used to establish the minimum hourly volume of gas required at each park lot. Tables also are available for calculating the required gas supply for buildings and other fuel gas-consuming appliances connected to a mobile home.

Gas piping that is installed below ground must have a minimum depth coverage of 18 inches. In other words, there must be at least 18 inches of dirt covering any portion of a buried gas pipe. Although it is acceptable to install gas piping above grade, no gas piping shall be installed aboveground when the piping is under a mobile home. This is basically what the code says, but the context of what is said can be confusing. If you simply read this one section of the code, it appears that gas piping can be installed under a mobile home if the piping is buried. Don't believe it, at least not as it first appears. Just after this section of the code is language that explains much more fully the rules for gas piping under a mobile home.

The code states clearly, but later than I feel it should, that gas piping must not be installed underground beneath buildings or any portion of a mobile home lot where a mobile home or accessory buildings, or structures, or concrete slabs, or automobile parking are intended to be placed. Gas piping can be buried under the locations discussed above, but the piping must be installed in a gastight conduit.

Conduit used for gas piping must be made of an approved material. No material with less strength than Schedule 40 pipe shall be used for a gas piping conduit. The diameter of any conduit used for buried gas piping must be at least one-half inch larger than the outside diameter of the gas pipe housed in the conduit.

Conduit required for underground gas piping must extend for a minimum of 12 inches beyond all areas where the conduit is required. In other words, if you are installing buried gas pipe in a conduit under a concrete patio, the conduit must extend for at least 12 inches beyond the edges of the concrete. This extension distance also applies to conduit used to sleeve gas piping that penetrates the outside walls of buildings. The outer end of the conduit is allowed to be sealed. If a conduit terminates within a building, the conduit must be readily accessible and the space between the conduit and the gas pipe being protected must be sealed to prevent any leakage of gas into the building. Does this confuse you?

You can't seal the end of the conduit, but you must seal the space between the pipe and the conduit. It is a little confusing, but it is also the code. The end of the conduit pipe that protrudes the outer wall of a building must not be sealed, but the end of the conduit that terminates within a building must be sealed between the conduit and the gas pipe. This prevents gas from entering a building and allows any gas leakage to vent out to open air.

There is another aspect of the code that can be confusing. All gas piping that is surrounded by concrete must be in a conduit, right? Wrong. A gas piping lateral that terminates in a park lot outlet riser that is surrounded by a concrete slab is not required to be installed in a conduit when certain conditions exist. If the concrete slab is entirely outside the wall line of a mobile home, a conduit is not needed when the concrete is not continuous with any other concrete slab and is used for stabilizing other utility connections.

SYSTEM SHUTOFF VALVES

A system shutoff valve is required for controlling the flow of gas to the entire gas piping system. The valve must be readily accessible, identified, and approved for its intended use. A system shutoff valve must be installed near the point of connection to the service piping or supply connection of the liquefied petroleum (LP) gas tank.

In addition to a system shutoff valve, every mobile home lot must have an approved gas shutoff valve installed upstream of the lot's gas outlet. The valve must be located on the outlet riser at a height of not less than 4 inches above ground. It is a violation of code to place the valve under any mobile home. When a gas outlet is not in use, the outlet must be capped or plugged with an approved device to prevent accidental discharge of gas.

GAS CONNECTOR

Every connection between the mobile home and the gas outlet must be made with an approved or listed gas connector. The connector must not be more than 6 feet long. When additional length is needed, approved pipe fittings may be used between the flexible connector and the lot gas outlet. The connector used must be sized adequately to supply the total gas demand of the mobile home to which it is connected.

MECHANICAL PROTECTION

Proper protection is essential for all gas piping and equipment. Gas outlet risers, regulators, meters, valves, and other exposed equipment must be protected from any form of mechanical damage. Various items can be used to provide such protection. Posts, fencing, and other permanent barriers all can be used to form protection for exposed gas devices.

Atmospherically controlled regulators must be installed so that moisture cannot enter the regulator vent and accumulate above the diaphragm. Regulator vents that may be obstructed by snow or ice must be protected. Shields usually are used to guard against the vent opening becoming closed or obstructed by such conditions as snow or ice.

Gas meters are not allowed to be installed in unvented or inaccessible locations. Gas meters installed in an area that might present a source of ignition must be set at least three feet from any source of ignition.

Proper support is required for gas meters. The pipe of a gas outlet riser is not considered sufficient support. A gas meter must be supported by a post or bracket that is placed on a firm footing, or some other means of providing equivalent support must be provided.

SIZING AND MAINTENANCE

The sizing of gas pipe for either natural gas or LP gas can be done in conjunction with sizing tables found in local code books. Sizing also may be done by suitable professionals, such as engineers. All sizing determinations must be in compliance with code requirements.

The operator of a mobile home park is responsible for maintaining all gas piping installations and equipment. The park operator must keep all systems and devices in good working order. If a gas system is found to be defective or outside code compliance, a code officer or a serving gas supplier has the authority to disconnect the gas service. If a disconnection is made, a notice will be attached to the gas piping or appliance, or both, that states the service has been disconnected and the reasons for disconnection. All inspections and tests are to be made in accordance with local code requirements.

RECREATIONAL VEHICLE PARKS

There are many types of recreational vehicles (RVs), including motor homes, travel trailers, fifth-wheel trailers, camping trailers, park trailers, and truck campers. These vehicular type units are designed primarily as temporary living quarters for recreational, camping, travel, or seasonal use. The units can have their own motive power, can be mounted on another vehicle, or can be towed by a vehicle.

Parks designed for RVs consist of a plot of land where at least two or more RV sites are located, established or maintained for occupancy by RVs of the general public as temporary living quarters for recreational or vacation purposes. An RV site is a plot of ground located within an RV park that is intended for the accommodation of an RV, tent, or other individual camping unit on a temporary basis. Plumbing installations for RV parks and sites must meet code requirements.

TOILET AND SHOWER FACILITIES

The toilet and shower facilities for RV parks are substantial. Every RV park is required to provide toilets and urinals at one or more locations. The facilities must be convenient and located within a 500-foot radius of any RV site that is not provided with an individual sewer connection. Facilities for males and females must be marked appropriately. There must be at least one toilet for each sex up to the first 25 sites. Each additional 25 sites not provided with sewer connections require the installation of at least one additional toilet.

Interior walls in restrooms and bathrooms must be moisture resistant up to a minimum height of 4 feet. This is done to make cleaning the walls easier and more effective. Floors have to be made of materials that are impervious to water and easily cleaned. Facilities housing toilets that are flushed with water must be equipped with suitable floor drains. The trap of the floor drain must be provided with a means, such as a trap primer, to protect the trap seal.

When toilets flushed with water are installed, an equal number of lavatories is required, up to the first six toilets. Additional toilets call for more lavatories. When two additional toilets are installed, one additional lavatory is needed. All lavatories must be piped with a supply of potable water and be provided with approved drainage into a drainage system.

Urinals can be used in place of toilets in facilities that are separated for male use. Urinals, however, cannot replace more than one-third of the number of toilets required. There is an exception. A urinal may be used to replace a toilet in a minimum park. Individual stall or wall-hung types of urinals are acceptable, but floor type trough units are not.

Toilets installed must be of an approved type, have an elongated bowl, and shall be provided with open-front seats. Every toilet must have a separate compartment and be provided with a door and a latch for privacy. A holder or dispenser for toilet tissue also is required. Any dividing walls or partitions must be at least five feet high and must be separated from the floor by a space not greater than 12 inches. The toilet enclosures cannot be less than 30 inches in width and a minimum of 30 inches of clear space is needed in front of each toilet. No toilet may be installed so that it is closer than 15 inches to any side wall. A special requirement is needed in toilet rooms for females. A receptacle of durable, impervious, and readily cleanable material, with a lid, is required for the disposal of sanitary napkins.

Showers, when provided, must have a minimum floor area of 36 inches by 36 inches and must be capable of encompassing a 30-inch diameter circle. The showers must be of an individual type. Every shower area must be visually screened from view. Individual dressing areas that are screened from view also must be provided. Each dressing area must contain at least one clothing hook and a stool or equivalent bench area.

Shower areas must be designed to minimize the flow of water into a dressing area. The shower area must be connected to the drainage system by means of a properly vented and trapped inlet. These areas must have an impervious, skid-resistant surface. Wooden racks (duck boards) over shower floors are prohibited.

Toilet buildings must have a minimum ceiling height of 7 feet. If artificial light is not provided, a window or skylight with an area equal to at least 10 percent of the floor area must be supplied. Doors to the exterior are required to open outward. They must be self-closing and shall be screened visually by means of a vestibule or wall to prevent direct view of the interior when the exterior doors are open. The screening is not required on single toilet units.

Toilet rooms must have permanent non-closable, screened opening, or openings, that has a total area not less than 5 percent of the floor area, opening directly to the exterior. The reason for this is to provide proper ventilation. Listed exhaust fans, vented to the exterior, the rating of which in cubic feet per minute is at least 25 percent of the total volume of the rooms served can be considered as meeting the requirements for ventilation. Any openable windows and vents to the outside must be provided with fly-proof screens of not less than number 16 mesh.

RV PARK POTABLE WATER

All water supply to RV parks must meet the criteria as set forth by local authorities. When an approved public water supply is available, the system shall be used. If a private supply is used, all components of the system must be approved. Any seasonal water supply must be equipped with a means for draining the system. There must never be a situation in which a potable water supply is connected to any non-potable or unapproved water supply or in which a connection might be subjected to any backflow or backsiphonage.

Figuring the water demand can be done by referring to your local code book. Park sites that don't have individual water connections are required to have a minimum of 25 gallons of water allocated in the total supply demand. If a site has an individual connection, the calculation demand rises to 50 gallons per day. When toilets that depend on water for flushing are used, the supply demand has to be a minimum of 50 gallons a day per site.

The minimum water pressure requirement for water outlets in which water is distributed under pressure to any individual site is 20 psi, with a minimum flow of two gallons per minute (gpm). Maximum allowable pressure is 80 psi. All water outlets must be convenient to access and, when not piped to individual recreational vehicle sites, shall not be located farther than 300 feet from any site. Standing water and muddy conditions around a water outlet must be prevented.

When water storage tanks are used, they must be made of impervious materials, protected from contamination, and provided with watertight covers that are locked. Any vents or overflows must be pointed downward. Overflows and vents must be fitted with corrosion-resistant screening that is not less than number 24 mesh. The reason for the mesh is to prevent insects and vermin from entering the containers. It is not allowable for a water storage tank to have a direct connection to a sewer.

INDIVIDUAL WATER CONNECTIONS

Individual water connections for RVs must be installed so that they are at the rear of the RV site, to the left side of the RV and within 4 feet of the RV location. Potable water connections must consist of a water riser pipe equipped with a threaded male spigot located at least 12 inches, but not more than 24 inches, above grade level. The riser must accept the threads of a standard hose connection. All risers must be protected from physical damage. The riser spigot must be equipped with a listed antisiphon backflow-prevention device. If a drinking fountain is supplied, it must conform to all code requirements.

RV PARK DRAINAGE SYSTEMS

All RV parks must have an adequate and approved drainage system for conveying and disposing of all sewage. When feasible, all RV parks are to be connected to a public sewer. Materials used in the creation of a park drainage system must be of an approved type. Installations must be done in accordance with the plumbing code.

Sizing drainage pipes for an RV park requires you to know the number of RVs to be served in the park. A 3-inch pipe can handle up to five RVs. To accommodate up to 36 units, you will need a 4-inch pipe. Five-inch pipes can handle up to 71 recreational vehicles. One-hundred-twenty RVs can be taken care of with a 6-inch pipe. An 8-inch pipe will accommodate up to 440 RV units. Sewer pipes have to be located to prevent damage from vehicular traffic. Cleanouts must be provided in accordance with the plumbing code.

DRAINAGE SYSTEM INLETS

Any drainage system provided in an RV park must be protected from damage. Drainage connections must consist of a sewer riser that extends vertically to the finished grade. A minimum pipe diameter for a sewer riser is three inches. The end of the riser must be equipped with a four-inch inlet or a minimum three-inch female fitting. Sewer inlets for individual RV lots must be on the left, rear-half of the site, to the left side of the RV, and within four feet of the RV location.

DISPOSAL STATIONS

A sanitary disposal station is required for each 100 RV sites that are not equipped with individual sewer inlets. Disposal stations have to be level, convenient for access from a service road, and must provide easy ingress and egress for the recreational vehicles using the facilities. Except where otherwise approved, disposal stations must have a concrete slab with drainage system inlet located so as to be on the left side of any RV unit. The slab must not be less than 3 feet by 3 feet in surface area. Thickness of the slab must be not less than 3½ inches and properly reinforced. A surface finish on the slab must be troweled to a smooth finish and sloped from each side inward to a drainage system inlet.

The drainage system inlet must consist of a 4-inch, self-closing foot-operated hatch of approved material with cover milled to fit tightly. The hatch body must be set in the concrete of the slab with the lip of the hatch opening flush with the slab surface. This is required to facilitate easy cleaning. The hatch must be connected properly to a drainage system inlet that shall discharge to an approved sanitary sewage disposal facility.

RV parks that are equipped with piped water supplies for flushing RV holding tanks and the sanitary disposal station slab must consist of a piped water supply under pressure, terminating in an outlet located and installed so as to prevent damage to the piping. Risers of this type must be properly supported and terminate at least 2 feet above the finished grade. A ¾-inch, valved outlet adaptable for a flexible hose is required. A suitable vacuum breaker or backflow preventer is required downstream from the last shutoff valve.

A flushing area must have a sign of durable material posted that indicates the following message: DANGER—NOT TO BE USED FOR DRINKING OR DOMESTIC PURPOSES. The sign must not be less than 2 feet by 2 feet in its dimensions.

RV PARK WATER SUPPLY STATIONS

Potable water stations in RV parks, when provided, must be located at least 50 feet from any sanitary disposal station. The water station, used for refilling the potable water tanks of RVs, must not be used for flushing waste tanks. A sign is required to be posted to prevent the improper use of the water station. The sign must be at least 2 feet by 2 feet in its dimensions, made of durable material, and inscribed in clear legible letters on a contrasting background with the following message: POTABLE WATER. NOT TO BE USED FOR FLUSHING WASTE TANKS. The potable water source must be protected from backflow by means of a listed vacuum breaker located downstream from the last shutoff valve.

FUEL-GAS

All fuel-gas equipment and installation in RV parks must be installed in a manner to meet the full criteria of the governing code. With this chapter completed, let's move on to Chapter 19 and discuss safety issues when working on a job.

CHAPTER 19
COMMON SENSE SAFETY ON THE JOB

Safety procedures and precautions don't get the attention they deserve. Far too many people are injured on jobs every year. Most of the injuries could be prevented, but they are not. Why is this? People are in a hurry to make a few extra bucks, so they cut corners. This happens with plumbing contractors and piece workers. It even affects hourly plumbers who want to shave 15 minutes off their work day so they can head back to the shop early.

Based on my field experience, most accidents occur as a result of negligence. Plumbers try to cut a corner, and they wind up getting hurt. This has proved true with my personal injuries. I've only suffered two serious on-the-job injuries, and both were a direct result of my carelessness. I knew better than to do what I was doing when I was hurt, but I did it anyway. Well, sometimes you don't get a second chance, and the life you affect may not be your own. So, let's look at some sensible safety procedures that you can implement in your daily activity.

VERY DANGEROUS

Plumbing can be a very dangerous trade. The tools of the trade have potential to be killers. Requirements of the job can place you in positions where a lack of concentration could result in serious injury or death. The fact that plumbing can be dangerous is no reason to rule out the trade as your profession. Driving can be extremely dangerous, but few people never set foot in an automobile out of fear.

Fear is generally a result of ignorance. When you have a depth of knowledge and skill, fear begins to subside. As you become more accomplished at what you do, fear is forgotten. While it is advisable to learn to work without fear, you should never work without respect. There is a huge difference between fear and respect.

If, as a plumber, you are afraid to climb up on a roof to flash a pipe, you are not going to last long in the plumbing trade. If you scurry up the roof recklessly, however, you could be injured severely, perhaps even killed. You must respect the position into which you are putting yourself. If you are using a ladder to get onto the roof, you must respect the outcome of what a mistake could have. Once you are on the roof, you must be conscious of footing conditions and the way you negotiate the pitch of the roof. If you pay attention, are properly trained, and don't get careless, you are not likely to get hurt.

Being afraid of a roof will limit or eliminate your plumbing career. Treating your trip to and from the roof like a walk into your living room could be deadly. Respect is the key. If you respect the consequences of your actions, you are aware of what you are doing and the odds for a safe trip improve.

Many young plumbers are fearless in the beginning. They think nothing of darting around on a roof or jumping down in a sewer trench. As their careers progress, they usually hear about or see on-the-job accidents. Someone gets buried in the cave-in of a trench. Somebody falls off a roof. A metal ladder being set up hits a power line. A careless plumber steps into a flooded basement and is electrocuted because of submerged equipment. The list of possible job-related injuries is a long one.

Millions of people are hurt every year in job-related accidents. Most of these people were not following solid safety procedures. Sure, some of them were victims of unavoidable accidents, but most were hurt by their own hand, in one way or another. You don't have to be one of these statistics.

In nearly 20 years of plumbing, I have only been hurt seriously on the job twice. Both times were my fault. I got careless. In one of the instances, I let loose clothing and a powerful drill work together to chew up my arm. In the other incident, I tried to save myself the trouble of repositioning my stepladder while drilling holes in floor joists. My desire to save a minute cost me torn stomach muscles and months of pain from a twisting drill.

My accidents were not mistakes; they resulted from stupidity. Mistakes are made through ignorance. I wasn't ignorant about what could happen to me. I knew the risk I was taking, and I knew the proper way to perform my job. Even with my knowledge, I slipped up and got hurt. Luckily, both of my injuries healed, and I didn't pay a lifelong price for my stupidity.

During my long plumbing career, I have seen a lot of people get hurt. Most of these people have been helpers and apprentices. Every one of the on-the-job accidents I have witnessed could have been avoided. Many of the incidents were not extremely serious, but a few were.

As a plumber, you will be doing some dangerous work. You will be drilling holes, running threading machines, snaking drains, installing roof flashings, and a lot of other potentially dangerous jobs. Hopefully, your employer will provide you with quality tools and equipment. If you have the right tool for the job, you are off to a good start in staying safe.

Safety training is another factor you should seek from your employer. Some plumbing contractors fail to tell their employees how to do their jobs safely. It is easy for someone, such as an experienced plumber who knows a job inside and out, to forget to inform an inexperienced person about potential danger.

For example, a plumber might tell you to break up the concrete around a pipe to allow the installation of a closet flange and never consider telling you to wear safety glasses. The plumber will assume you know the concrete is going to fly up in your face as it is chiseled. As a rookie, however, you might not know about the reaction concrete has when hit with a cold chisel. One swing of the hammer could cause extreme damage to your eyesight.

Simple jobs, like the one in the example, are all it takes to ruin a career. You might be really on your toes when asked to scoot across an I-beam, but how much thought are you going to give to tightening the bolts on a toilet? The risk of falling off the I-beam is obvious. Having chips of china, from a broken toilet where the nuts were turned one time too many, flying into your eyes is not so obvious. Either way, you can have a work-stopping injury.

Safety is a serious issue. Some job sites are very strict in maintaining safety requirements. But a lot of jobs have no written safety rules. If you are working on a commercial job, supervisors are likely to make sure you abide by the rules of the Occupational Safety and Health Administration (OSHA). Failure to comply with OSHA regulations can result in stiff financial penalties. If you are working in residential plumbing, however, you may never set foot on a job where OSHA regulations are observed.

In all cases, you are responsible for your own safety. Your employer and OSHA can help you remain safe, but in the end, it is up to you. You are the one who has to know what to do and how to do it. And not only do you have to take responsibility for your own actions, you also have to watch out for the actions of others. It is not unlikely that you could be injured by someone else's carelessness. Now that you have had the primer course, let's get down to the specifics of job-related safety.

You will find the suggestions in this chapter broken down into various categories. Each category will deal with specific safety issues related to the category. For example, in the section on tool safety, you will learn procedures for working safely with tools. As you move from section to section, you might notice some overlapping of safety tips. For example, in the section on general safety, you will see that it is wise to work without wearing jewelry. Then you will find jewelry mentioned again in the tool section. The duplication is done to pinpoint definite safety risks and procedures. We will start with general safety.

GENERAL SAFETY

General safety covers a lot of territory. It starts from the time you get into the company vehicle and carries you right through to the end of the day. Many of the general safety recommendations involve the use of common sense. Now, let's get started.

VEHICLES

Many plumbers are given company trucks to get to and from jobs. You probably will spend a lot of time loading and unloading company trucks. And, of course, you will spend time either riding in or driving them. All of these areas can threaten your safety.

If you will be driving the truck, take the time to get used to how it handles. Loaded plumbing trucks don't drive like the family car. Remember to check the vehicle's fluids, tires, lights, and related equipment. Many plumbing trucks are old and have seen better days. Failure to check the vehicle's equipment could result in unwanted headaches. Also, remember to use the safety belts; they do save lives.

Apprentices usually are charged with the duty of unloading the truck at the job site. There are a lot of ways to get hurt while doing this job. Many plumbing trucks use roof racks to haul pipe and ladders. If you are unloading these items, make sure they will not come into contact with low-hanging electrical wires. Copper pipe and aluminum ladders make very good electrical conductors, and they will carry the power surge through you on the way to the ground. If you are unloading heavy items, don't put your body in awkward positions. Learn the proper ways for lifting, and never lift objects inappropriately. If the weather is wet, be careful climbing on the truck. Step bumpers get slippery, and a fall can impale you on an object or bang up your knee.

When it is time to load the truck, observe the same safety precautions you did in unloading. In addition to these considerations, always make sure your load is packed evenly and is well secured. Be especially careful of any load you attach to the roof rack, and always double check the cargo doors on trucks with utility bodies.

It will not only be embarrassing to lose your load going down the road, it could be deadly. I have seen a one-piece fiberglass tub-shower unit fly out of the back of a pickup truck as the truck was rolling up an interstate highway. As a young helper, I lost a load of pipe in the middle of a busy intersection. In that same year, the cargo doors on the utility body of my truck flew open as I came off a ramp onto a major highway. Tools were scattered across two lanes of traffic. These types of accidents don't have to happen. It's your job to make sure they don't.

CLOTHING

Clothing is responsible for a lot of on-the-job injuries. Sometimes it is the lack of clothing that causes the accidents, and there are many times when too much clothing creates the problem. Generally, it is wise not to wear loose-fitting clothes. Shirt tails should be tucked in, and short-sleeve shirts are safer than long-sleeved shirts when operating some types of equipment.

Caps can save you from minor inconveniences, such as getting glue in your hair, and hard hats provide some protection from potentially damaging accidents, like having a steel fitting dropped on your head. If you have long hair, keep it up and under a hat.

Good footwear is essential in the trade. A strong pair of hunting-style boots usually will be best. The thick soles provide some protection from nails and other sharp objects you might step on. Boots with steel toes can make a big difference in your physical well-being. If you are going to be climbing, wear footgear with a flexible sole that grips well.

Gloves can keep your hands warm and clean, but they also can contribute to serious accidents. Wear gloves sparingly, depending upon the job you are doing.

JEWELRY

On the whole, jewelry should not be worn in the workplace. Rings can inflict deep cuts in your fingers. They also can work with machinery to amputate fingers. Chains and bracelets are equally dangerous, probably more so.

EYE AND EAR PROTECTION

Eye and ear protection often is overlooked. An inexpensive pair of safety glasses can prevent you from being blind the rest of your life. Ear protection reduces the effect of loud noises, such as jackhammers and drills. You might not notice much benefit now, but in later years, you will be glad you wore it. If you don't want to lose your hearing, wear ear protection when subjected to loud noises.

PADS

Knee pads not only make a plumber's job more comfortable, they help to protect the knees. Some plumbers spend a lot of time on their knees, and pads should be worn to ensure you can continue to work for many years.

The embarrassment factor plays a significant role in job-related injuries. People, especially young people, feel the need to fit in and to make a name for themselves. Plumbing is sort of a macho trade. There is no secret that plumbers often fancy themselves as strong human specimens. Many plumbers are strong. The work can be hard and in doing it, becoming strong is a side benefit. But you can't allow safety to be pushed aside for the purpose of making a macho statement.

All too many people believe that working without safety glasses, ear protection, and so forth makes them tough. That's just not true; it might get them hurt, but it does not make them look tough. If anything, it makes them look stupid or inexperienced.

Don't fall into the trap so many young plumbers do. Never let people goad you into bad safety practices. Some plumbers are going to laugh at your knee pads. Let them laugh, you still will have good knees when they are hobbling around on canes. I'm dead serious about this issue. There is nothing sissy about safety. Wear your gear in confidence, and don't let the few jokesters get to you.

TOOL SAFETY

Tool safety is a big issue in plumbing. Anyone in the plumbing trade will work with numerous tools, all of which are potentially dangerous. Some of them are especially hazardous. This section is broken down by the various tools used on the job. You cannot afford to start working without the basics in tool safety. The more you can absorb about tool safety, the better off you will be.

The best starting point is reading all literature available from the manufacturers of your tools. The people that make the tools provide some good safety suggestions with them. Read and follow the manufacturers' recommendations.

The next step in working safely with your tools is to ask questions. If you don't understand how a tool operates, ask someone to explain it to you. Don't experiment on your own; the price you pay could be much too high.

Common sense is irreplaceable in the safe operation of tools. If you see an electrical cord with cut insulation, you should have enough common sense to avoid using it. In addition to this type of simple observation, you will learn some interesting facts about tool safety. Now, let me tell you what I've learned about tool safety over the years.

Some basic principles apply to all of your work with tools. We will start with the basics, and then we will move on to specific tools. Here are the basics:

- Keep body parts away from moving parts.
- Don't work with poor lighting conditions.
- Be careful of wet areas when working with electrical tools.
- If special clothing is recommended for working with your tools, wear it.
- Use tools only for their intended purposes.
- Get to know your tools well.
- Keep your tools in good condition.

Now, let's take a close look at the tools you are likely to use. Plumbers use a wide variety of hand tools and electrical tools. They also use specialty tools. So let's see how you can use all these tools without injury.

TORCHES

Plumbers often use torches on a daily basis. Some plumbers use propane torches, and others use acetylene torches. In either case, the containers of fuel for these torches can be very dangerous. The flames produced from torches also can do a lot of damage.

When working with torches and tanks of fuel, you must be very careful. Don't allow your torch equipment to fall on a hard surface; the valves may break. Check all your connections closely; a leak allowing fuel to fill an area could result in an explosion when you light the torch.

Always pay attention to where your flame is pointed. Carelessness with the flame could start unwanted fires. This is especially true when working near insulation and other flammable substances. If the flame is directed at concrete, the concrete might explode. Since moisture is retained in concrete, intense heat can cause the moisture to force the concrete to explode.

It's not done often enough, but you always should have a fire extinguisher close by when working with a torch. If you have to work close to flammable substances, use a heat shield on the torch. When your flame is close to wood or insulation, try to remove the insulation or wet the flammable substance before applying the flame. When you are done using the torch, make sure the fuel tank is turned off and the hose is drained of all fuel. Use a striker to light your torch. The use of a match or cigarette lighter puts your hand too close to the source of the flame.

LEAD POTS AND LADLES

Lead pots and ladles offer their own style of potential danger. Plumbers today don't use molten lead as much as they used to, but hot lead still is used to make joints with cast-iron pipe and fittings.

When working with a small quantity of lead, many plumbers heat the lead in a ladle. They melt the lead with their torch and pour it straight from the ladle. When larger quantities of lead are needed, lead pots are used. The pots are filled with lead and set over a flame. All of these types of work are dangerous.

Never put wet materials in hot lead. If the ladle is wet or cold when it is dipped into the pot of molten lead, the hot lead can explode. Don't add wet lead to the melting pot if it contains molten lead. Before you pour hot lead in a waiting joint, make sure the joint is not wet. Another word of caution: Don't leave a working lead pot where rain dripping off a roof can fall into it.

Obviously, molten lead is hot, so you shouldn't touch it. Be very careful to avoid overturning the pot of hot lead. I remember one accident in which a pot of hot lead was knocked over onto a plumber's foot. Let me just say the scene was terrifying.

DRILLS AND BITS

Drills have been my worst enemy in plumbing. The two serious injuries I have received both were related to my work with a right-angle drill. The drills used most by plumbers are not little pistol-grip, handheld types of drills. The day-to-day drilling done by plumbers involves the use of large, powerful right-angle drills. These drills have enormous power when they get in a bind. Hitting a nail or a knot in the wood being drilled can do a lot of damage. You can break fingers, lose teeth, suffer head injuries, and a lot more. As with all electrical tools, you always should check the electrical cord before using your drill. If the cord is not in good shape, don't use the drill.

Always know what you are drilling into. If you are doing new construction work, it is fairly easy to look before you drill. Drilling in a remodeling job, however, can be much more difficult. You cannot always see what you are getting into. If you are unfortunate enough to drill into a hot wire, you can get a considerable electrical shock.

The bits you use in a drill are part of the safe operation of the tool. If your drill bits are dull, sharpen them. Dull bits are much more dangerous than sharp ones. When you are using a standard plumber's bit to drill through thin wood, like plywood, be careful. Once the worm driver of the bit penetrates the plywood fully, the big teeth on the bit can bite and jump, causing you to lose control of the drill. If you will be drilling metal, be aware that the metal shavings will be sharp and hot.

POWER SAWS

Plumbers don't use power saws as much as carpenters, but they do use them. The most common type of power saw used by plumbers is the reciprocating saw. These saws are used to cut pipe, plywood, floor joists, and a whole lot more. In addition to reciprocating saws, plumbers use circular saws and chop saws. All of the saws have the potential for serious injury.

Reciprocating saws are reasonably safe. Most models are insulated to help avoid electrical shocks if a hot wire is cut. The blade is typically a safe distance from the user, and the saws are pretty easy to hold and control. However, the brittle blades do break, which could result in an eye injury.

Plumbers occasionally use circular saws. The blades on these saws can bind and cause the saws to kick back. Chop saws are sometimes used to cut pipe. If you keep your body parts out of the way and wear eye protection, chop saws are not unusually dangerous.

HANDHELD DRAIN CLEANERS

Handheld drain cleaners don't get a lot of use from plumbers who do new construction work, but they are a frequently used tool of service plumbers. Most of these drain cleaners resemble, to some extent, a straight, handheld drill. Some models sit on stands, but most small snakes are handheld. These small-diameter snakes are not nearly as dangerous as their big brothers, but they do deserve respect. These units carry all the normal hazards of an electric tool, but there is more.

The cables used for small drain-cleaning jobs are usually very flexible. They are basically springs. The heads attached to the cables take on different shapes and looks. When you look at these thin cables, they don't look dangerous, but they can be. When the cables are being fed down a drain and turning, they can hit hard stoppages and go out of control. The cable can twist and kink. If your finger, hand, or arm is caught in the cable, injury can be the result. To avoid this, don't allow excessive cable to exist between the drainpipe and the machine.

LARGE SEWER MACHINES

Large sewer machines are much more dangerous than small ones. These machines have tremendous power and their cables are capable of removing fingers. Broken bones, severe cuts, and assorted other injuries are also possible with big snakes.

One of the most common conflicts with large sewer machines is with the cables. When the cutting heads hit roots or similar hard items in the pipe, the cable goes wild. The twisting, thrashing cable can do a lot of damage. Again, limiting excess cable is one of the best protections possible. Special sleeves are also available to contain unruly cables.

Most big machines can be operated by one person, but it is wise to have someone standing by in case help is needed. Loose clothing results in many drain machine accidents. The use of special mitts can help reduce the risk of hand injuries. Electrical shocks are also possible when doing drain cleaning.

POWER PIPE THREADERS

Power pipe threaders are very nice to have if you are doing much work with threaded pipe. These threading machines, however, can grind body parts as well as they thread pipe. Electric threaders are very dangerous in the hands of untrained people. It is critical to keep fingers and clothing away from the power mechanisms. The metal shavings produced by pipe threaders can be very sharp, and burrs left on the threaded pipe can slash your skin. The cutting oil used to keep the dies from getting too hot can make the floor around the machine slippery.

AIR-POWERED TOOLS

Plumbers do not often use air-powered tools. Jackhammers are probably the most-often used air-powered tools for plumbers. When using tools with air hoses, check all connections carefully. If you experience a blowout, the hose can spiral wildly out of control.

POWDER-ACTUATED TOOLS

Plumbers use powder-actuated tools to secure objects to hard surfaces, such as concrete. If the user is properly trained, these tools are not too dangerous. However, good training, eye protection, and ear protection are all necessary. Misfires and chipping hard surfaces are the most common problems with these tools.

LADDERS

Plumbers frequently use ladders, both stepladders and extension ladders. Many ladder accidents are possible. You must always be aware of what is around you when handling a ladder. If you brush against a live electrical wire with a ladder you are carrying, your life could be over. Ladders often fall over when the people using them are not careful. Reaching too far from a ladder can be all it takes to cause a fall.

When you set up a ladder, or rolling scaffold, make sure it is set up properly. The ladder should be on firm footing and all safety braces and clamps should be in place. When using an extension ladder, many plumbers use a rope to tie rungs together where the sections overlap. The rope provides an extra guard against the ladder's safety clamps failing and the ladder collapsing. When using an extension ladder, be sure to secure both the base and the top. I had an unusual accident on a ladder that I would like to share with you.

I was on a tall extension ladder, working near the top of a commercial building. The top of my ladder was resting on the edge of the flat roof. There was metal flashing surrounding the edge of the roof, and the top of the ladder was leaning against the flashing. There was a picket fence behind me and electrical wires entering the building to my right. The entrance wires were a good distance away, so I was in no immediate danger. As I worked on the ladder, a huge gust of wind blew around the building. I don't know where it came from; it hadn't been very windy when I went up the ladder.

The wind hit me and pushed me and the ladder sideways. The top of the ladder slid easily along the metal flashing, and I couldn't grab anything to stop me. I knew the ladder was going to go down, and I didn't have much time to make a decision. If I pushed off the ladder, I probably would be impaled on the fence. If I rode the ladder down, it might hit the electrical wires and fry me. I waited until the last minute and jumped off the ladder.

I landed on the wet ground with a thud, but I missed the fence. The ladder hit the wires and sparks flew. Fortunately, I wasn't hurt and electricians were available to take care of the electrical problem. This was a case where I wasn't really negligent, but I could have been killed. If I had secured the top of the ladder, my accident wouldn't have happened.

SCREWDRIVERS AND CHISELS

Eye injuries and puncture wounds are common when working with screwdrivers and chisels. When the tools are used properly and safety glasses are worn, few accidents occur. The key to avoiding injury with most hand tools is simply to use the right tool for the job. If you use a wrench as a hammer or a screwdriver as a chisel, you are asking for trouble.

There are, of course, other types of tools and safety hazards found in the plumbing trade. However, this list covers the ones that result in the most injuries. In all cases, observe proper safety procedures and utilize safety gear, such as eye and ear protection.

COWORKER SAFETY

Coworker safety is the last segment of this chapter. I am including it because workers frequently are injured by the actions of co-workers. This section is meant to protect you from others and to make you aware of how your actions might affect your co-workers.

Most plumbers find themselves working around other people. This is especially true on construction jobs. When working around other people, you must be aware of their actions, as well as your own. If you are walking out of a house to get something off the truck and a roll of roofing paper gets away from a roofer, you could get an instant headache.

If you don't pay attention to what is going on around you, it is possible to wind up in all sorts of trouble. Cranes lose their loads sometimes, and such a load landing on you is likely to be fatal. Equipment operators don't always see the plumber kneeling down for a piece of pipe. It's not hard to have a close encounter with heavy equipment. While we are on the subject of equipment, let me bore you with another war story.

One day I was in a sewer ditch, connecting the sewer from a new house to the sewer main. The section of ditch I was working in was only about four feet deep. There was a large pile of dirt near the edge of the trench; it had been created when the ditch was dug. The dirt wasn't laid back like it should have been; it was piled up. As I worked in the ditch, a backhoe came by. The operator had no idea I was in the ditch. When he swung the backhoe around to make a turn, the small scorpion-type bucket on the back of the equipment hit the dirt pile.

I had stood up when I heard the hoe approaching, and it was a good thing I had. When the equipment hit the pile of dirt, part of the mound caved in on me. I tried to run, but it caught both of my legs and the weight drove me to the ground. I was buried from just below my waist. My head was all right, and my arms were free. I was still holding my shovel.

I yelled, but nobody heard me. I must admit, I was a little panicked. I tried to get up and couldn't. After a while, I was able to move enough dirt with the shovel to crawl out from under the dirt. I was lucky. If I had been on my knees making the connection or grading the pipe, I might have been smothered. As it was, I came out of the ditch no worse for the wear. But, boy, was I mad at the careless backhoe operator. I won't go into the details of the little confrontation I had with him.

That accident is a prime example of how other workers can hurt you and never know they did it. You have to watch out for yourself at all times. As you gain field experience, you will develop a second nature for detecting impending coworker problems. You will learn to sense when something is wrong or is about to go wrong. But you have to stay alive and healthy long enough to get that experience.

Always be aware of what is going on over your head. Avoid working under other people and hazardous overhead conditions. Let people know where you are, so you won't get stranded on a roof or in an attic when your ladder is moved or falls over.

You also must remember that your actions could harm coworkers. If you are on a roof to flash a pipe and your hammer gets away from you, somebody could get hurt. Open communication between workers is one of the best ways to avoid injuries. If everyone knows where everyone else is working, injuries are less likely. Primarily, think and think some more. There is no substitute for common sense. Try to avoid working alone, and remain alert at all times.

CHAPTER 20
FIRST-AID BASICS

Everyone should invest some time in learning the basics of first aid. You never know when having skills in first-aid treatments might save your life. Plumbers live what can be a dangerous life. On-the-job injuries are not uncommon. Most injuries are fairly minor, but they often require treatment. Do you know the right way to get a sliver of copper out of your hand? If your helper suffers from an electrical shock when a drill cord goes bad, do you know what to do? Well, many plumbers don't possess good first-aid skills.

Before we get too far into this chapter, there are a few points I want to make. First of all, I'm not a medical doctor or any type of trained medical professional. I've taken first-aid classes, but I'm certainly not an authority about medical issues. The suggestions that I will give you in this chapter are for informational purposes only. This book is not a substitute for first-aid training offered by qualified professionals.

My intent here is to make you aware of some basic first-aid procedures that can make life on the job much easier. But I want you to understand that I'm not suggesting you use my advice to administer first aid. Hopefully, this chapter will show you the advantages you can gain from taking first-aid classes. Before you attempt first aid on anyone, including yourself, you should attend a structured, approved first-aid class. I'm going to give you information that is as accurate as I can make it, but don't assume that my words are enough. Take a little time to seek professional training in the art of first aid. You might never use what you learn, but the one time it is needed, you will be glad you made the effort to learn what to do. With this said, let's jump right into some tips about first aid.

OPEN WOUNDS

Open wounds are a common problem for plumbers. Many tools and materials used by plumbers can create open wounds. What should you do if you or one of your workers is cut?

- Stop the bleeding as soon as possible.
- Disinfect and protect the wound from contamination.
- You might have to take steps to avoid shock symptoms.
- Once the patient is stable, seek medical attention for severe cuts.

When a bad cut is sustained, the victim might slip into shock. Loss of consciousness could result from a loss of blood. Death from extreme bleeding is also a risk. As a first-aid provider, you must act quickly to reduce the risk of serious complications.

BLEEDING

To stop bleeding, direct pressure is usually a good tactic. This may be as crude as clamping your hand over the wound, but a cleaner compression is desirable. Ideally, a sterile material should be placed over the wound and secured, usually with tape (even if it's duct tape). Thick gauze used as a pressure material can absorb blood and allow the clotting process to begin.

Bad wounds might bleed right through the compress material. If this happens, don't remove the blood-soaked material. Add a new layer of material over it. Keep pressure on the wound. If you are not prepared with a first-aid kit, you could substitute gauze and tape with strips cut from clothing that can be tied in place over the wound.

ELEVATE IT

When you are dealing with a bleeding wound, it is usually best to elevate it. If you suspect a fractured or broken bone in the area of the wound, elevation might not be practical. When we talk about elevating a wound, it simply means to raise the wound above the level of the victim's heart. This helps the blood flow to slow, because of gravity.

SUPER SERIOUS

Super-serious bleeding might not stop even after a compression bandage is applied and the wound is elevated. When this is the case, you must resort to putting pressure on the main artery producing the blood. Constricting an artery is not an alternative for the steps that we have discussed previously.

Putting pressure on an artery is serious business. First, you must be able to locate the artery, and you should not keep the artery constricted any longer than necessary. You might have to apply pressure for a while, release it, and then apply it again. It's important that you do not restrict the flow of blood in arteries for long periods of time. I hesitate to go into too much detail about this process, because I believe it is a method you should be taught in a controlled, classroom situation. However, I will hit the high spots. But remember, these words are not a substitute for professional training from qualified instructors.

Open arm wounds are controlled with the brachial artery, which is in the area between the biceps and triceps, on the inside of the arm. It's about halfway between the armpit and the elbow. Pressure is created with the flat parts of your fingertips. Basically, you are holding the victim's wrist with one hand and closing off the artery with your other hand. Pressure exerted by your fingers pushes the artery against the arm bone and restricts blood flow. Again, don't attempt this type of first aid until you have been trained properly in the execution of the procedure.

Severe leg wounds might require the constriction of the femoral artery, which is in the pelvic region. Bleeding victims usually are placed on their backs for this procedure. The heel of a hand is placed on the artery to restrict blood flow. In some cases, fingertips are used to apply pressure. I'm uncomfortable with going into great detail about these procedures, because I don't want you to rely solely on what I'm telling you. It's enough that you understand that knowing when and where to apply pressure to arteries can save lives and that you should seek professional training in these techniques.

TOURNIQUETS

Tourniquets get a lot of attention in movies, but they can do as much harm as they do good if not used properly. A tourniquet should be used only in a life-threatening situation. When a tourniquet is applied, there is a risk of losing the limb to which the restriction is applied. This is obviously a serious decision and one that must be made only when all other means of stopping blood loss have been exhausted.

Unfortunately, plumbers might run into a situation where a tourniquet is the only answer. For example, if a worker allowed a power saw to get out of control, a hand might be severed or some other type of life-threatening injury could occur. This would be cause for the use of a tourniquet. Let me give you a basic overview of what's involved when a tourniquet is used.

Tourniquets should be at least two inches wide. A tourniquet should be placed at a point that is above a wound, between the bleeding and the victim's heart. The binding, however, should not encroach directly on the wound. Tourniquets can be fashioned out of many materials. If you are using strips of cloth, wrap the cloth around the limb that is wounded and tie a knot in the material. Use a stick, screwdriver, or whatever else you can lay your hands on to tighten the binding.

After you have made a commitment to apply a tourniquet, the wrapping should be removed only by a physician. It's a good idea to note the time that a tourniquet is applied, as this will help doctors later in assessing their options. As an extension of the tourniquet treatment, you most likely will have to treat the patient for shock.

INFECTION

Infection is always a concern with open wounds. When a wound is serious enough to require a compression bandage, don't attempt to clean it. Keep pressure on the wound to stop bleeding. In cases of severe wounds, be on the lookout for shock symptoms and be prepared to treat them. Your primary concern with a serious open wound is to stop the bleeding and gain professional medical help as soon as possible.

Lesser cuts, which are more common than deep ones, should be cleaned. Regular soap and water can be used to clean a wound before applying a bandage. Remember, we are talking about minor cuts and scrapes at this point. Flush the wound generously with clean water. A piece of sterile gauze can be used to pat the wound dry. Then a clean, dry bandage can be applied to protect the wound while in transport to a medical facility.

SPLINTERS AND SUCH

Splinters and similar foreign objects often invade the skin of plumbers. Getting these items out cleanly is best done by a doctor, but there are some on-the-job methods that you might want to try. A magnifying glass and a pair of tweezers work well together when removing embedded objects, such as splinters and slivers of copper tubing. Ideally, tweezers being used should be sterilized either over an open flame, such as the flame of your torch, or in boiling water.

Splinters and slivers that are submerged beneath the skin often can be lifted out with the tip of a sterilized needle. The use of a needle in conjunction with a pair of tweezers is very effective in the removal of most simple splinters. If you are dealing with something that has gone extremely deep into tissue, it is best to leave the object alone until a doctor can remove it.

EYE INJURIES

Eye injuries are very common on construction and remodeling jobs. Most of these injuries could be avoided if proper eye protection were worn, but far too many workers don't wear safety glasses and goggles. This sets the stage for eye irritations and injuries.

Before you attempt to help someone who is suffering from an eye injury, you should wash your hands thoroughly. I know this is not always possible on construction sites, but cleaning your hands is advantageous. In the meantime, keep the victim from rubbing the injured eye. Rubbing can make matters much worse.

Never attempt to remove a foreign object from someone's eye with the use of a rigid device, such as a toothpick. Cotton swabs that have been wetted can serve well as a magnet to remove some types of invasion objects. If the person you are helping has something embedded in an eye, get the person to a doctor as soon as possible. Don't attempt to remove the object yourself.

When you are investigating the cause of an eye injury, you should pull down the lower lid of the eye to determine if you can see the object causing the trouble. A floating object, such as a piece of sawdust trapped between an eye and an eyelid can be removed with a tissue, a damp cotton swab, or even a clean handkerchief. Don't allow dry cotton material to come into contact with an eye.

If looking under the lower lid doesn't reveal the source of discomfort, check under the upper lid. Clean water can be used to flush out many eye contaminants without much risk of damage to the eye. Objects that cannot be removed easily should be left alone until a physician can take over.

SCALP INJURIES

Scalp injuries can be misleading. What looks like a serious wound can be a fairly minor cut. On the other hand, what appears to be only a cut can involve a fractured skull. If you or someone around you sustains a scalp injury, such as having a hammer fall on your head from an overhead worker, take it seriously. Don't attempt to clean the wound. Expect profuse bleeding.

If you don't suspect a skull fracture, raise the victim's head and shoulders to reduce bleeding. Try not to bend the neck. Put a sterile bandage over the wound, but don't apply excessive pressure. If there is a bone fracture, pressure could worsen the situation. Secure the bandage with gauze or some other material that you can wrap around it. Seek medical attention immediately.

FACIAL INJURIES

Facial injuries can occur on plumbing jobs. I've seen helpers let their right-angle drills get away from them with the result being hard knocks to the face. On one occasion, I remember a tooth being lost, and split lips and tongues that have been bitten are common when a drill goes on the rampage.

Extremely bad facial injuries can cause a blockage of the victim's air passages. This, of course, is a very serious condition. It's critical that air passages be open at all times. If the person's mouth contains broken teeth or dentures, remove them. Be careful not to jar the individual's spine if you have reason to believe there might be injury to the back or neck.

Conscious victims should be positioned, when possible, so that secretions from the mouth and nose will drain out. Shock is a potential concern in severe facial injuries. For most on-the-job injuries, plumbers should be treated for comfort and sent for medical attention.

NOSE BLEEDS

Nose bleeds are not usually difficult to treat. Typically, pressure applied to the side of the nose where bleeding is occurring will stop the flow of blood. Applying cold compresses also can help. If external pressure is not stopping the bleeding, use a small, clean pad of gauze to create a dam on the inside of the nose. Then, apply pressure on the outside of the nose. This almost always will work. If it doesn't, get to a doctor.

BACK INJURIES

There is really only one thing that you need to know about back injuries: Don't move the injured person. Call for professional help and see that the victim remains still until help arrives. Moving someone who has suffered a back injury can be very risky. Don't do it unless there is a life-threatening cause for your action, such as a person trapped in a fire or some other type of deadly situation.

LEGS AND FEET

Legs and feet sometimes become injured on job sites. The worst case of this type that I can remember was when a plumber knocked over a pot of molten lead onto his foot. It sends shivers up my spine just to recall that incident. Anyway, when someone suffers a minor foot or leg injury, you should clean and cover the wound. Bandages should be supportive without being constrictive. The appendage should be elevated above the victim's heart level when possible. Prohibit the person from walking. Remove boots and socks so that you can keep an eye on the person's toes. If the toes begin to swell or turn blue, loosen the supportive bandages.

BLISTERS

Blisters might not seem like much of an emergency, but they sure can take the steam out of a helper or plumber. In most cases, blisters can be covered with a heavy gauze pad to reduce pain. It generally is recommended to leave blisters unbroken. When a blister breaks, the area should be cleaned and treated as an open wound. Some blisters tend to be more serious than others. For example, blisters in the palm of a hand or on the sole of a foot should be examined by a doctor.

HAND INJURIES

Hand injuries are common in the plumbing trade. Little cuts are the most frequent complaint. Getting flux in a cut is an eye-opening experience, so even the smallest break in the skin should be covered. Serious hand injuries should be elevated. This tends to reduce swelling. You should not try to clean really serious hand injuries. Use a pressure bandage to control bleeding. If the cut is on the palm of a hand, the victim can squeeze a roll of gauze to slow the flow of blood. Pressure should stop the bleeding, but if it doesn't, seek medical assistance. As with all injuries, use common sense when determining if professional attention is needed after first aid is applied.

SHOCK

Shock is a condition that can be life-threatening even when the injury responsible for a person going into shock is not otherwise fatal. We are talking about traumatic shock, not electrical shock. Many factors can lead to a person going into shock. A serious injury is a common cause, but many other causes exist. There are certain signs of shock you can look for.

If a person's skin turns pale or blue and is cold to the touch, it's a likely sign of shock. Skin that becomes moist and clammy can indicate shock is present. A general weakness is also a sign of shock. When a person is going into shock, the individual's pulse is likely to exceed 100 beats per minute. Breathing usually is increased, but it may be shallow, deep, or irregular. Chest injuries usually result in shallow breathing. Victims who have lost blood might thrash about as they enter into shock. Vomiting and nausea also can signal shock.

As a person slips into deeper shock, the individual might become unresponsive. Look at the eyes, they may be dilated widely. Blood pressure can drop, and in time, the victim will lose consciousness. Body temperature will fall, and death will be likely if treatment is not rendered.

There are three main goals when treating someone for shock. Get the person's blood circulating well. Make sure an adequate supply of oxygen is available to the individual, and maintain the person's body temperature.

When you have to treat a person for shock, you should keep the victim lying down. Cover the individual so that the loss of body heat will be minimal. Get medical help as soon as possible. The reason it's best to keep a person lying down is so the blood will circulate better. Remember, if you suspect back or neck injuries, don't move the person.

People who are unconscious should be placed on one side so that fluids will run out of the mouth and nose. It's also important to make sure that air passages are open. A person with a head injury can be laid out flat or propped up, but the head should not be lower than the rest of the body. It is sometimes advantageous to elevate a person's feet when they are in shock. If the person has any difficulty in breathing or if pain increases when the feet are raised, lower them.

Body temperature is a big concern with shock patients. You want to overcome or avoid chilling. However, don't attempt to add additional heat to the surface of the person's body with artificial means. This can be damaging. Use only blankets, clothes, and other similar items to regain and maintain body temperature.

Avoid the temptation to offer the victim fluids, unless medical care is not going to be available for a long time. Avoid fluids completely if the person is unconscious or is subject to vomiting. Under most job-site conditions, fluids should not be administered.

BURNS

Burns are not common among plumbers, but they can occur in the workplace. There are three types of burns that you may have to deal with. First-degree burns are the least serious. These burns typically come from overexposure to the sun, which construction workers often suffer from, quick contact with a hot object, such as the tip of a torch, and scalding water, which could be the case when working with a boiler or water heater.

Second-degree burns are more serious. They can come from a deep sunburn or from contact with hot liquids and flames. A person who is affected by a second-degree burn might have a red or mottled appearance, blisters, and a wet appearance of the skin within the burn area. This wet look comes from a loss of plasma through the damaged layers of skin.

Third-degree burns are the most serious. They can be caused by contact with open flames, hot objects, or immersion in very hot water. Electrical injuries also can result in third-degree burns. This type of burn can look similar to a second-degree burn, but the difference will be the loss of all layers of skin.

TREATMENT

Treatment for most job-related burns can be administered on the job site and will not require hospitalization. First-degree burns should be washed with or submerged in cold water. A dry dressing can be applied if necessary. These burns are not too serious. Eliminating pain is the primary goal with first-degree burns.

Second-degree burns should be immersed in cold (but not ice) water. The soaking should continue for at least one hour and up to two hours. After soaking, the wound should be layered with clean cloths that have been dipped in ice water and wrung out. Then the wound should be dried by blotting, not rubbing. A dry, sterile gauze then should be applied. Don't break open any blisters. It also is not advisable to use ointments and sprays on severe burns. Burned arms and legs should be elevated, and medical attention should be acquired.

Bad burns, the third-degree type, need quick medical attention. First, don't remove a burn victim's clothing, skin might come off with it. A thick, sterile dressing can be applied to the burn area. Personally, I would avoid this if possible. A dressing might stick to the mutilated skin and cause additional skin loss when the dressing is removed. When hands are burned, keep them elevated above the victim's heart. The same goes for feet and legs. You should not soak a third-degree burn in cold water, it could induce more shock symptoms. Don't use ointments, sprays, or other types of treatments. Get the burn victim to competent medical care as soon as possible.

HEAT-RELATED PROBLEMS

Heat-related problems can include heat stroke and heat exhaustion. Cramps are also possible when working in hot weather. Some people don't consider heat stroke to be serious. They are wrong. Heat stroke can be life-threatening. People affected by heat stroke can develop body temperatures in excess of 106°F. Their skin is likely to be hot, red, and dry.

You might think sweating would take place, but it doesn't. Pulse is rapid and strong, and victims can sink into an unconscious state.

If you are dealing with heat stroke, you need to lower the person's body temperature quickly. There is a risk, however, of cooling the body too quickly after the victim's temperature is below 102°F. You can lower body temperature with rubbing alcohol, cold packs, cold water on clothes or in a bathtub of cold water. Avoid the use of ice in the cooling process. Fans and air-conditioned space can be used to achieve your cooling goals. Get the body temperature down to at least 102° and then go for medical help.

CRAMPS

Cramps are not uncommon among workers during hot spells. A simple massage can be all it takes to cure this problem. Saltwater solutions are another way to control cramps. Mix one teaspoonful of salt per glass of water and have the victim drink half a glass about every 15 minutes.

EXHAUSTION

Heat exhaustion is more common than heat stroke. A person affected by heat exhaustion is likely to maintain a fairly normal body temperature, but the person's skin might be pale and clammy. Sweating might be very noticeable, and the individual probably will complain of being tired and weak. Headaches, cramps, and nausea can accompany the other symptoms. In some cases, fainting might occur.

The saltwater treatment described for cramps usually will work with heat exhaustion. Victims should lie down and elevate their feet about a foot off the floor or bed. Clothing should be loosened, and cool, wet cloths can be used to add comfort. If vomiting occurs, get the person to a hospital for intravenous fluids.

We could continue talking about first aid for a much longer time, but the help I can give you here for medical procedures is limited. You owe it to yourself, your family, and the people you work with to learn first-aid techniques. This can be done best by attending formal classes in your area. Most towns and cities offer first-aid classes on a regular basis. I strongly suggest that you enroll in one. Until you have some hands-on experience in a classroom and gain the depth of knowledge needed, you are not prepared for emergencies. Don't get caught short. Prepare now for the emergency that might never happen.

CHAPTER 21
CHARTS AND TABLES

RECOMMENDED TABULAR ARRANGEMENT FOR USE IN SOLVING PIPE SIZING PROBLEMS

COLUMN	1	2	3	4	5	6	7	8	9	10
Line	Description	Lbs. per square inch (psi)	Gal. per min. through section	Length of section (feet)	Trial pipe size (inches)	Equivalent length of fittings and valves (feet)	Total equivalent length Col. 4 and Col. 6 (100 feet)	Friction loss per 100 feet of trial size pipe (psi)	Friction loss in equivalent length Col. 8 × Col. 7 (psi)	Excess pressure over friction losses (psi)
a	Minimum pressure available at main	55.00								
b	Highest pressure required at a fixture (Section 604.3)	15.00								
c	Meter loss 2" meter	11.00								
d	Tap in main loss 2" tap (Table E103A) 1.61									
e	Static head loss 21 × 0.43 psi	9.03								
f	Special fixture loss backflow preventer	9.00								
g	Special fixture loss—Filter	0.00								
h	Special fixture loss—Other	0.00								
i	Total overall losses and requirements (sum of Lines b through h)	45.64								
j	Pressure available to overcome pipe friction (Line a minus Lines b to h)	9.36								
						FU				
	Designation — Pipe section (from diagram) — Cold water distribution piping	AB	294	108.0	54	2½	12	0.66	3.3	2.18
		BC	264	108.0	8	2½	2.5	0.105	3.2	0.34
		CD	132	77.0	13	2½	8	0.21	1.9	0.40
		CF	132	77.0	150	2½	12	1.62	1.9	3.08
		DE	132	77.0	150	2½	14.5	1.645	1.9	3.12
k	Total pipe friction losses (cold)					9.36	6.24	6.24		
l	Difference (Line j minus Line k)									3.12
	Pipe section (from diagram) — Hot water distribution piping	A'B'	294	108.0	54	2½	9.6	0.64	3.3	2.1
		B'C'	24	38.0	8	2	9.0	0.17	1.4	0.24
		C'D'^b	12	28.6	13	1½	5	0.18	3.2	0.58
		C'F'^b	12	28.6	150	1½	14	1.64	3.2	5.25
		D'E'^b	12	28.6	150	1½	7	1.57	3.2	5.02
k	Total pipe friction losses (hot)					9.36	7.94	7.94		
l	Difference (Line j minus Line k)									1.42

Line a–f left label: Service and cold water distribution piping^a

For SI: 1 inch = 25.4 mm, 1 foot = 304.8 mm, 1 psi = 6.895 kPa, 1 gpm = 3.785 L/m.

a. To be considered as pressure gain for fixtures below main (to consider separately, omit from "i" and add to "j").

b. To consider separately, in k use C-F only if greater loss than above.

FIGURE 21.1 This table can help resolve any problems that might arise in relation to pipe sizing. (*Courtesy of International Code Council, Inc. and International Plumbing Code 2000.*)

LOAD VALUES ASSIGNED TO FIXTURES[a]

FIXTURE	OCCUPANCY	TYPE OF SUPPLY CONTROL	LOAD VALUES, IN WATER SUPPLY FIXTURE UNITS (wsfu)		
			Cold	Hot	Total
Bathroom group	Private	Flush tank	2.7	1.5	3.6
Bathroom group	Private	Flush valve	6.0	3.0	8.0
Bathtub	Private	Faucet	1.0	1.0	1.4
Bathtub	Public	Faucet	3.0	3.0	4.0
Bidet	Private	Faucet	1.5	1.5	2.0
Combination fixture	Private	Faucet	2.25	2.25	3.0
Dishwashing machine	Private	Automatic		1.4	1.4
Drinking fountain	Offices, etc.	3/8" valve	0.25		0.25
Kitchen sink	Private	Faucet	1.0	1.0	1.4
Kitchen sink	Hotel, restaurant	Faucet	3.0	3.0	4.0
Laundry trays (1 to 3)	Private	Faucet	1.0	1.0	1.4
Lavatory	Private	Faucet	0.5	0.5	0.7
Lavatory	Public	Faucet	1.5	1.5	2.0
Service sink	Offices, etc.	Faucet	2.25	2.25	3.0
Shower head	Public	Mixing valve	3.0	3.0	4.0
Shower head	Private	Mixing valve	1.0	1.0	1.4
Urinal	Public	1" flush valve	10.0		10.0
Urinal	Public	3/4" flush valve	5.0		5.0
Urinal	Public	Flush tank	3.0		3.0
Washing machine (8 lbs.)	Private	Automatic	1.0	1.0	1.4
Washing machine (8 lbs.)	Public	Automatic	2.25	2.25	3.0
Washing machine (15 lbs.)	Public	Automatic	3.0	3.0	4.0
Water closet	Private	Flush valve	6.0		6.0
Water closet	Private	Flush tank	2.2		2.2
Water closet	Public	Flush valve	10.0		10.0
Water closet	Public	Flush tank	5.0		5.0
Water closet	Public or private	Flushometer tank	2.0		2.0

For SI: 1 inch = 25.4 mm, 1 pound = 0.454 kg.

a. For fixtures not listed, loads should be assumed by comparing the fixture to one listed using water in similar quantities and at similar rates. The assigned loads for fixtures with both hot and cold water supplies are given for separate hot and cold water loads and for total load, the separate hot and cold water loads being three-fourths of the total load for the fixture in each case.

FIGURE 21.2 Every fixture involved in plumbing has a load value. They are determined here. (*Courtesy of International Code Council, Inc. and International Plumbing Code 2000.*)

Table for Estimating Demand

SUPPLY SYSTEMS PREDOMINANTLY FOR FLUSH TANKS			SUPPLY SYSTEMS PREDOMINANTLY FOR FLUSH VALVES		
Load	Demand		Load	Demand	
(Water supply fixture units)	(Gallons per minute)	(Cubic feet per minute)	(Water supply fixture units)	(Gallons per minute)	(Cubic feet per minute)
1	3.0	0.04104			
2	5.0	0.0684			
3	6.5	0.86892			
4	8.0	1.06944			
5	9.4	1.256592	5	15.0	2.0052
6	10.7	1.430376	6	17.4	2.326032
7	11.8	1.577424	7	19.8	2.646364
8	12.8	1.711104	8	22.2	2.967696
9	13.7	1.831416	9	24.6	3.288528
10	14.6	1.951728	10	27.0	3.60936
11	15.4	2.058672	11	27.8	3.716304
12	16.0	2.13888	12	28.6	3.823248
13	16.5	2.20572	13	29.4	3.930192
14	17.0	2.27256	14	30.2	4.037136
15	17.5	2.3394	15	31.0	4.14408
16	18.0	2.90624	16	31.8	4.241024
17	18.4	2.459712	17	32.6	4.357968
18	18.8	2.513184	18	33.4	4.464912
19	19.2	2.566656	19	34.2	4.571856
20	19.6	2.620128	20	35.0	4.6788
25	21.5	2.87412	25	38.0	5.07984
30	23.3	3.114744	30	42.0	5.61356
35	24.9	3.328632	35	44.0	5.88192
40	26.3	3.515784	40	46.0	6.14928
45	27.7	3.702936	45	48.0	6.41664
50	29.1	3.890088	50	50.0	6.684
60	32.0	4.27776	60	54.0	7.21872
70	35.0	4.6788	70	58.0	7.75344
80	38.0	5.07984	80	61.2	8.181216
90	41.0	5.48088	90	64.3	8.595624
100	43.5	5.81508	100	67.5	9.0234
120	48.0	6.41664	120	73.0	9.75864
140	52.5	7.0182	140	77.0	10.29336
160	57.0	7.61976	160	81.0	10.82808
180	61.0	8.15448	180	85.5	11.42964
200	65.0	8.6892	200	90.0	12.0312
225	70.0	9.3576	225	95.5	12.76644
250	75.0	10.0260	250	101.0	13.50168
275	80.0	10.6944	275	104.5	13.96956
300	85.0	11.3628	300	108.0	14.43744
400	105.0	14.0364	400	127.0	16.97736
500	124.0	16.57632	500	143.0	19.11624
750	170.0	22.7256	750	177.0	23.66136
1,000	208.0	27.80544	1,000	208.0	27.80544
1,250	239.0	31.94952	1,250	239.0	31.94952
1,500	269.0	35.95992	1,500	269.0	35.95992
1,750	297.0	39.70296	1,750	297.0	39.70296

FIGURE 21.3 This table will let a user estimate demand. *(Courtesy of International Code Council, Inc. and International Plumbing Code 2000.)*

Table for Estimating Demand—cont'd

| SUPPLY SYSTEMS PREDOMINANTLY FOR FLUSH TANKS | | | SUPPLY SYSTEMS PREDOMINANTLY FOR FLUSH VALVES | | |
| Load | Demand | | Load | Demand | |
(Water supply fixture units)	(Gallons per minute)	(Cubic feet per minute)	(Water supply fixture units)	(Gallons per minute)	(Cubic feet per minute)
2,000	325.0	43.446	2,000	325.0	43.446
2,500	380.0	50.7984	2,500	380.0	50.7984
3,000	433.0	57.88344	3,000	433.0	57.88344
4,000	535.0	70.182	4,000	525.0	70.182
5,000	593.0	79.27224	5,000	593.0	79.27224

For SI: 1 gpm = 3.785 L/m, 1 cfm = 0.4719 L/s.

FIGURE 21.4 The table for estimating demand for flush tanks and valves continues. (*Courtesy of International Code Council, Inc. and International Plumbing Code 2000.*)

LOSS OF PRESSURE THROUGH TAPS AND TEES IN POUNDS PER SQUARE INCH (psi)

GALLONS PER MINUTE	SIZE OF TAP OR TEE (inches)						
	5/8	3/4	1	1 1/4	1 1/2	2	3
10	1.35	0.64	0.18	0.08	0.14		
20	5.38	2.54	0.77	0.31	0.33	0.10	
30	12.1	5.72	1.62	0.69	0.58	0.18	
40		10.2	3.07	1.23	0.58	0.28	
50		15.9	4.49	1.92	0.91	0.40	
60			6.46	2.76	1.31	0.55	0.10
70			8.79	3.76	1.78	0.72	0.13
80			11.5	4.90	2.32	0.91	0.16
90			14.5	6.21	2.94	1.12	0.21
100			17.94	7.67	3.63	1.61	0.30
120			25.8	11.0	5.23	2.20	0.41
140			35.2	15.0	7.12	2.52	0.47
150				17.2	8.16	2.92	0.54
160				19.6	9.30	3.62	0.68
180				24.8	11.8	4.48	0.84
200				30.7	14.5	5.6	1.06
225				38.8	18.4	7.00	1.31
250				47.9	22.7	7.70	1.59
275					27.4	10.1	1.88
300					32.6		

For SI: 1 inch = 25.4 mm, 1 psi = 6.895 kPa, 1 gpm = 3.785 L/m.

FIGURE 21.5 Pressure can be lost in taps and tees. This examines the numbers. *(Courtesy of International Code Council, Inc. and International Plumbing Code 2000.)*

ALLOWANCE IN EQUIVALENT LENGTH OF PIPE FOR FRICTION LOSS IN VALVES AND THREADED FITTINGS (feet)

FITTING OR VALVE	PIPE SIZES (inches)							
	½	¾	1	1¼	1½	2	2½	3
45-degree elbow	1.2	1.5	1.8	2.4	3.0	4.0	5.0	6.0
90-degree elbow	2.0	2.5	3.0	4.0	5.0	7.0	8.0	10.0
Tee, run	0.6	0.8	0.9	1.2	1.5	2.0	2.5	3.0
Tee, branch	3.0	4.0	5.0	6.0	7.0	10.0	12.0	15.0
Gate valve	0.4	0.5	0.6	0.8	1.0	1.3	1.6	2.0
Balancing valve	0.8	1.1	1.5	1.9	2.2	3.0	3.7	4.5
Plug-type cock	0.8	1.1	1.5	1.9	2.2	3.0	3.7	4.5
Check valve, swing	5.6	8.4	11.2	14.0	16.8	22.4	28.0	33.6
Globe valve	15.0	20.0	25.0	35.0	45.0	55.0	65.0	80.0
Angle valve	8.0	12.0	15.0	18.0	22.0	28.0	34.0	40.0

For SI: 1 inch = 25.4 mm, 1 foot = 304.8 mm, 1 degree = 0.0175 rad.

FIGURE 21.6 This chart examines the allowances involved in friction loss in valves and threaded fittings. *(Courtesy of International Code Council, Inc. and International Plumbing Code 2000.)*

PRESSURE LOSS IN FITTINGS AND VALVES EXPRESSED AS EQUIVALENT LENGTH OF TUBE[a] (feet)

NOMINAL OR STANDARD SIZE (inches)	FITTINGS					VALVES			
	Standard Ell		90-Degree Tee		Coupling	Ball	Gate	Butterfly	Check
	90 Degree	45 Degree	Side Branch	Straight Run					
3/8	0.5	—	1.5	—	—	—	—	—	1.5
1/2	1	0.5	2	—	—	—	—	—	2
5/8	1.5	0.5	2	—	—	—	—	—	2.5
3/4	2	0.5	3	—	—	—	—	—	3
1	2.5	1	4.5	—	—	0.5	—	—	4.5
1 1/4	3	1	5.5	0.5	0.5	0.5	—	—	5.5
1 1/2	4	1.5	7	0.5	0.5	0.5	—	—	6.5
2	5.5	2	9	0.5	0.5	0.5	0.5	7.5	9
2 1/2	7	2.5	12	0.5	0.5	—	1	10	11.5
3	9	3.5	15	1	1	—	1.5	15.5	14.5
3 1/2	9	3.5	14	1	1	—	2	—	12.5
4	12.5	5	21	1	1	—	2	16	18.5
5	16	6	27	1.5	1.5	—	3	11.5	23.5
6	19	7	34	2	2	—	3.5	13.5	26.5
8	29	11	50	3	3	—	5	12.5	39

For SI: 1 inch = 25.4 mm, 1 foot = 304.8 mm, 1 degree = 0.0175 rad.

a. Allowances are for streamlined soldered fittings and recessed threaded fittings. For threaded fittings, double the allowances shown in the table. The equivalent lengths presented above are based on a C factor of 150 in the Hazen-Williams friction loss formula. The lengths shown are rounded to the nearest half-foot.

FIGURE 21.7 You can determine pressure losses as equivalent lengths from this table. (*Courtesy of International Code Council, Inc. and International Plumbing Code 2000.*)

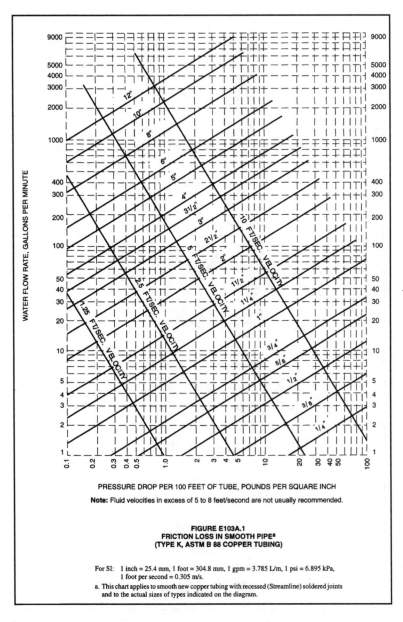

PRESSURE DROP PER 100 FEET OF TUBE, POUNDS PER SQUARE INCH

Note: Fluid velocities in excess of 5 to 8 feet/second are not usually recommended.

FIGURE E103A.1
FRICTION LOSS IN SMOOTH PIPE[a]
(TYPE K, ASTM B 88 COPPER TUBING)

For SI: 1 inch = 25.4 mm, 1 foot = 304.8 mm, 1 gpm = 3.785 L/m, 1 psi = 6.895 kPa,
1 foot per second = 0.305 m/s.

a. This chart applies to smooth new copper tubing with recessed (Streamline) soldered joints
and to the actual sizes of types indicated on the diagram.

FIGURE 21.8 This is one of several tables that determine friction loss. *(Courtesy of International Code Council, Inc. and International Plumbing Code 2000.)*

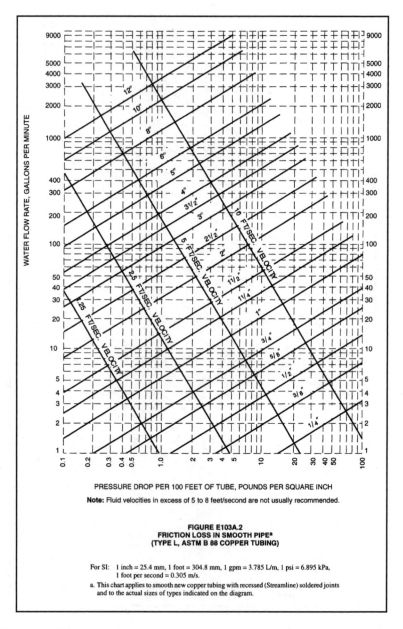

PRESSURE DROP PER 100 FEET OF TUBE, POUNDS PER SQUARE INCH

Note: Fluid velocities in excess of 5 to 8 feet/second are not usually recommended.

FIGURE E103A.2
FRICTION LOSS IN SMOOTH PIPE[a]
(TYPE L, ASTM B 88 COPPER TUBING)

For SI: 1 inch = 25.4 mm, 1 foot = 304.8 mm, 1 gpm = 3.785 L/m, 1 psi = 6.895 kPa,
1 foot per second = 0.305 m/s.

a. This chart applies to smooth new copper tubing with recessed (Streamline) soldered joints
and to the actual sizes of types indicated on the diagram.

FIGURE 21.9 This is one of several tables that determine friction loss. *(Courtesy of International Code Council, Inc. and International Plumbing Code 2000.)*

21.10

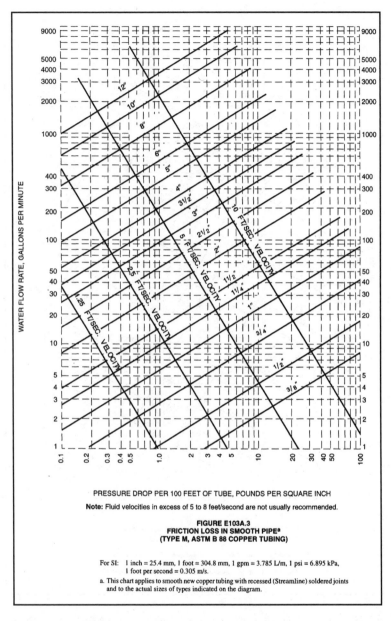

PRESSURE DROP PER 100 FEET OF TUBE, POUNDS PER SQUARE INCH

Note: Fluid velocities in excess of 5 to 8 feet/second are not usually recommended.

**FIGURE E103A.3
FRICTION LOSS IN SMOOTH PIPE[a]
(TYPE M, ASTM B 88 COPPER TUBING)**

For SI: 1 inch = 25.4 mm, 1 foot = 304.8 mm, 1 gpm = 3.785 L/m, 1 psi = 6.895 kPa,
1 foot per second = 0.305 m/s.

a. This chart applies to smooth new copper tubing with recessed (Streamline) soldered joints
and to the actual sizes of types indicated on the diagram.

FIGURE 21.10 This is one of several tables that determine friction loss. *(Courtesy of International Code Council, Inc. and International Plumbing Code 2000.)*

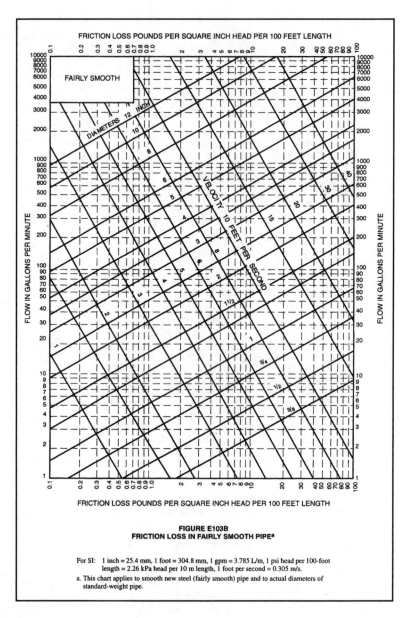

FIGURE E103B
FRICTION LOSS IN FAIRLY SMOOTH PIPE[a]

For SI: 1 inch = 25.4 mm, 1 foot = 304.8 mm, 1 gpm = 3.785 L/m, 1 psi head per 100-foot
length = 2.26 kPa head per 10 m length, 1 foot per second = 0.305 m/s.

a. This chart applies to smooth new steel (fairly smooth) pipe and to actual diameters of
standard-weight pipe.

FIGURE 21.11 This is one of several tables that determine friction loss. *(Courtesy of International Code Council, Inc. and International Plumbing Code 2000.)*

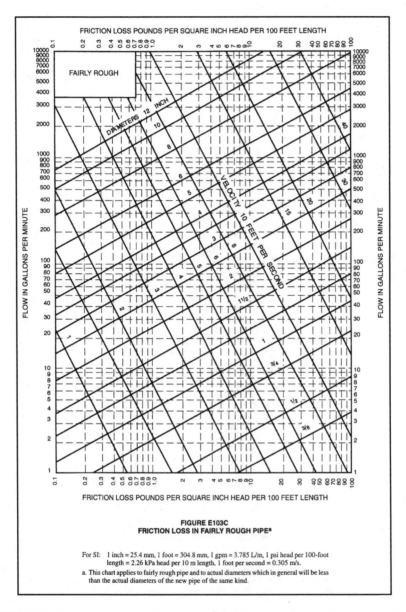

FIGURE E103C
FRICTION LOSS IN FAIRLY ROUGH PIPE[a]

For SI: 1 inch = 25.4 mm, 1 foot = 304.8 mm, 1 gpm = 3.785 L/m, 1 psi head per 100-foot length = 2.26 kPa head per 10 m length, 1 foot per second = 0.305 m/s.

a. This chart applies to fairly rough pipe and to actual diameters which in general will be less than the actual diameters of the new pipe of the same kind.

FIGURE 21.12 This is one of several tables that determine friction loss. *(Courtesy of International Code Council, Inc. and International Plumbing Code 2000.)*

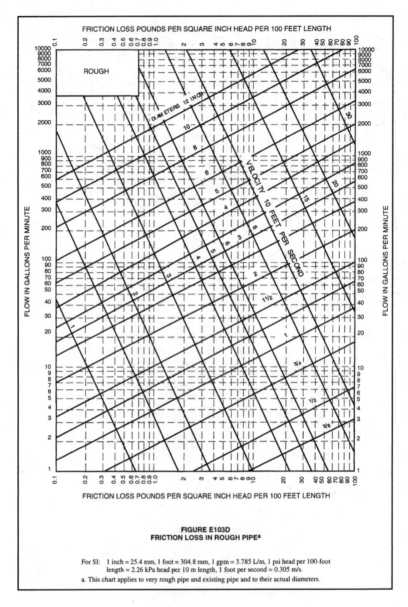

FIGURE E103D
FRICTION LOSS IN ROUGH PIPE[a]

For SI: 1 inch = 25.4 mm, 1 foot = 304.8 mm, 1 gpm = 3.785 L/m, 1 psi head per 100-foot
length = 2.26 kPa head per 10 m length, 1 foot per second = 0.305 m/s.

a. This chart applies to very rough pipe and existing pipe and to their actual diameters.

FIGURE 21.13 This is one of several tables that determine friction loss. *(Courtesy of International Code Council, Inc. and International Plumbing Code 2000.)*

	Inch	mm
	1/2	15
	3/4	20
	1	25

Water Supply Fixture Units (WSFU) and Minimum Fixture Branch Pipe Sizes[3]

Appliances, Appurtenances or Fixtures[2]	Minimum Fixture Branch Pipe Size[1,4]	Private	Public	Assembly[5]
Bathtub or Combination Bath/Shower (fill)	1/2"	4.0	4.0	
3/4" Bathtub Fill Valve	3/4"	10.0	10.0	
Bidet	1/2"	1.0		
Clotheswasher	1/2"	4.0	4.0	
Dental Unit, cuspidor	1/2"		1.0	
Dishwasher, domestic	1/2"	1.5	1.5	
Drinking Fountain or Watercooler	1/2"	0.5	0.5	0.75
Hose Bibb	1/2"	2.5	2.5	
Hose Bibb, each additional[7]	1/2"	1.0	1.0	
Lavatory	1/2"	1.0	1.0	1.0
Lawn Sprinkler, each head[5]		1.0	1.0	
Mobile Home, each (minimum)		12.0		
Sinks				
Bar	1/2"	1.0	2.0	
Clinic Faucet	1/2"		3.0	
Clinic Flushometer Valve				
with or without faucet	1"		8.0	
Kitchen, domestic	1/2"	1.5	1.5	
Laundry	1/2"	1.5	1.5	
Service or Mop Basin	1/2"	1.5	3.0	
Washup, each set of faucets	1/2"		2.0	
Shower	1/2"	2.0	2.0	
Urinal, 1.0 GPF	3/4"	3.0	4.0	5.0
Urinal, greater than 1.0 GPF	3/4"	4.0	5.0	6.0
Urinal, flush tank	1/2"	2.0	2.0	3.0
Washfountain, circular spray	3/4"		4.0	
Water Closet, 1.6 GPF Gravity Tank	1/2"	2.5	2.5	3.5
Water Closet, 1.6 GPF Flushometer Tank	1/2"	2.5	2.5	3.5
Water Closet, 1.6 GPF Flushometer Valve	1"	5.0	5.0	6.0
Water Closet, greater than 1.6 GPF Gravity Tank	1/2"	3.0	5.5	7.0
Water Closet, greater than 1.6 GPF Flushometer Valve	1"	7.0	8.0	10.0

Notes:

1. Size of the cold branch outlet pipe, or both the hot and cold branch outlet pipes.

2. Appliances, Appurtenances or Fixtures not included in this Table may be sized by reference to fixtures having a similar flow rate and frequency of use.

3. The listed fixture unit values represent their total load on the cold water service. The separate cold water and hot water fixture unit value for fixtures having both cold and hot water connections shall each be taken as three-quarters (3/4) of the listed total value of the fixture.

4. The listed minimum supply branch pipe sizes for individual fixtures are the nominal (I.D.) pipe size.

5. For fixtures or supply connections likely to impose continuous flow demands, determine the required flow in gallons per minute (GPM) and add it separately to the demand (in GPM) for the distribution system or portions thereof.

6. Assembly [Public Use (See Table 4-1)].

7. Reduced fixture unit loading for additional hose bibbs as used is to be used only when sizing total building demand and for pipe sizing when more than one hose bibb is supplied by a segment of water distributing pipe. The fixture branch to each hose bibb shall be sized on the basis of 2.5 fixture units.

FIGURE 21.14 This shows the relationship between fixture units and fixture branch pipes for a water supply. (*Reprinted from the* 2000 *Uniform Plumbing Code (UPC)* with the permission of the International Association of Plumbing and Mechanical Officials (IAPMO).)

Allowance in equivalent length of pipe for friction loss in valves and threaded fittings.*

Equivalent Length of Pipe for Various Fittings

Diameter of fitting Inches	90° Standard Elbow Feet	45° Standard Elbow Feet	Standard Tee 90° Feet	Coupling or Straight Run of Tee Feet	Gate Valve Feet	Globe Valve Feet	Angle Valve Feet
3/8	1.0	0.6	1.5	0.3	0.2	8	4
1/2	2.0	1.2	3.0	0.6	0.4	15	8
3/4	2.5	1.5	4.0	0.8	0.5	20	12
1	3.0	1.8	5.0	0.9	0.6	25	15
1-1/4	4.0	2.4	6.0	1.2	0.8	35	18
1-1/2	5.0	3.0	7.0	1.5	1.0	45	22
2	7.0	4.0	10.0	2.0	1.3	55	28
2-1/2	8.0	5.0	12.0	2.5	1.6	65	34
3	10.0	6.0	15.0	3.0	2.0	80	40
4	14.0	8.0	21.0	4.0	2.7	125	55
5	17.0	10.0	25.0	5.0	3.3	140	70
6	20.0	12.0	30.0	6.0	4.0	165	80

FIGURE 21.15 Pipe length of various fittings is described. *(Reprinted from the 2000 Uniform Plumbing Code (UPC) with the permission of the International Association of Plumbing and Mechanical Officials (IAPMO).)*

Equivalent Length of Pipe for Various Fittings

Diameter of fitting	90° Standard Elbow	45° Standard Elbow	Standard Tee 90°	Coupling or Straight Run of Tee	Gate Valve	Globe Valve	Angle Valve
mm	mm	mm	mm	mm	mm	mm	mm
10	305	183	457	91	61	2438	1219
15	610	366	914	183	122	4572	2438
20	762	457	1219	244	152	6096	3658
25	914	549	1524	274	183	7620	4572
32	1219	732	1829	366	244	10668	5486
40	1524	914	2134	457	305	13716	6706
50	2134	1219	3048	610	396	16764	8534
65	2438	1524	3658	762	488	19812	10363
80	3048	1829	4572	914	610	24384	12192
100	4267	2438	6401	1219	823	38100	16764
125	5182	3048	7620	1524	1006	42672	21336
150	6096	3658	9144	1829	1219	50292	24384

*Allowances are based on non-recessed threaded fittings. Use one-half (1/2) the allowances for recessed threaded fittings or streamline solder fittings.

FIGURE 21.16 This is the metric version of the previous table. *(Reprinted from the 2000 Uniform Plumbing Code (UPC) with the permission of the International Association of Plumbing and Mechanical Officials (IAPMO).)* *(Continued)*

Example

Fixture Units and Estimated Demands

	Building Supply Demand				Branch to Hot Water System		
Kind of Fixtures	No. of Fixtures	Fixture Unit Demand	Total Units	Building Supply Demand in gpm (L per sec)	No. of Fixtures	Fixture Unit Demand Calculation	Demand in gallons per minute (L per sec)
Water Closets	130	8.0	1040	–	–	–	–
Urinals	30	4.0	120	–	–	–	–
Shower Heads	12	2.0	24	–	12	12 x 2 x 3/4 = 18	–
Lavatories	100	1.0	100	–	100	100 x 1 x 3/4 = 75	–
Service Sinks	27	3.0	81	–	27	27 x 3 x 3/4 = 61	–
Total			1365	252 gpm (15.8 L/s)		154	55 gpm (3.4 L/s)

Allowing for 15 psi (103.4 kPa) at the highest fixture under the maximum demand of 252 gallons per minute (15.8 L/sec.), the pressure available for friction loss is found by the following:

$$55 - [15 + (45 \times 0.43)] = 20.65 \text{ psi}$$

Metric: $379 - [103.4 + (13.7 \times 9.8)] = 142.3$ kPa

The allowable friction loss per 100 feet (30.4 m) of pipe is therefore:

$$100 \times 20.65 \div 200 = 10.32 \text{ psi}$$

Metric: $30.4 \times 142.3 \div 60.8 = 71.1$ kPa

FIGURE 21.17 Estimated demand can be figured with this information. (*Reprinted from the 2000 Uniform Plumbing Code (UPC) with the permission of the International Association of Plumbing and Mechanical Officials (IAPMO).*)

Friction Losses for Disk Type Water Meters

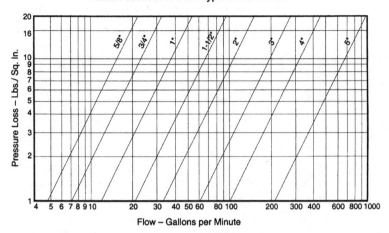

Flow – Gallons per Minute

Friction Losses for Disk Type Water Meters

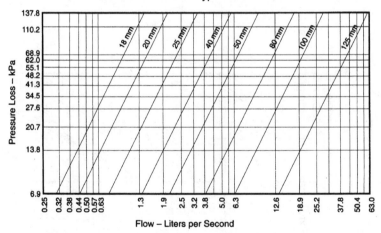

Flow – Liters per Second

FIGURE 21.18 English and metric unit information about friction loss is above. *(Reprinted from the* 2000 Uniform Plumbing Code (UPC) *with the permission of the International Association of Plumbing and Mechanical Officials (IAPMO).)*

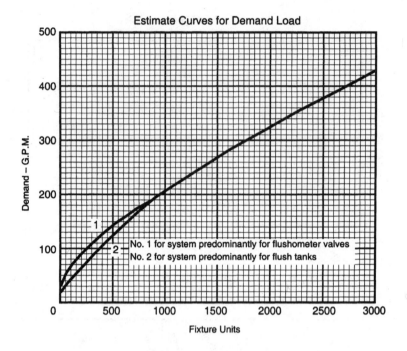

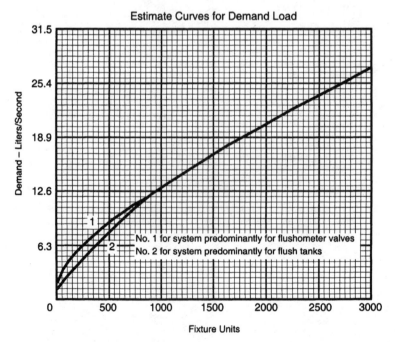

FIGURE 21.19 English and metric lengths are shown. *(Reprinted from the 2000 Uniform Plumbing Code (UPC) with the permission of the International Association of Plumbing and Mechanical Officials (IAPMO).)*

21.20

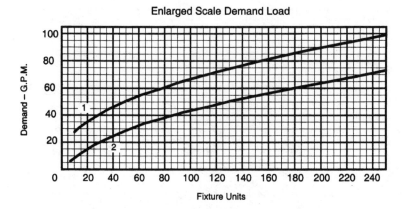

Enlarged Scale Demand Load

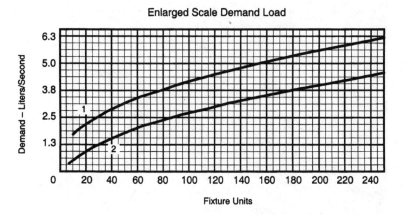

Enlarged Scale Demand Load

FIGURE 21.20 These are an enlarged scale of demand loads. *(Reprinted from the* 2000 Uniform Plumbing Code (UPC) *with the permission of the International Association of Plumbing and Mechanical Officials (IAPMO).)*

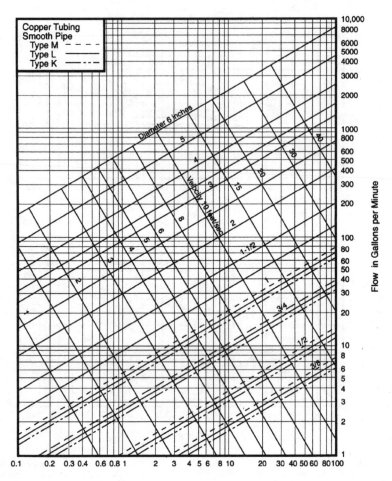

FIGURE 21.21 This is one of several graphs about friction loss. *(Reprinted from the* 2000 Uniform Plumbing Code *(UPC) with the permission of the International Association of Plumbing and Mechanical Officials (IAPMO).)*

21.22

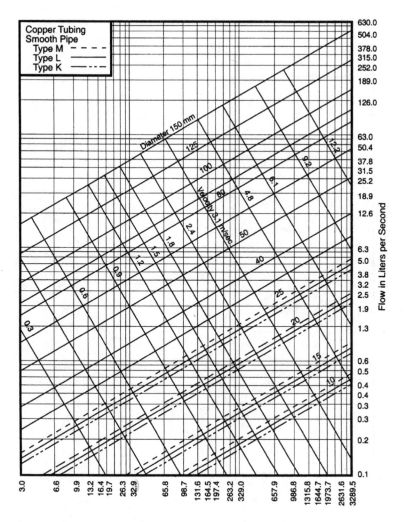

FIGURE 21.22 This is one of several graphs about friction loss. *(Reprinted from the 2000 Uniform Plumbing Code (UPC) with the permission of the International Association of Plumbing and Mechanical Officials (IAPMO).)*

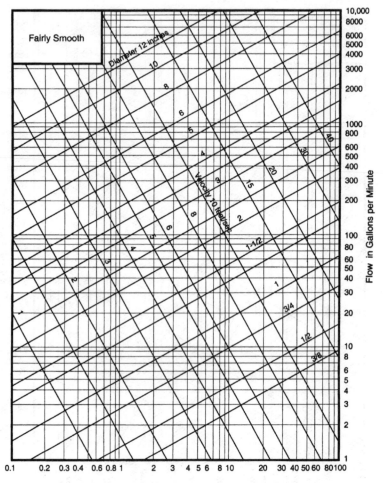

Flow in Gallons per Minute

Friction Loss – Lbs. per Square Inch Head per 100 Foot Length

FIGURE 22.23 This is one of several graphs about friction loss. *(Reprinted from the* 2000 Uniform Plumbing Code *(UPC) with the permission of the International Association of Plumbing and Mechanical Officials (IAPMO).)*

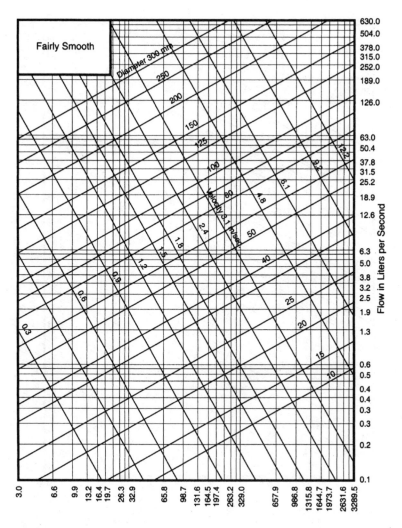

FIGURE 21.24 This is one of several graphs about friction loss. *(Reprinted from the 2000 Uniform Plumbing Code (UPC) with the permission of the International Association of Plumbing and Mechanical Officials (IAPMO).)*

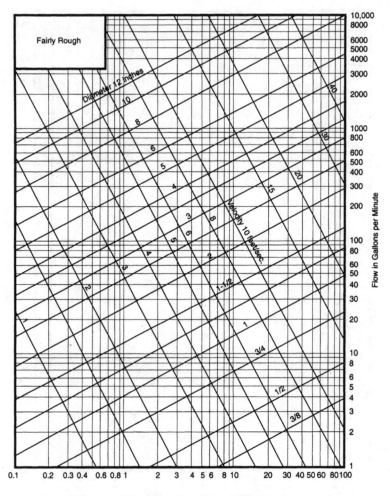

Fairly Rough

Diameter 12 inches

Velocity 10 feet/sec.

Flow in Gallons per Minute

Friction Loss – Lbs. per Square Inch Head per 100 Foot Length

FIGURE 21.25 This is one of several graphs about friction loss. *(Reprinted from the* 2000 Uniform Plumbing Code *(UPC) with the permission of the International Association of Plumbing and Mechanical Officials (IAPMO).)*

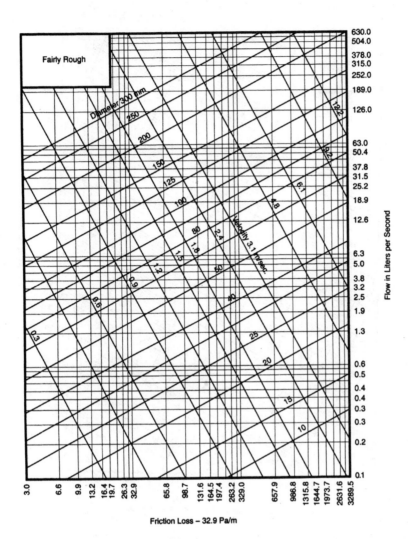

FIGURE 21.26 This is one of several graphs about friction loss. *(Reprinted from the 2000 Uniform Plumbing Code (UPC) with the permission of the International Association of Plumbing and Mechanical Officials (IAPMO).)*

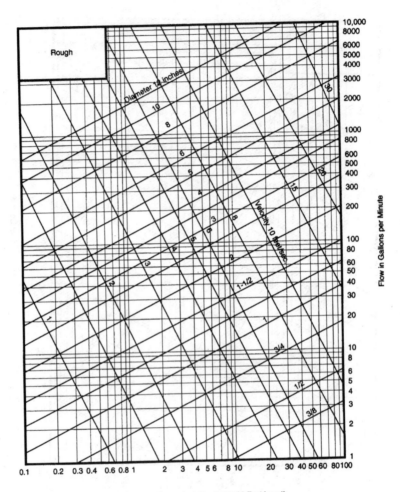

FIGURE 21.27 This is one of several graphs about friction loss. *(Reprinted from the* 2000 Uniform Plumbing Code *(UPC) with the permission of the International Association of Plumbing and Mechanical Officials (IAPMO).)*

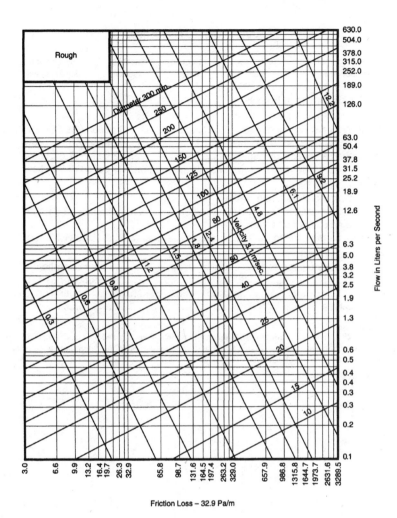

FIGURE 21.28 This is one of several graphs about friction loss. *(Reprinted from the* 2000 Uniform Plumbing Code (UPC) *with the permission of the International Association of Plumbing and Mechanical Officials (IAPMO).)*

Sizing of Grease Interceptors

Number of meals per peak hour[1]	x	Waste flow rate[2]	x	retention time[3]	x	storage factor[4]	=	Interceptor size (liquid capacity)

1. **Meals Served at Peak Hour**

2. **Waste Flow Rate**
   a. With dishwashing machine .... 6 gallon (22.7 L) flow
   b. Without dishwashing machine .... 5 gallon (18.9 L) flow
   c. Single service kitchen .... 2 gallon (7.6 L) flow
   d. Food waste disposer .... 1 gallon (3.8 L) flow

3. **Retention Times**
   Commercial kitchen waste
      Dishwasher .... 2.5 hours
   Single service kitchen
      Single serving .... 1.5 hours

4. **Storage Factors**
   Fully equipped commercial kitchen .... 8 hour operation: 1
   .... 16 hour operation: 2
   .... 24 hour operation: 3
   Single Service Kitchen 1.5

FIGURE 21.29   This chart shows information about grease interceptors and how to size them. *(Reprinted from the 2000 Uniform Plumbing Code (UPC) with the permission of the International Association of Plumbing and Mechanical Officials (IAPMO).)*

APPROVED CONSTRUCTION OF TILE-LINED SHOWER RECEPTORS

STANDARD SPECIFICATION FOR THE INSTALLATION OF TILE-LINED SHOWER RECEPTORS

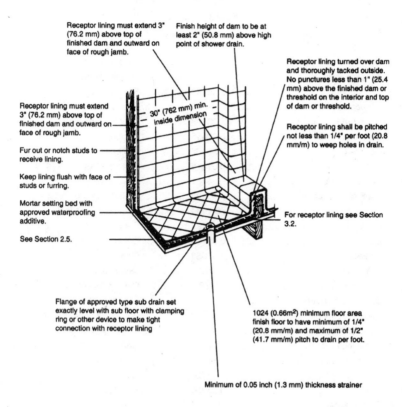

Receptor lining must extend 3" (76.2 mm) above top of finished dam and outward on face of rough jamb.

Finish height of dam to be at least 2" (50.8 mm) above high point of shower drain.

Receptor lining turned over dam and thoroughly tacked outside. No punctures less than 1" (25.4 mm) above the finished dam or threshold on the interior and top of dam or threshold.

Receptor lining shall be pitched not less than 1/4" per foot (20.8 mm/m) to weep holes in drain.

Receptor lining must extend 3" (76.2 mm) above top of finished dam and outward on face of rough jamb.

Fur out or notch studs to receive lining.

Keep lining flush with face of studs or furring.

Mortar setting bed with approved waterproofing additive.

See Section 2.5.

30" (762 mm) min. inside dimension

For receptor lining see Section 3.2.

Flange of approved type sub drain set exactly level with sub floor with clamping ring or other device to make tight connection with receptor lining

1024 (0.66m²) minimum floor area finish floor to have minimum of 1/4" (20.8 mm/m) and maximum of 1/2" (41.7 mm/m) pitch to drain per foot.

Minimum of 0.05 inch (1.3 mm) thickness strainer

FIGURE 21.30 Tile-lined shower receptors have specific requirements. (*Reprinted from the* 2000 *Uniform Plumbing Code (UPC)* with the permission of the International Association of Plumbing and Mechanical Officials (IAPMO).)

21.31

Conversion Table

MULTIPLY	BY	TO OBTAIN
Acres	43,560	Square feet
Acre-feet	43,560	Cubic feet
Acre-feet	325,851	Gallons
Atmospheres	76.0	Cms of mercury
Atmospheres	29.92	Inches of mercury
Atmospheres	33.90	Feet of water
Atmospheres	14.70	Pounds/square inch
Btu/minute	12.96	Foot-Pounds/second
Btu/minute	0.02356	Horse-power
Centimeters	0.3937	Inches
Centimeters of mercury	0.01316	Atmospheres
Centimeters of mercury	0.4461	Feet of water
Centimeters of mercury	27.85	Pounds/square feet
Centimeters of mercury	0.1934	Pounds/square inch
Cubic feet	1728	Cubic inches
Cubic feet	0.03704	Cubic yards
Cubic feet	7.48052	Gallons
Cubic feet	29.92	Quarts (liquid)
Cubic feet/minute	472.0	Cubic cms/second
Cubic feet/minute	0.1247	Gallons/second
Cubic feet/minute	62.43	Pounds of water/minute
Cubic feet/second	0.0646317	Million gallons/day
Cubic feet/second	448.831	Gallons/minute
Cubic yards	27	Cubic feet
Cubic yards	202.0	Gallons
Feet of water	0.02950	Atmospheres
Feet of water	0.8826	Inches of mercury
Feet of water	62.43	Pounds/square feet
Feet of water	0.4335	Pounds/square inch
Feet/minute	0.01667	Feet/second
Feet/minute	0.01136	Miles/hour
Feet/second	0.6818	Miles/hour
Feet/second	0.01136	Miles/minute
Gallons	3785	Cubic centimeters
Gallons	0.1337	Cubic feet
Gallons	231	Cubic inches
Gallons	4	Quarts (liquid)
Gallons water	8.3453	Pounds of water
Gallons/minute	0.002228	Cubic feet/second

FIGURE 21.31A A useful set of tables to keep on hand. (*Reprinted from the* 2000 Uniform Plumbing Code *(UPC) with the permission of the International Association of Plumbing and Mechanical Officials (IAPMO).*)

21.32

MULTIPLY	BY	TO OBTAIN
Gallons/minute	8.0208	Cubic feet/hour
Gallons water/minute	6.0086	Tons of water/24 hours
Inches	2.540	Centimeters
Inches of mercury	0.03342	Atmospheres
Inches of mercury	1.133	Feet of water
Inches of mercury	0.4912	Pounds/square inch
Inches of water	0.002458	Atmospheres
Inches of water	0.07355	Inches of mercury
Inches of water	5.202	Pounds/square feet
Inches of water	0.03613	Pounds/square inch
Liters	1000	Cubic centimeters
Liters	61.02	Cubic inches
Liters	0.2642	Gallons
Miles	5280	Feet
Miles/hour	88	Feet/minute
Miles/hour	1.467	Feet/second
Millimeters	0.1	Centimeters
Millimeters	0.03937	Inches
Million gallon/day	1.54723	Cubic feet/second
Pounds of water	0.01602	Cubic feet
Pounds of water	27.68	Cubic inches
Pounds of water	0.1198	Gallons
Pounds/cubic inch	1728	Pounds/cubic feet
Pounds/square foot	0.01602	Feet of water
Pounds/square inch	0.06804	Atmospheres
Pounds/square inch	2.307	Feet of water
Pounds/square inch	2.036	Inches of mercury
Quarts (dry)	67.20	Cubic inches
Quarts (liquid)	57.75	Cubic inches
Square feet	144	Square inches
Square miles	640	Acres
Square yards	9	Square feet
Temperature (°C) + 273	1	Abs. temperature (°C)
Temperature (°C) + 17.28	1.8	Temperature (°F)
Temperature (°F) + 460	1	Abs. temperature (°F)
Temperature (°F) - 32	5/9	Temperature (°C)
Tons (short)	2000	Pounds
Tons of water/24 hours	83.333	Pounds water/hour
Tons of water/24 hours	0.16643	Gallons/minute
Tons of water/24 hours	1.3349	Cubic feet/hour

FIGURE 21.31B A useful set of tables to keep on hand. *(Reprinted from the* 2000 Uniform Plumbing Code (UPC) *with the permission of the International Association of Plumbing and Mechanical Officials (IAPMO).) (Continued)*

AREAS AND CIRCUMFERENCE OF CIRCLES

Diameter		Circumference		Area	
Inches	mm	Inches	mm	Inches^2	mm^2
1/8	6	0.40	10	0.01227	8.0
1/4	8	0.79	20	0.04909	31.7
3/8	10	1.18	30	0.11045	71.3
1/2	15	1.57	40	0.19635	126.7
3/4	20	2.36	60	0.44179	285.0
1	25	3.14	80	0.7854	506.7
1-1/4	32	3.93	100	1.2272	791.7
1-1/2	40	4.71	120	1.7671	1140.1
2	50	6.28	160	3.1416	2026.8
2-1/2	65	7.85	200	4.9087	3166.9
3	80	9.43	240	7.0686	4560.4
4	100	12.55	320	12.566	8107.1
5	125	15.71	400	19.635	12,667.7
6	150	18.85	480	28.274	18,241.3
7	175	21.99	560	38.485	24,828.9
8	200	25.13	640	50.265	32,428.9
9	225	28.27	720	63.617	41,043.1
10	250	31.42	800	78.540	50,670.9

EQUAL PERIPHERIES

$S = 0.7854 D$

$D = 1.2732 S$

$S = 0.8862 D$

$D = 1.1284 S$

$S = 0.2821 C$

EQUAL AREAS

Area of square (S') =
1.2732 x area of circle

Area of square (S) =
0.6366 x area of circle

$C = \pi D = 2\pi R$

$C = 3.5446 \sqrt{area}$

$D = 0.3183 C = 2R$

$D = 1.1283 \sqrt{area}$

$Area = \pi R^2 = 0.7854 D^2$

$Area = 0.07958 C^2 = \dfrac{\pi D^2}{4}$

$\pi = 3.1416$

FIGURE 21.32 More useful information. *(Reprinted from the* 2000 Uniform Plumbing Code *(UPC) with the permission of the International Association of Plumbing and Mechanical Officials (IAPMO).)*

METRIC SYSTEM
(INTERNATIONAL SYSTEM OF UNITS – SI)

TO CONVERT	INTO	MULTIPLY BY
Atmospheres	Cms of mercury	76.0
Btu	Joules	1,054.8
Btu/hour	Watts	0.2931
Btu/minute	Kilowatts	0.01757
Btu/minute	Watts	17.57
Centigrade	Fahrenheit	(°C x 9/5) + 32°
Circumference	Radians	6.283
Cubic centimeters	Cubic inches	0.06102
Cubic feet	Cubic meters	0.02832
Cubic feet	Liters	28.32
Cubic feet/minute	Cubic cms/second	472.0
Cubic inches	Cubic cms	16.39
Cubic inches	Liters	0.01639
Cubic meters	Gallons (U.S. liquid)	264.2
Feet	Centimeters	30.48
Feet	Meters	0.3048
Feet	Millimeters	304.8
Feet of water	Kgs/square cm	0.03048
Foot-Pounds	Joules	1.356
Foot-pounds/minute	Kilowatts	2.260×10^{-5}
Foot-pounds/second	Kilowatts	1.356×10^{-3}
Gallons	Liters	3.785
Horsepower	Watts	745.7
Horsepower-hours	Joules	2.684×10^{6}
Horsepower-hours	Kilowatt-hours	0.7457
Joules	Btu	9.480×10^{-4}
Joules	Foot-pounds	0.7376
Joules	Watt-hours	2.778×10^{-4}
Kilograms	Pounds	2.205
Kilograms	Tons (short)	1.102×10^{-3}
Kilometers	Miles	0.6214
Kilometers/hour	Miles/hour	0.6214
Kilowatts	Horsepower	1.341
Kilowatt-hours	Btu	3,413
Kilowatt-hours	Foot-pounds	2.655×10^{6}
Kilowatt-hours	Joules	3.6×10^{6}
Liters	Cubic feet	0.03531

FIGURE 21.33 Metric conversions are important. *(Reprinted from the* 2000 Uniform Plumbing Code (UPC) *with the permission of the International Association of Plumbing and Mechanical Officials (IAPMO).)*

METRIC SYSTEM
(INTERNATIONAL SYSTEM OF UNITS – SI)
(Continued)

TO CONVERT	INTO	MULTIPLY BY
Liters	Gallons (U.S. liquid)	0.2642
Meters	Feet	3.281
Meters	Inches	39.37
Meters	Yards	1.094
Meters/second	Feet/second	3.281
Meters/second	Miles/hr	2.237
Miles (statute)	Kilometers	1.609
Miles/hour	Meters/minute	26.82
Millimeters	Inches	0.03937
Ounces (fluid)	Liters	0.02957
Pints (liquid)	Cubic centimeters	473.2
Pounds	Kilograms	0.4536
PSI	Pascals	6,895
Quarts (liquid)	Liters	0.9463
Radians	Degrees	57.30
Square inches	Square millimeters	645.2
Square meters	Square inches	1,550
Square millimeters	Square inches	1.550×10^{-3}
Watts	Btu/hour	3.4129
Watts	Btu/minute	0.05688
Watts	Foot-pounds/second	0.7378
Watts	Horsepower	1.341×10^{-3}

FIGURE 21.34 Metric conversions. *(Reprinted from the 2000 Uniform Plumbing Code (UPC) with the permission of the International Association of Plumbing and Mechanical Officials (IAPMO)*

CHAPTER 22

TROUBLESHOOTING WATER-DISTRIBUTION SYSTEMS

Troubleshooting water-distribution systems can be very simple. Plumbers often can look for the source of a leak and find it by following the sound of rushing water or a trail of water stains. If it were always this easy, you wouldn't need to read this chapter. However, troubleshooting a water system can require a lot of time and thought. Some problems disguise themselves well and are difficult to find.

Plumbers who have developed strong troubleshooting skills usually can guess what's wrong on a job before they ever arrive at the site. This type of knowledge, however, doesn't come quickly or easily. To become a proficient troubleshooter, you must learn the principles of using a methodic approach in your problem-solving situation. Experience is one of the best teachers, but this chapter can help you jump start your learning experience.

What's the first type of problem you think of when considering trouble with a water system? Is it water leaks? Problems with leaks are common. Fortunately, the source of leaks in a water system often are easy to locate. And correcting such a problem is not usually a major problem. But, how about water pressure? Many homeowners complain about their water pressure, usually because it is not adequate to satisfy them. What would you look for if a customer complained about low water pressure? There are many possible causes of such a problem. Let's look at some of them.

LOW WATER PRESSURE

Low water pressure often becomes a problem in some plumbing systems. Trying to take a shower in a home with low water pressure is very frustrating, and waiting several minutes for a water closet to refill its tank can be nerve-racking. There are many reasons why a lack of water pressure will induce customers to call in professional plumbers.

What causes low water pressure? Many factors can be found guilty of causing water pressure to be less than what is desirable. Buildings served by water pumps can be plagued by low water pressure when there is a problem with a pressure tank or the pressure switch. Homes that are connected to city water supplies might have pressure-reducing valves that need to be adjusted. Properties that are plumbed with old galvanized steel piping could be suffering from rust obstructions in the pipes.

PRESSURE TANKS

Properties that obtain their water from private water sources depend on pressure tanks to give them the working pressure they want from their plumbing systems. If the pressure tanks become waterlogged, they cannot produce the type of water pressure they are designed to provide.

A waterlogged pressure tank will give a symptom that is hard for a serious troubleshooter to miss. When a pressure tank is waterlogged, the well pump will cut on frequently, often every time a faucet is opened. These tanks sometimes provide adequate pressure, even though they are forcing the pump to work much harder than it is intended. At other times, a loss in pressure will be noticeable.

If you are faced with a building that has unsatisfactory water pressure and a well pump that cuts on frequently, you should take a close look at the pressure tank. When a tank is suspected of being waterlogged, it should be drained, recharged with air pressure, and then refilled with water. This is a simple process, and it can solve your pressure problems.

PRESSURE SWITCHES

Properties that are served by private wells depend not only on pressure tanks for their water pressure, but also on pressure switches. If the cut-in pressure on a well system is not set properly, it is possible for pressure to drop to unacceptable levels before the pump will produce more water.

A fast visual inspection of the pressure gauge on the well system will tell you if the system is maintaining a satisfactory working pressure. It is important, however, to make sure there is demand from a plumbing fixture while you are observing the pressure gauge. If no plumbing fixtures are being called upon for water, the pressure gauge will remain static at its highest level. To be sure the system is maintaining an acceptable working pressure, you must put a demand on the system.

A pressure gauge that shows a sharp fall in pressure before the pump cuts on indicates that the cut-in pressure is set too low. Going into the box of the pressure switch and adjusting the spring-loaded nut will correct this problem.

PRESSURE-REDUCING VALVES

Pressure-reducing valves that are not set properly can cause trouble in the form of low water pressure. If these valves are not adjusted to the proper settings, they can reduce water to little more than a trickle.

Sometimes the adjustment screws on pressure-reducing valves are turned down too tightly. If you are dealing with a low-pressure situation in which a pressure-reducing valve is involved, check the level of the screw setting. Turning the adjustment screw counterclockwise will increase the water pressure on the system.

Pressure-reducing valves are not often the cause of a pressure problem, but they can be. It might be that the adjustment screw is set improperly or it could be that the entire valve is bad. If you cut the water off at the street connection, you can check the building pressure by removing the pressure-reducing valve and installing a pressure gauge on the piping. This will eliminate, or confirm, any doubts you might have about the pressure-reducing valve.

GALVANIZED PIPE

If you have been in the plumbing trade long enough to have worked with much galvanized pipe, you know how badly it can rust, corrode, and build up obstructions. Any time you are working with a water-distribution system that is made up of galvanized pipe, you shouldn't be surprised to find low water pressure.

I have cut out sections of galvanized pipe in which the open pipe diameter wasn't large enough to allow the insertion of a common drinking straw. When the open diameter of the pipe is constricted, water pressure must drop. Any longtime plumber will tell you that galvanized water pipe is a nightmare waiting to happen.

Most systems plumbed with galvanized pipe are equipped with unions on various sections of the piping. If you cut off the water and loosen the unions, you are likely to find the cause of your low pressure. A quick look at the interior of the pipe probably will be all it will take to convince you.

The only real solution to low pressure from blocked galvanized pipe is the replacement of the water piping. This can be a big and expensive job, but it is the only way to solve the problem positively.

UNDERSIZED PIPING

Undersized piping can create problems with water pressure. If the pipes delivering water to fixtures are too small, the fixtures will not receive an adequate volume of water at a desired pressure.

HARD WATER

Hard water can be a cause of reduced water pressure. The scale that hard water allows to build up on the inside of water pipes, fixtures, and tanks can reduce water pressure by a noticeable amount. When you have a job that is giving you trouble with low pressure and no sign of a cause, inspect the interior of some piping to see if it is being blocked by a scale buildup.

CLOGGED FILTERS

Clogged in-line filters are notorious for their ability to restrict water flow. People have these little filters put in to trap sediment, and they do. They trap the sediment so well that they eventually clog up and block the normal flow of water, thereby reducing water pressure.

When you respond to a low-pressure call, ask the property owner if any filters are installed on the water distribution system. If one is, check the filter to see if it needs to be replaced.

STOPPED-UP AERATORS

One of the most common, and simplest to fix, causes of low water pressure is stopped-up aerators. The little screens in the aerators stop up frequently when a house is served by a private water supply. Iron particles and mineral deposits are usually the cause for stoppages.

Aerators can be removed and will sometimes clean up, but many times they must be replaced. As long as you have an assortment of aerators on your service truck, this is one problem that is both easy to find and to fix.

EXCESSIVE WATER PRESSURE

Excessive water pressure in a water distribution system can be dangerous and destructive. When fixtures are under too much pressure, they do not perform well over extended periods of time. The O rings, stems, and other components of the plumbing system are not usually meant to work with pressures exceeding 80 pounds per square inch (psi).

Extreme water pressure can be dangerous for the users of the plumbing system. For example, I once worked for a homeowner who had cut herself badly at the kitchen sink as a result of high water pressure. She was holding a drinking glass under the kitchen faucet when she turned on the water to fill the glass. The water rushed out with such force that it knocked the glass out of her hand. Glass shattered on impact with the sink and pieces of the broken glass sliced into the woman's hand and arm.

High water pressure is easy to control. The pressure can be reduced with a pressure-reducing valve, and in the case of well systems, with adjustments to the pressure switch.

Troubleshooting high water pressure is simple, all you have to do is look at the pressure gauge on the well system or install a pressure gauge on a faucet, usually a hose bibb. If the pressure is higher than 60 psi, it usually should be lowered. Most residential properties operate well with a water pressure of between 40 and 50 psi.

If you're working with a well system, all you have to do is go into the pressure switch and adjust the cut-out setting. In the case of city water service, you will have to install a pressure-reducing valve. If a pressure-reducing valve is already in place, you can lower the system pressure by turning the adjustment screw clockwise.

NOISY PIPES

Noisy pipes, resulting from water hammer, can make living around a plumbing system very annoying. Sometimes the pipes bang, sometimes they squeak, and sometimes they chatter. This condition can be so severe that living with the noise is almost unbearable. What causes noisy pipes? Water hammer is one reason pipes play their unfavorable tunes, but it is not the only reason.

Not all pipe noises are the same. The type of noise you hear gives a strong indication about the type of action that will be required to solve the problem. You must listen closely to the sounds being made in order to diagnose the problems properly.

What will you be listening for? The tone and type of noise being made might be all you have to solve the problem of a water hammer. Let's look at the various noises individually and see what they mean and how you can correct the problem.

WATER HAMMER

Water hammer is the most common cause of noisy pipes in a water distribution system. Banging pipes are a sure sign of water hammer. Can water hammer be stopped? Yes, there are ways to eliminate its actions and effects, but implementing the procedure is not always easy.

What causes water hammer? Water hammer occurs most often with quick-closing valves, like ballcocks and washing-machine valves, but it can be a problem with other fixtures. The condition can be worsened when the water distribution pipes are installed with long, straight runs. When the water is shut off quickly, it bangs into the fittings at the end of the pipe run or at the fixture and produces the hammering or banging noise. The shock wave can produce some loud noises.

If a plumbing system is under higher than average pressure, it can be more likely to suffer from water hammer. There are several ways to approach the problem to eliminate the banging. To illustrate these options, I would like to put the problems into the form of real-world situations. Allow me to give you an example from one of my company's past service calls.

A TOILET

This first example of a water hammer problem involves a troublesome toilet. The residents of the home where the toilet was located were annoyed by the banging noise their water pipes made on most occasions when the toilet was flushed. The toilet was located in the second-floor bathroom, and when it was flushed, it would rattle the pipes all through the home. There came a day when the homeowners decided they could not live with the problem any longer.

The couple called in a plumber to troubleshoot their problem. The plumber looked over the situation and went about his work in trying to locate the cause of the problem. He knew they were suffering from a water-hammer situation, but he wasn't sure how to handle the call. In fact, the plumber gave up, washed his hands of the job and left.

Distraught, the homeowners called my company. One of my plumbers responded to the call and quickly assessed that the upstairs toilet was the culprit. He recommended that the couple allow him to install an air chamber in the wall, near the toilet.

At first, the homeowners were reluctant to give their permission to cut into their bathroom wall. Under the circumstances, however, there were not many viable options, so they gave their consent.

My plumber made a modest cut in the bathroom wall, around and above the closet supply, and installed an air chamber. This particular homeowner was one who wanted to know every move that was being made, and he was not too sure the plumber was doing anything that would really improve the situation.

After the air chamber was installed, my plumber activated the system and began to flush the toilet. After several sequences of flushing the pipes remained quiet; there was no banging. The homeowner had been doubtful about my plumber's decision, but he did admit the work solved the problem, and he was happy.

Air chambers are frequently all that are needed to reduce or eliminate banging pipes. My plumber was correct in his diagnosis of the problem and in effecting an efficient cure for the problem. It is possible, however, that the problem could have been solved by installing a master unit for a water-hammer arrestor in the basement of the home, eliminating the need to cut into the bathroom wall. There is no guarantee, however, that a master unit would have controlled the situation on the second floor.

The plumber made a wise and prudent decision. By installing the air chamber at the fixture, satisfaction was practically guaranteed. While the master unit might have worked, there was some risk that it wouldn't. If the homeowner had objected to opening the bathroom wall, I'm sure the plumber would have installed a master unit in an attempt to control the problem.

EXISTING AIR CHAMBERS

Sometimes existing air chambers become waterlogged and fail to function properly. This problem is not uncommon, but it is a bother. If you have a plumbing system equipped with air chambers that is being affected by water hammer, you must recharge the existing air chambers with a fresh supply of air. This is a simple process.

To recharge air chambers, you first must drain the water pipes of most of their reserve water. This can be done by opening a faucet or valve that is at the low end of the system and one that is at the upper end of the system.

After the water has drained out of the water distribution pipes, you can close the faucets or valves and turn the water supply back on. As the new water fills the system, air will be replenished in the air chambers. As easy as this procedure is, it is all that is required to recharge waterlogged air chambers.

UNSECURED PIPES

Unsecured pipes are prime targets when you are having a noise problem. If the pipes that make up the water distribution system are not secured properly, they are likely to vibrate and make all sorts of noise. Not only is this annoying to the ears, it can be damaging to the pipes.

Pipes that are not secure in their hangers can vibrate to the point that they wear holes in themselves. If enough stress is present, the connection joints might even be broken. Local plumbing codes dictate how far apart the hangers for pipe may be. If the hangers fall within the guidelines of the local code and secure the pipes tightly, no problems should exist. Trouble can develop, however, when the hangers are farther apart than they should be, are the wrong size, or are not attached firmly.

Pipes that are not secured in the manner described by the plumbing code can create loud banging sounds. This noise imitates the noise made by a system that is experiencing a water hammer. Squeaking and chattering noises also can be present when pipes are not secured properly.

The problem of poorly secured pipes is easy to fix, if you have access to the pipes. Unfortunately, the pipes and their hangers often are concealed by finished walls. Sometimes the problem pipes can be accessed in basements or crawl spaces, but they often are inaccessible unless walls and ceilings are destroyed.

After you have access to the pipes that are not secured properly, all you have to do is add hangers or tighten the existing hangers. This is fine if you can get to the pipes easily, but it is a hard-sell to a homeowner who doesn't want the walls and ceilings of a home cut open. There is, however, no other way to eradicate the problem.

THAT SQUEAKING NOISE

That squeaking noise you often hear in pipes is almost always a hot-water pipe. Whether the water distribution pipe is copper or plastic, it tends to expand when it gets hot. When hot water moves through the pipe, the pipe expands and creates friction against the pipe hanger.

The expansion aspect of hot-water pipes is not practical to eliminate, but you can do something about the squeaking noise. The simple solution, when you have access to the pipe and hanger, is to install an insulator between the pipe and the support. This can be a piece of foam, a piece of rubber, or any other suitable insulator. Your only goal is to prevent the pipe from rubbing against the hanger when the pipe expands.

CHATTERING

The chattering heard in plumbing pipes is not from cold temperatures, it is from problems in the faucets of fixtures. While this is more of a faucet and valve problem than a water distribution problem, we will cover it here, since the noise often is associated with water pipes.

When you hear a chattering sound in a plumbing system, you should inspect the faucet stems and washers at nearby faucets. You should find that the faucet washer has become loose and is being vibrated, or fluttered as some plumbers say, by the water pressure between it and the faucet seat.

To solve the problem of chattering pipes, all you should have to do is tighten the washers in the faucets. After the washers are tight, the noise should disappear.

MUFFLING THE SYSTEM AT THE WATER SERVICE

When you have a plumbing system that is banging because of water hammer, you can try muffling the system at the water service. This procedure doesn't always work, but sometimes it does.

Piping in many buildings is not readily accessible. Under these conditions, the only easy way to attack a water hammer is at the water-service or main water-distribution pipe. By installing air chambers or water-hammer arrestors on sections of the available pipe, you might be able to eliminate the symptoms of water hammer throughout the building.

To install an air chamber or water arrestor on the water-service or main water-distribution pipe is not a big job. After the water is cut off, you simply have to cut in some tee fittings and install the air chambers or arrestors. The risers that accept these devices should be at least two feet high, when possible.

Depending upon the severity of the water-hammer problem, several devices might need to be installed at different locations along the water piping. Typically, one at the beginning of the piping, one at the end of the piping, and devices installed at the ends of branches will greatly reduce, if not eliminate, the problems associated with water hammers.

ADDING OFFSETS

Adding offsets to long runs of straight piping is another way to reduce the effects of water hammer. Long runs of piping invite the slamming and banging noise of a water hammer. If you cut out the straight sections and rework them with some offsets, you increase your odds of beating the banging noises.

LEAKS IN COPPER PIPING

Copper leaks are the most common type of leak found in the piping of water distribution systems. Copper tubing and pipe has been used for many years, and it is still quite popular for new installations.

While any experienced plumber knows how to solder joints on copper pipe, there are times when the job doesn't go by the book. A little water trapped in a copper pipe can make soldering a joint a hair-pulling experience if you don't know how to handle the situation. There are also those times when the fittings that must be soldered are located in places where the flame from a torch poses some potentially serious problems.

Finding leaks in copper pipe and tubing is not usually difficult. By the time plumbers get called in to fix a leak, someone usually knows where the leak is. There are times, though, when the location of the leak is not known.

Big leaks are easy to find; you can either see or hear them. It is the tiny leak that does its damage over an extended period of time that is difficult to put your finger on. With these leaks it is sometimes necessary to start at the evidence of the leak and work your way back to the origin of the water. This can mean cutting out walls and ceilings, but there are times when there just isn't any other way to pinpoint the problem.

Copper leaks often spray water in many directions. This also can make finding the leak difficult. If a solder joint weakens to the point that water can spray out of it, the water might travel quite a distance before it splashes down. These leaks are not hard to find when they are exposed, but if they are concealed, you can find yourself several feet away from the leak when you cut into a wall or ceiling that is showing water damage.

The good thing about spraying leaks is that you usually can hear them after you have made an access hole. This is a big advantage over trying to find a mysterious drainage leak. The spraying water will lead you to the leak in a hurry.

Copper pipe and the joints made on it can leak for a number of reasons. Pinholes in the pipe can occur from acidic water. The pipe can swell and split or blow out of a fitting if it has frozen. Stress can break joints loose, and sometimes joints that were never soldered properly will blow completely out of a fitting. Bad solder joints also can begin to drip slowly.

Fixing copper water lines is usually not a complex procedure, but what will you do if there is water in the lines? Water in the piping can be of two types; it can be standing water that is trapped, and it can be moving water that is leaking past a closed valve. In either case, the water makes it hard to get a good solder joint.

Inexperienced plumbers often don't know how to overcome the problem of water in the piping they need to solder. Some inexperienced plumbers will try to make a solder joint and be fooled by the actions of the solder. This type of situation will result in a new leak, but the leak might not show up immediately; let me explain.

When water is present in the area being soldered, several things can happen. If there is enough water in the pipe, the joint will not get hot enough to melt solder. This is frustrating, but at least the plumber is aware that the solder joint can't be made without some type of action being taken against the standing water.

In some cases, the pipe and fitting will get hot enough to allow solder to melt, but the fitting will not obtain a temperature suitable for a solid joint. When this happens, solder will melt and roll out around the fitting, but it is not being sucked into the fitting as it should be. To inexperienced eyes, this type of joint can look all right, but it's not.

When the water is turned back on, the defective joint might leak immediately, if you're lucky. If the joint leaks right away, the plumber knows the job is not done. Sometimes, however, these fouled joints will not leak immediately, and this means trouble.

When the weak joint doesn't leak during the initial inspection, it might be left and possibly concealed by a wall or ceiling repair. In time, and it usually won't be long, the bad joint will begin to leak. It might drip, or it might blow out. Either way, the plumber's insurance company won't be happy.

When the bad joint fails and is inspected by the next plumber, it will be obvious the joint was not made properly. There will not be evidence of solder deep in the fitting, because it was never there. This usually will be considered neglectful on the professional plumber's part.

A third way that water in the piping will drive an inexperienced plumber crazy is with the results of steam building up in the pipe. As the area around the joint is heated, the water will turn to steam. The steam will vent itself, usually through some portion of the fitting being soldered. This steam might escape without being seen. When this happens, solder runs around the fitting as it should, except that a void is created where the steam is blowing out. If the plumber can't see, hear, or sense the steam, the joint will look good. After the water is turned on, the joint no longer will look so good; it will leak and the process will have to be repeated.

Inexperienced plumbers will think they just had a leak because of poor soldering skills. They will go back through the same process and wind up with the same result. Until they figure out what is happening, and what to do about it, they will just spin their wheels trying to solder around the steam. There are ways to beat all of these problems, and I'm about to show them to you.

TRAPPED WATER

Trapped water is easier to overcome than moving water. Trapped water sometimes can be removed by bending down the cut ends of horizontal pipes. Opening fixtures above and below the work area often will remove trapped water from pipes. When the problem pipe is installed vertically, a regular drinking straw can be inserted into the pipe and the water blown out. If you have the equipment and time, you can use an air compressor to blow trapped water from pipes. When none of these options works, you have to get a bit more creative.

The easiest way to beat trapped water in pipes that just won't drain is the use of bleed fittings, which are cast fittings made with removable drain caps. Heating mechanics call them vent fittings, I call them bleed fittings, and many people just call them drain fittings.

Drain fittings are available as couplings and ells. The vent on a coupling is on the side of the fitting, at about the center point. Drain ells have their weep holes on the back of the ell, where the ell makes its turn.

If you install a drain fitting in close proximity to the trapped water, you can steam the water out of the pipe as you solder the fitting. Remove the cover cap and the little black seal that covers the drain opening. Make sure the drain opening is not pointing toward you.

As you heat the pipe and fitting, the trapped water will turn to steam and vent through the weep hole in the fitting. The hot water can spit out of the hole and the steam can come out fast and hot, which is why you don't want the hole pointed in your direction.

With the water and steam coming out of the vent hole, you can solder the joint with minimal problems. Unless you have a huge amount of water in the pipe, the soldering process shouldn't take long, and it should go about the same as it would if the water weren't in the pipe.

Once the joints have cooled, you can replace the black seal and the cap to finish your watertight joint. If you get caught without bleed fittings on your truck, a stop-and-waste valve can be substituted for the drain fitting. The little weep drain on the side of the valve will work in the same way described for the drain fittings. Make sure, however, that the valve is in its open position before you begin to solder.

Bread is an old standby for seasoned plumbers who are troubled by water in the pipes they are trying to solder. Inserting bread into the pipe will block the water long enough for a good joint to be soldered. When the water pressure is returned to the pipe, the bread will break down and come out of a faucet. This procedure works well, a few traps are involved with it.

I've used bread countless times to control water that was inhibiting my soldering. On occasion, I've regretted it. Sometimes I've created more problems for myself by using bread. I would recommend that you use drain fittings rather than bread, but I'll tell you what to look out for when bread is used, just in case you don't have a choice.

First, always remove the crust from the bread before stuffing the pipe. The crust is much more dense than the heart of the bread, and it doesn't dissolve as well. When bread has been used in a pipe, try to remove it through the spout of a bathtub. If you must get it out through a faucet, remove the aerator before you turn on the water. The broken-down bread will clog the screen of the aerator in the blink of an eye.

Don't flush a toilet to remove bread from the line. The bread particles might become lodged in the fill valve and cause more trouble. Avoid packing the bread in the pipes too tightly. I once used a pencil to push and pack bread into a pipe that contained moving water. The bread blocked the water and allowed me to make my solder joint, but it didn't come out after the water was turned on.

MOVING WATER

Moving water makes the job of soldering joints an adventure. Old valves don't always hold back water completely. There will be many times when a trickle, or more, of water will creep past the valve and make soldering a bad experience. When you have moving water in a pipe, you will have to use some special way to get the soldering job done. Bleed fittings will let you do your work if the water is not too abundant.

There is a special tool that I've seen advertised, but I can't remember its name, that is designed to made soldering wet pipes possible. As I recall, the tool has special fittings that are inserted into the pipe and expanded. The plug holds back the water while a valve is installed on the pipe. When the valve is soldered onto the end of the pipe, the tool is removed and the valve is closed, allowing you to work from the valve without the threat of water.

The tool is nifty, but not needed. You can do the same thing with the old standby, bread. Stuff the pipe with bread, and pack it if you have to. After the water is stopped, solder a gate valve onto the end of the pipe. Don't use a stop-and-waste valve, because you might need access to the bread in order to get it out.

After the valve is soldered properly and cooled, close it and turn on the water. Open the valve to let the bread out of the system; remember that water will be coming out right behind the bread. If the bread is packed too tightly for the water to come through, poke holes in it with a piece of wire, like a coat hanger. The bread will break up and come out of the pipe. After the pipe is clear, close the valve and go on about your business.

PB LEAKS

PB leaks are not common, except at connections that were not made properly. Because of its flexibility and durability, PB pipe rarely gives plumbers problems with leaks. It doesn't even usually burst under freezing conditions. There are, however, several times when the pipe is not put together properly, and this can result in leaks.

Bad crimps account for some leaks with PB pipe. If a crimping tool gets out of adjustment, and they do after extended use, the crimp ring might not be seated properly. If this is the case, the bad connection must be removed and a new one made.

It is not often that an insert fitting is not inserted far enough, but I've seen it happen. This type of problem is obvious and usually not difficult to repair. The replacement of the bad connection is all that is required.

Some plumbers are not familiar with PB pipe, and they sometimes use stainless steel clamps to hold their connections together. This rarely works for long, if at all. If you have a leak at a PB connection where a stainless steel clamp has been used, remove the clamp and crimp in a proper connection.

Compression ferrules account for most of the after-installation leaks in PB piping systems. Compression fittings can be used on PB pipe, but brass ferrules should be avoided.

Nylon ferrules are the proper type to use with compression fittings and PB pipe.

If you have a PB pipe or supply tube leaking around the point of connection in which a compression fitting has been used, inspect the ferrules. If brass ferrules have been used, replace them with nylon ferrules. The brass ferrules can cut into the PB pipe if the compression nut is tightened too much.

PE LEAKS

PE leaks are not much of a consideration when talking about water distribution systems. PE pipe is used frequently for water services, but its use is very limited in water distribution systems.

Cracked fittings and loose clamps are the most common causes of leaks in PE piping. The fittings can be replaced and the clamps can be tightened or replaced. On the occasions when PE pipe develops pinhole leaks, there are a few easy ways to fix the problem. The best way to deal with the situation is to cut out the bad section of the piping and replace it. When this isn't convenient or possible, you can use repair clamps.

Repair clamps can be used to patch small holes in all types of piping. You can buy the clamps or in some cases you make your own. In the case of PE pipe, rubber tape and a stainless steel pipe clamp will make a good repair on small holes. All you have to do is wrap rubber tape around the hole and clamp it into place.

CPVC LEAKS

CPVC leaks are common, and in my opinion, are a real pain to deal with. I've worked with CPVC off and on for about 20 years, and I've never liked it. The one house that I plumbed with CPVC was enough to turn me off from using it in future jobs. Since that first house, early in my career, my only association with CPVC has been in repairing it.

When you are troubleshooting a water-distribution system made from CPVC piping, there is a lot to look for. You have to keep your eyes open for little cracks in the pipe. Because of its makeup, CPVC will crack without a lot of provocation. These hairline cracks can be difficult to find.

In addition to cracks in the piping, you have to look for cracks in the threaded fittings. Fittings that have been cross-threaded are also common in CPVC systems. And to top it off, CPVC pipe that is not supported properly can vibrate itself to the point of weakening joints that will leak.

When you find a leak in CPVC pipe or fittings, don't attempt to reglue the fitting. Cut it out and make a good connection with new materials. Attempting to patch old CPVC more often than not will result in frustration and continued problems.

You don't have to solder CPVC, but water in the line still will mess up your new joints. The water will seep into the cement and create a void that will leak. I know some plumbers just cut the pipe, slap a little glue on it and stick it together with its fitting, but I don't think you should operate this way. CPVC is so finicky that I believe you should make your connections by the book.

Cut the ends of the pipe squarely and rough them up with some sandpaper. Using a cleaner and a primer prior to applying the cement is also a good idea. When you glue and connect the pipe with the fitting, turn it, if you can, to make sure the cement gets good coverage.

Don't turn the pipe loose right away. If you release your grip too soon, the pipe is likely to push out of the fitting to some extent, weakening the joint. Hold the connection in place as long as your patience will allow. CPVC takes a long time to set up, so don't move it or turn on the water too soon. New joints should set for at least an hour before being subjected to water pressure.

GALVANIZED STEEL AND BRASS

Galvanized steel and brass pipe isn't used much for water distribution these days, but it still does exist in some buildings. Leaking threads are a common problem with old piping. The threads are the weakest point in the piping, and they tend to be the first part of the system to deteriorate. Acidic and corrosive water can work on these threads to make them leak prematurely.

When you have a leak at the threads going into a fitting, a repair clamp is not going to help you. Under these conditions, the leaking section of pipe must be removed and replaced. This can become quite a job. What starts out as a single leak at one fitting quickly can become a plumber's nightmare. As you cut and turn on the bad section of pipe, you are likely to weaken or break the threads at some other connection. This chain reaction can go on and on until you practically have to replace an entire section of the water-distribution system.

Many young plumbers see leaks at threads and believe they can correct the problem by tightening the pipe. In old piping, this is only a pipe dream, no pun intended. Twisting the old pipe tighter is not likely to solve problem, but it might worsen it.

If you have leaks at the threads, you might as well come to terms with the fact that the section is going to have to be replaced. Unlike drain pipe, you can't use rubber couplings to put water piping back together. You have to remove sections of the pipe until you can get to good threads. Then you can convert to some type of modern piping to replace the bad section. Be prepared to have to replace more than you plan to.

Pinhole leaks in steel and brass pipes can be repaired with repair clamps. Rubber tape and a pipe clamp will work, but a real repair clamp is best. Brass pipe imitates copper to the point that some plumbers think they are working with copper. In fact, I hate to admit it, but I've even mistaken brass for copper. It is easy to do in poor lighting.

Brass water pipe can be cut with copper tubing cutters. The cut takes a little longer, but unless you suspect the pipe is brass, it can fool you. When you try to get a copper fitting on the end of the pipe, however, you'll know something is wrong; the fitting won't fit. If you think you're working with copper and can't figure out why your fittings won't work, look for an existing joint and see if it is a threaded connection. If it is, you're probably dealing with brass pipe.

If the fittings are soldered and your standard fittings won't fit the pipe, somebody plumbed the job with refrigeration tubing. When this is the case, you're going to need refrigeration fittings to get the job done. This one tip can save you a lot of frustration.

COMPRESSION LEAKS

Compression leaks are frequent problems with water distribution systems. The compression fittings usually are easy to fix. All that usually is required is the tightening of the compression nut. These leaks happen when the nut was not tight to begin with or when the pipe has vibrated enough to loosen the fitting. It is also possible that the connection was hit with something and loosened.

It might be necessary to remove the existing ferrule and replace it with a new one. This usually requires you to replace the section of pipe or tubing to which the old ferrule is attached. In any event, correcting compression leaks is one of the easier jobs that plumbers face.

FROZEN PIPES

Frozen pipes can be a plumber's bread-and-butter money in the winter. They also can be troublesome to work with, hard to find, difficult to fix, and are potentially dangerous if they are not worked with in the proper manner.

Steel pipe, and sometimes copper pipe, can be thawed with the use of a welding machine or a special pipe-thawing machine. If the leads are attached at opposite ends of the pipe, with the frozen section in the middle, the machines can produce enough warmth to thaw the pipe. I've seen plumbers do this many times over the years, but the process can be dangerous; fires can be started.

When I was a plumber in Virginia, I used to get numerous calls for thawing and repairing frozen pipes. The job usually entailed only a single pipe, often that of an outside hose bibb. A heat gun, a hair dryer, or a torch made quick work of thawing these individual sections of pipe.

In Maine, the freeze-ups often are considerably larger than the ones I dealt with in Virginia. Here it is not uncommon for entire systems to freeze. To thaw these pipes, I usually use a portable heater. A large space heater, such as those used on construction sites, will bring a building up to a thawing temperature quickly and safely. After the building is warm and the pipes are thawed, necessary repairs can be made.

If you haven't worked with many frozen pipes, you might not be aware of the way the freezing action can swell copper pipe. Even though the split resulting from the freeze-up is in one place, the pipe might be swollen for several feet on either side of the split. This makes it impossible to get a fitting on the end of the swollen pipe. To handle this dilemma, you will have to keep moving back on the pipe until you find a piece that has not swollen from the cold.

Common sense and field experience are a plumber's two best tools when troubleshooting. The more calls you handle, the more you learn. As your experience grows, the time it takes you to identify and correct a problem decreases. This chapter should shave a considerable amount of time off your learning curve.

CHAPTER 23
TROUBLESHOOTING DWV SYSTEMS

Troubleshooting and repairing DWV systems is work that can age plumbers before their time. Finding the causes of problems with DWV systems can be extremely frustrating. Even after a problem is found, figuring out how to get to it or fix it can be enough to make you want to pull out your hair. Service work is a job that many plumbers enjoy, but dealing with DWV systems can be all it takes to make a service plumber consider going into new construction.

Is working with existing DWV systems really so bad? It can be. Some of the potential problems that plumbers have to identify and solve are considerably aggravating. Finding a leak in a water-distribution system is usually pretty easy. Since the leak is under constant pressure, it leaks all the time. This is not the case with a drain. There are a number of situations that can make troubleshooting and repairing DWV systems somewhat less than fun.

One reason so many plumbers get frustrated with DWV systems is that they have limited experience. Like most things in life, troubleshooting gets easier as experience is gained. I can't accelerate your career and give you ten years of field experience. But I can give you the benefit of my experience, which spans more than 20 years as a plumber. When you read this chapter, keep in mind that it is not based on research or theory. The information contained here is a direct result of my hands-on experience. Let's get into it and see how much you can learn before your next service call.

SLOW DRAINS

Slow drains, drains that empty slowly, are one of the most frequent complaints with the DWV system. Slow-moving drains can be the result of a number of possible problems. The type of drain that has the problem has a lot to do with how you troubleshoot it. Knowing the type of pipe with which you are working can influence the troubleshooting steps you take. Let me give you a brief rundown about the various types of pipes and their most common problems.

COPPER

Copper drainage systems are usually not much trouble. The copper provides many years of good service and it is not usually a contributor to stoppages. In fact, in 20 years I can't recall a time when a copper DWV system was at fault for any problems. Oh sure, they get stopped up, but all drains can. In general, copper is a good aboveground drainage material that is very dependable. It is possible for copper drains to build up scale and slow down, but this is rare. Most problems with copper systems will be the result of a user putting something in the drain that shouldn't be in there.

PLASTIC PIPE

Plastic pipe, as you probably know, is the most common pipe used for DWV systems today. It may be ABS or PVC, but either is good, and both are dependable. PVC is more brittle than ABS and is more likely to be cracked or broken, but under normal conditions, neither pipe gives much cause for trouble.

I have cut out sections of plastic pipe where a layer of crud was building up on the pipe walls. This type of situation could lead to a slow drain, but I've never seen a plastic pipe with a gradual buildup sufficient to render the pipe useless or even unsatisfactory. Like copper, plastic won't usually slow down or stop up unless something is put in the drain that shouldn't be there.

GALVANIZED PIPE

Galvanized pipe has to be one of the worst materials ever used in plumbing. This pipe is famous for its ability to rust and catch every imaginable thing that goes down it. Grease and hair are especially common stoppages to find in galvanized drains.

As galvanized pipes age, they begin to develop build-ups that slowly restrict their openings. Eventually, the build-up will block the pipe completely. Snaking the drain will punch holes in the obstructions, but the stoppage will reoccur in a matter of weeks or months.

Drains that go down very slowly and get worse and worse as time goes on are likely to be galvanized drains. Of all the various types of drain pipes in use, galvanized drains are the most likely to produce symptoms of growing slower and slower over a period of time.

Old galvanized drains also are known to rust out at their threads, causing leaks at their joints. The best solution to galvanized drain problems is the replacement of the old piping with more modern plumbing materials. Rubber couplings make it easy to adapt new drainage materials to the old piping.

CAST-IRON PIPE

Cast-iron pipe has been used for DWV systems for a very long time. This pipe material has proved itself dependable. There are, however, some drawbacks to cast-iron drains. The interior of cast-iron pipe can be rough, especially as it starts to rust. These rough surfaces catch a lot of debris as it is being drained down the pipes. This leads to stoppages, but they are not as bad as the ones found in galvanized pipes and they don't recur as quickly.

One of the biggest problems with cast-iron drainage systems is the removal of cleanout plugs. The old brass plugs that have been in the cleanouts for years can be next to impossible to remove.

I've had occasions when a 24-inch pipe wrench with a two-foot cheater bar wouldn't turn the cleanout plug. If you've done much work with pipe wrenches and cheater bars, you know the kind of leverage that is being applied under these conditions. Even so, some cleanout plugs just won't budge, and some of them are in locations where you just can't get enough leverage on them to make the turn.

When this happens, your options are limited. Some plumbers drill out the brass plugs. Others use cold chisels and hammers to knock out the plugs. And others just cut the pipe and put it back together with a rubber coupling to make cleaning the drain the next time easier. I usually opt for the latter.

While cast-iron pipe will last for decades, it can be susceptible to slow-downs. The use of an electric snake that is equipped with a cutting head can scour the pipe walls and solve the problem. New stoppages should not occur for years, unless there is something more wrong with the pipe, such as inadequate grading.

BATHTUB DRAINS

Bathtub drains can be running slowly for several reasons. The waste and overflow might need to be adjusted. There might be an accumulation of hair in the tub waste. The trap could be obstructed by foreign objects. The drain's vent might be obstructed. It's possible that the drain is not vented at all. These reasons, and a few others, can make a bathtub drain slowly.

If you are having trouble with a newly installed bathtub, the problem is most likely in the tub waste and overflow. There are several types of waste and overflows. The type where you flip a lever to hold and release water in the tub might need to be adjusted. The length of the rod on these units is adjustable for tubs with various heights. If the rod has been set at a length too long for the tub, the fixture might drain slowly.

To determine if the tub waste needs to be adjusted, remove the cover from the overflow pipe. Remove the assembly with the rod and plunger on it. Next, remove the pop-up plug from the tub's drain, if it has one. Block the tub's drain with a rag and fill the tub with water. Remove the rag or stopper and see if the tub drains properly. If it does, you need to adjust the tub waste. If it does not, your problem is farther along in the drainage system.

Adjusting the tub waste is simple. The rod that you have removed will have threads and nuts on it. The housing it screws into usually will be marked with various numbers. These numbers represent the height of the tub. If you are adjusting the waste to allow water to flow more freely, you will want to shorten the rod. To do this, loosen the set-nuts and screw the rod farther into the housing. You might have to achieve the proper setting through a trial-and-error method. Just keep adjusting the waste rod until the tub drains satisfactorily. You have proved that the problem is in the tub waste; all you have to do is reach the proper setting and the tub drain will work fine.

If the tub waste does not have a trip lever, it most likely is controlled by a stopper plug. These plugs usually are depressed to open and close the tub drain. At times, the rubber gasket on the stopper impedes the flow of the draining water. When this is the case, turn the stopper counterclockwise for a few turns. With continued adjustments, you should reach a point where the drain functions properly.

Bathtub traps are usually P traps. If you have determined that your problem is not associated with the tub waste and overflow, the trap is the next logical place to look. With new installations, there is rarely a problem with the trap, but there could be. You can run a small spring snake down the overflow pipe of the tub and into the trap. If the snake meets no resistance, your trap should not be causing your problem.

If you are dealing with new plumbing, there is no reason to have trouble with the bathtub's drainpipe. If you connected to an old piece of galvanized pipe to install your new tub or are working on existing plumbing, however, the drain pipe could very well be your problem. You can continue to put your snake through the trap and into the drainpipe. For some blockages, a hand-snake will clear the pipe. But if you have a strong blockage in a galvanized pipe, you will need an electric drain cleaner to eliminate the clog.

When galvanized pipes become blocked, the stoppage often consumes the entire pipe. Small drain snakes will only punch holes in the clog. The drain will work better for a while, but then it will stop up again.

To remove the typical stoppage from a galvanized pipe, you will need an electric drain cleaner with a cutting head. As the head rotates its way through the pipe, it cuts away the debris from the sides of the pipe.

Unfortunately, even a thorough cutting of the blockage is often only a temporary repair. The rusted interior of galvanized pipe will not waste time in catching grease, hair, and other particles to create a new clog.

SHOWER DRAINS

Showers do not have a waste and overflow arrangement to adjust. When shower drains move slowly, you know your problem is in the trap or the drain line. The same methods described for clearing blockages in tub traps and drains will apply to showers.

SINK DRAINS

If you have a sink that is draining slowly, you can assume the problem is in the trap or the drainage piping. Unlike tub and shower traps, the traps for most sinks are readily accessible. These traps frequently are joined with nuts that allow the trap to be disassembled. When a sink drains slowly, disassemble the trap and inspect for any obstruction that might cause the slow draining of the sink.

If the trap is clear, run a snake down the drain pipe. It is unusual for a sink drain to operate poorly following a new installation, but sink drains in older homes often fill with grease and slow down. In a lavatory, you might have a problem with a pop-up assembly. The pop-up might need to be adjusted to allow water to drain faster. It is also possible that you have a hair clog on the pop-up rod. To test these possibilities, you must remove the pop-up plug.

If you have a hair clog, you will see it when you look down the drain assembly. To determine if the pop-up needs adjusting, place a rag or removable stopper over the lavatory drain. Fill the lavatory bowl to capacity and remove the rag or stopper. If the water drains well under these conditions, the pop-up assembly needs to be adjusted. If the water continues to leave the bowl slowly, your problem is in the trap or drainage pipe.

TOILET DRAINS

Toilets with a lazy flush can be affected by a number of afflictions. The wax seal might not be seated properly. The flush-valve might not be operating properly, or the flush holes might be clogged. If all of these inspections pass muster, the problem is in the drainage pipe.

FLUSH HOLES

When you have a toilet that is swirling and flushing slowly, check the flush holes. Even if a toilet is new, don't rule out the flush valve or the flush holes.

The flush holes are located inside the toilet bowl, under the rim. When you flush the toilet, water should wash down the sides of the toilet bowl. If it doesn't, check the flush holes. The holes may be plugged with minerals and sediment. It also is possible that the holes were not formed properly during the making of the toilet. To check the flush holes, you will need a stiff piece of wire. A wire coat hanger will work fine for checking the holes.

You can locate the holes with a mirror or by running your hand around the underside of the toilet's rim.

Push the wire into the holes. If the wire goes in easily, you can rule out the flush holes. If you hit resistance, work the wire back and forth. If white crystallized particles fall out, you have a mineral build-up. If the wire will not go in at all, you might have a defective toilet bowl.

TANK PARTS

If flush holes are not at fault, check out the tank parts. Inspect the water level in the tank. If the water is at the fill-line, etched into the tank, you have adequate water for a normal flush.

While you have the cover off the tank, flush the toilet. Watch the flush valve when the tank empties. The flapper or tank ball should stay up long enough to allow most of the water to leave the tank. If the flapper or tank ball closes the flush valve early, there will not be enough water for a satisfactory flush. When this is the case, you must adjust the interior parts of the toilet tank.

WAX RING

If all the aspects of the toilet pass your inspection, you must look to the wax ring or the drain pipe. Wax rings that are not installed properly can cause a toilet to flush slowly. In new installations, wax rings might be the primary cause of lazy flushes in the toilet. The problem occurs when the toilet is set on the flange. If the toilet bowl is not positioned directly over the wax ring, the wax might be compressed and spread out over the drain pipe. Since you cannot see the wax seal when the toilet is set, the bad seal goes unnoticed.

As the toilet is used, toilet paper catches on the wax extending over the drain. After several flushes, the build-up can hinder the flushing action of the toilet. To confirm the condition of the wax ring, you must remove the toilet from its flange. As an alternative to pulling the toilet, you can use a closet auger to check the toilet's trap and drain. By running the auger through the toilet, you might retrieve wax on the end of the auger. If you do, you can bet the wax ring did not seat properly during the installation.

While the closet auger might reveal evidence of a faulty wax seal, the only way to be certain of the seal is to remove the toilet. If the hole at the bottom of the toilet has wax spread over it, you have found your problem. Scrape the old wax off and reset the toilet using a new wax ring. Be diligent to achieve a good seal when you reset the toilet.

DRAINPIPES

When you have completed all of these troubleshooting procedures and still have a slow-flushing toilet, the problem is in the drain pipes. You will have to use a snake to locate and clear the blockage. During construction and remodeling, some strange objects find their way into open drain pipes. This is especially true of the drains for water closets. Since these pipes are at floor level, it is easy for foreign objects to fall down the drain.

In addition to objects accidentally falling into the drains, sometimes someone will purposefully stick something in the pipe. When a construction job is in progress, all sorts of people have access to the job. Anyone from juveniles to irresponsible workers might stuff undesirable objects in the pipes. Rental tenants and children are often at fault for putting foreign objects in toilet bowls. During my career, I have found beer cans, nails, rubber balls, blocks of wood, and countless other objects in pipes. When these objects get in the pipes, they don't go down the drain well. Sometimes the pipe has to be cut to remove the blockage. A snake will not cut through a block of wood.

To determine if a drain pipe is stopped up, you can use a snake. Feed the snake into the drain until you come into contact with the blockage. For small drains, a spring-snake will negotiate the turns in the pipe better than a flat-tape snake. When you connect with the blockage, attempt to push it through the pipe with the snake. If the blockage will not clear, you will have to try an electric drain cleaner or cut the pipe.

For drains with a diameter of two inches or less, you will want to use a small electric drain cleaner. These units are handheld and often resemble an electric drill with a drum housing attached. For drains with diameters of three inches or more, you will want a larger machine. These large machines are potentially dangerous. When a clog is encountered, these powerful machines can cause severe damage to your body parts. The cable can kink and do serious damage to your appendages. Before using an electric drain cleaner, observe proper safety precautions.

Electric drain cleaners can be rented from most rental centers. When you rent the machine, request full instructions about how to safely operate the machine. In general, wear gloves, don't wear loose or dangling clothes, and never allow excess cable to build up outside of the drain. Slack cable can become very dangerous when you make contact with a solid blockage. Choose a flexible spring head for the drain cleaner. These heads are easier to feed down the pipe and are less likely to cause uncontrollable torque than solid cutting heads.

FROZEN DRAINS

Drain pipes should not freeze, because they are not supposed to hold water. In the drainage system, only the traps should hold water at all times. Even though drains shouldn't freeze, they sometimes do. If drains are not working and resist the passage of a snake, you could have a frozen pipe on your hands. Obviously, this will only be the case in freezing weather.

It is difficult to diagnose a frozen drain, unless you can see the pipe. When drains are run under a home, in a basement or crawlspace, you have access to the pipes. By putting your hands on the drains, you should be able to tell if they are frozen. When you cannot get to the pipes to touch them, you will have to make assumptions. If the pipe might be frozen, you can try a couple of different remedies.

If the pipe is draining, but doing so slowly, pouring very hot water down the pipe might melt the ice. If you have access to the pipe, a heat gun or a hand-held hair dryer might be sufficient to melt the ice. When the ice is far down the pipe, you can try filling the pipe with hot water. By attaching a garden hose to the hot water connection where a washing machine is, you can run hot water through the hose.

Keep an eye on your water hose. The hot water might cause the hose to overheat. Feed the hose into the drain that you suspect is frozen. Push it as far into the pipe as you reasonably can. When the hose reaches a stopping point, turn on the hot water. The hot water will force the water standing in the pipe out of the open end of the drain where the hose is inserted. As the hot water melts the ice, you can push the hose farther into the pipe. Continue this cycle until the drain is no longer clogged.

OVERLOADED DRAINS

When new plumbing has been added to an old system, it is possible the old drains are not adequate for the increased load of drainage. For example, the existing sewer might work fine when one bathtub is draining, but might run slowly with a second tub draining at the same time. The plumbing code is designed to prevent this type of problem. But things don't always work out the way they should.

If you are able to perform a visual inspection of all the drains, you can determine if they are large enough to handle a specified number of fixture units. The plumbing code assigns a fixture-unit rating to each fixture. Doing simple math will tell you if the pipe should work properly. Even when the numbers work, the drain might not. How could this be? Well, the drain might not be working to its full capacity for many reasons.

When roots grow into the sewer, the open diameter of the pipe is reduced. A sewer that is partially crushed will reduce the flow of water in the pipe. If the sewer is not installed with proper pitch, the pipe will not carry the waste as quickly as it should. All of these reasons could cause an old sewer to resist the demands of additional plumbing. These defects are hard to detect. You might have to expose all of the old piping to discover the cause for the slow-moving drains. In plumbing, you can never be sure of anything until you can see and verify all aspects of the system.

MATERIAL MISTAKES

Material mistakes account for some strange problems in the DWV systems of buildings. Good plumbers tend to think along the lines of the plumbing code when they create a mental picture of a plumbing problem. They basically assume the job was done to code, but this is sometimes far from the truth.

Whether you are troubleshooting a DWV system or any other plumbing problem, don't assume anything. I know it can be hard to think outside of the code, but there are times when you must. It is kind of like how police officers have to think like criminals to catch the bad guys. You have to think like an irresponsible plumber or a rank amateur sometimes to figure out plumbing problems. Let me give you a few examples of what I'm talking about.

THE SAND TRAP

This first example aptly could be dubbed the sand trap. The story has to do with a shower drain. A friend of mine who owns his own plumbing business was in my office recently. We were discussing a project that we will be working on together, and he told me this story.

The plumber was called in to work on a shower in the basement bathroom. During the course of his work, he noticed that something was not quite right with the drain in the shower. He removed the strainer plate for a closer look and was amazed at what he saw.

Looking through the drain of the shower, he saw sand. His first thought was that the trap had somehow become filled with sand. During his work with the drain, he discovered that it turned freely in the shower base. With a little more investigation, the plumber found that the trap had not been filled with sand. It turns out that the sand was the trap.

Whoever installed the shower never bothered to connect it to a trap or a drain. The fixture simply emptied its waste water into the sand beneath the concrete slab. This is clearly a situation few plumbers would ever imagine.

THE BUILDING DRAIN THAT WOULDN'T HOLD WATER

This next case history is about the building drain that wouldn't hold water. The job began when the customer called my office and complained about strong odors in the corner bedroom of the home. It was summer, and I suspected that the customer's problem was an overworked septic field, but I couldn't be sure without an on-site inspection.

I went to the house and met with the homeowner. She started to explain her problem to me, but I could already smell the odor and I hadn't even gone into the house.

The house was on a pier foundation. I walked around the left side of the home and saw two young children playing in the yard. They wore shorts and no shoes, and were splashing around in a puddle. This seemed odd since it hadn't rained in days. The closer I got to the back corner of the house and the children, the worse the smell became. I was starting to get a feeling that made me uncomfortable.

A bathroom recently had been added to the house, near the corner bedroom. Whoever installed the plumbing did so in a way that I'd never seen before. Looking under the house, I could see the problem, but I could hardly believe what I was seeing. The drainage hanging from the floor joists was piped with schedule-40 PVC. The PVC dropped straight down to the sewer that ran from under the house to the septic tank.

The sewer pipe was creating the problem. You see, someone installed slotted drainpipe for the sewer between the house and the septic tank. Much of the sewer had never been buried in the ground. The holes were on top, but the septic tank was full, and raw sewage was seeping out of the slotted sewer pipe. Sewage was puddled under the house, under the bedroom window, and, you guessed it, where the children were playing.

I asked the homeowner who had installed the plumbing, and she claimed the builder who had built the home some years back had done the work. Can you envision anyone using slotted pipe for a sewer? I couldn't either.

DUCT TAPE

Duct tape is a plumber's friend, and it is capable of many good uses, but it is not meant to be used as a repair clamp on a drainage line.

Many years ago, I was called to a house by a homeowner who said he could hear water dripping under his home after flushing the toilet. He thought perhaps the wax seal under his toilet had deteriorated. The house was built on a crawl space foundation. My first step in troubleshooting the situation was to flush the toilet. No water seeped out around the base of the toilet, but I could hear water dripping under the home, just as the homeowner had described.

I took my flashlight and went under the home to inspect the problem. As I worked my way back to the bathroom area, I got a strong whiff of sewage. Once I was close enough to the problem to see the piping and the ground, I noticed a rather large puddle on the ground. Shining my light on the pipe, I could see the remains of duct tape wrapped around the cast-iron drain. After a closer inspection, I found that the cast-iron pipe had a large hole in the side of it. The pipe appeared to have been hit with a hammer at sometime in the past. Apparently, whoever attempted to repair the hole used duct tape to seal the pipe. Maybe they had done this as a temporary measure until they could get materials to do the job right, I don't know.

In any event, the duct tape had long since lost its ability to retain the contents of the pipe, and raw sewage was blowing out the side of the pipe every time the toilet was flushed. I cut out the damaged pipe and replaced it with a section of ABS and some rubber repair couplings. The repair wasn't a big job, but it was a messy one.

When jobs are not installed with the proper materials, your troubleshooting skills are put to the ultimate test. As plumbers, we are trained to look for normal problems, not problems created by someone's use of illegal materials or installation methods.

If a drain from a lavatory is piped with one-inch PE pipe, as some I've found have been, it is not difficult to understand why the pipe is stopped up or why a snake is so difficult to feed into the drain. Until you see the pipe, it is natural to assume it is of a legal size and material. You cannot, however, assume anything when you're troubleshooting plumbing.

VENT PROBLEMS

Most vent problems will appear as a drainage problem. With the exception of escaping sewer gas, vent failures will affect the drainage system. When a vent is obstructed, drains will not drain as quickly as they should. This is the most common fault found in the vent system. Vents provide air to the drains, enabling the drains to flow freely. When vents become clogged, the drains don't work as well as they should. If there is a break in the vent piping, sewer gas can enter the home. The unpleasant odor associated with sewer gas will be your best indicator of a broken vent of a dry trap. Usually, these two potential failures are the only ones encountered with the venting system.

SEWER GAS

Sewer gas is potentially dangerous. It can be harmful to your health when inhaled, and it is explosive when contained in unvented conditions. Sewer gas is essentially methane gas. The gas is formed in septic tanks and sewers. When the traps and vents in your home are working properly, you will not notice any sewer gas. If there is a defect in these systems, you might notice an unpleasant smell. Some people fail to identify the odor and develop health problems because of the escaping sewer gas. In small quantities, sewer gas offers minimal risk, but in large quantities, especially in confined conditions, the gas is potentially fatal.

Sewer gas can invade a home from many sources. When a trap on a fixture goes dry, sewer gas can infiltrate a dwelling. If there are holes or bad connections in vent piping, sewer gas can escape. If a vent is terminating near a window, door, or ventilating opening, sewer gas can be pulled into the home through the opening.

CLOGGED VENTS

If vents become obstructed, drains will not drain well. When vents become clogged, leaves are a prime suspect for the clog. If leaves fall into the vent pipe, they can become lodged in fittings and can build up. With enough leaves, the vent can become useless.

Occasionally, a bird will build a nest on, or in, a vent pipe. The nest blocks the flow of air and disables the vent. These two situations do not occur often, but if vents fail, check for these types of stoppages in the pipe.

FROZEN VENTS

In cold climates, vent pipes sometimes freeze. The moisture accumulated in a vent pipe will begin to freeze on the walls of the pipe. If the temperature remains cold for an extended period, the ice will build up until it blocks the pipe. Cast- iron and galvanized pipes are more likely to freeze than plastic pipe. Unless the temperatures are brutal for a long time, it is unlikely that vents will freeze to an extent to hamper the operation of a plumbing system.

Persistence and patience are both important qualities when troubleshooting DWV systems. Look for the unexpected. Never underestimate how far from compliance with the plumbing code a plumbing system may be. It often is helpful to have a helper along when you are looking for problems with DWV systems. One of you might need to be upstairs flushing a toilet while the other one is under the home watching the piping. A planned troubleshooting approach is your best weapon against defects in DWV systems.

CHAPTER 24
STANDARD FIXTURE LAYOUT

Standard fixture layouts are dictated by local plumbing codes, which require certain amounts of space to be provided in front of and beside plumbing fixtures. The rules for standard fixtures are different than those used to control the installation of handicap fixtures. We will use this chapter to cover the essentials of standard fixtures and address the topic of handicap fixtures in the next chapter. For now, just concentrate on typical fixture installations when you review the information in this chapter. Before we get into specific details, I want to remind you to consult your local plumbing code for requirements specific to your region. The numbers I give you here are based on code requirements, but they may not be from the code that is enforced in your area.

If you work mostly with new construction, you probably work from blueprints. When this is the case, fixture locations are usually indicated and approved before a job is started. But remodeling jobs can require plumbers to make on-site determinations for fixture placement. A contractor might ask you to provide spacing requirements for small jobs. Knowing how to do this is important. For example, if a builder showed you a sketch, like

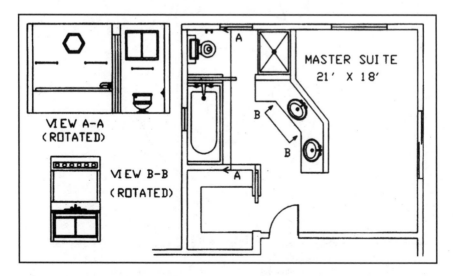

FIGURE 24.1 A typical bathroom layout. *(Courtesy of McGraw-Hill.)*

the one in figure 24.1, would you be able to assign numbers to the areas around the fixtures? How wide would the compartment where the toilet is housed be required to be? The answer is 30 inches. This is common knowledge for many plumbers, and code books define the distance. So even if you don't know the spacing requirements off the top of your head, you can always consult your local code for the answers.

A general rule for toilets is that there must be at least 15 inches of clear space on either side of the center of the drain for the toilet. As illustrated in figure 24.2, this equates to a total space of 30 inches. Now how much clearance is needed in front of a toilet? The normal answer is 18 inches, as shown in figure 24.3. Some bathrooms are small. This can create a problem for plumbers, especially when remodeling the bathroom with new fixtures or possibly different types of fixtures. Getting your rough-in for the fixtures right is crucial to the job. If you install a drain for a toilet and find out when you go to set fixtures that there is inadequate space for the toilet to comply with the plumbing code, there could be a lot of work and expense required to correct the situation.

When you are laying out plumbing fixtures, you should concentrate on what you are doing. Get to know your code requirements and check the fixture placement in all directions. Figure 24.4 shows how a legal layout might look. In contrast, figure 24.5 shows what would result in an illegal layout. Notice that the distance from the edge of the vanity is only 12 inches from the center of the toilet. To meet code, the distance must be at

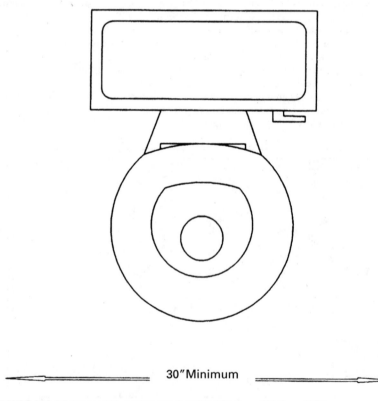

30″ Minimum

FIGURE 24.2 Minimum width requirements for toilet. *(Courtesy of McGraw-Hill.)*

least 15 inches. A problem like this might be avoided by using a smaller vanity. If the potential problem is caught on paper before pipes are installed, it is much easier and less costly to correct.

Another consideration when setting fixtures is the overall alignment. The plumbing codes not only require certain defined standards, they also deal with topics such as workmanship. This means that a job could be rejected if the fixtures are installed in a sloppy manner. Figure 24.6 shows a toilet where the flush tank is not installed with equal distance from the back wall. A proper installation would have the toilet tank set evenly, with equal distance from the back wall, as is indicated in figure 24.7.

CLEARANCES RELATED TO WATER CLOSETS

Let's talk about clearances related to water closets. There's not a lot to go over, so this moves along quickly. Remember that we are talking about standard plumbing fixtures here, not handicap fixtures. The minimum distance required from the center of a toilet drain to any obstruction on either side is 15 inches. Measuring from the front edge of a toilet to the nearest obstruction must prove a minimum of 18 inches of clear space. When toilets are installed in privacy stalls, you must make sure that the compartments are at least

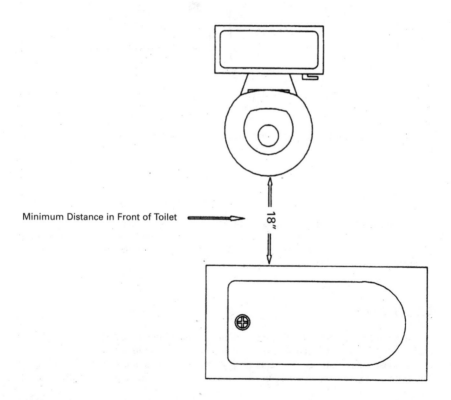

FIGURE 24.3 Minimum distance in front of toilet. *(Courtesy of McGraw-Hill.)*

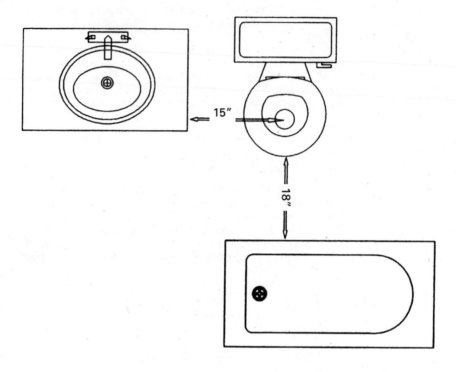

FIGURE 24.4 Minimum distance for legal layout. *(Courtesy of McGraw-Hill.)*

30 inches wide and at least 60 inches deep. That's all there is to a typical toilet layout (Fig. 24.7b).

URINALS

Urinals must have a minimum distance of 15 inches from the center of the drain to the nearest obstruction to either side. If multiple urinals are mounted side by side, there must be a minimum of 30 inches between the two urinal drains. The required clearance in front of a urinal is 18 inches (Fig. 24.8).

LAVATORIES

Lavatories are not affected by side measurements unless other types of plumbing fixtures are involved. The minimum distance in front of a lavatory should not be less than 18 inches. Obviously, minimum requirements are just that, minimums. It is best when more space can be dedicated to a bathroom in order to make the fixtures more user-friendly.

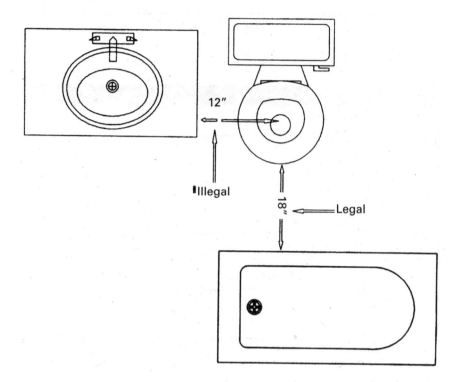

FIGURE 24.5 Illegal fixture spacing. *(Courtesy of McGraw-Hill.)*

KEEPING THE NUMBERS STRAIGHT

Keeping the numbers straight for standard plumbing fixtures doesn't require a lot of brain space. There are very few numbers to commit to memory. The process gets somewhat more complicated when you are dealing with handicap or "accessible" fixtures. We've finished discussing standard fixtures, so let's move on to the next chapter and learn about handicap fixture placement.

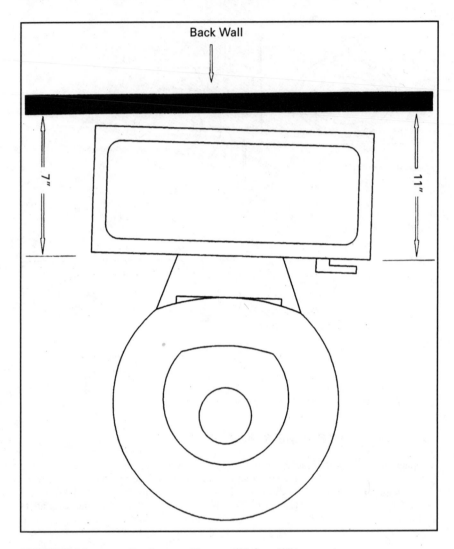

FIGURE 24.6 Improper toilet alignment. *(Courtesy of McGraw-Hill.)*

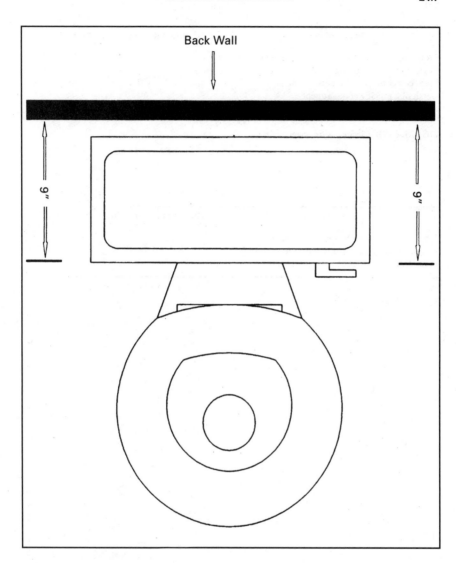

FIGURE 24.7A Proper toilet alignment. *(Courtesy of McGraw-Hill.)*

Measurement	Minimum distance (in inches)
From center of drain to any object on either side	15
From front of fixture to any object in front of it	18
From center to center of adjacent units	30
Depth of a privacy compartment	60

FIGURE 24.7B Clearances for water closets.

Measurement	Minimum distance (in inches)
From center of drain to any object on either side	15
From front of fixture to any object in front of it	18
From center to center of adjacent urinals	30

FIGURE 24.8 Clearances for urinals.

CHAPTER 25
HANDICAP FIXTURE LAYOUT

Layouts for handicap plumbing fixtures require more space than would be needed for standard plumbing fixtures. When you are planning the installation of accessible fixtures, you must take many factors into consideration. There are regulations pertaining to door widths, compartment sizes, fixture locations, and so forth. These rules and regulations generally are provided in local plumbing codes. It's not necessary for you to commit all the measurements to memory, but you need to be aware of them and know where to find the figures when they are needed for design issues.

When you are dealing with a building that requires the installation of handicap plumbing fixtures, you find yourself spread between the local building code and the local plumbing code. The two codes overlap when it comes to handicap facilities. Are you, as a plumber, responsible for the carpentry work? It depends on how you look at it. You probably have no responsibility to the width of a door used for access to a bathroom where handicap fixtures will be installed. But the width of a toilet compartment will affect you and your work. Most trades work well together on most jobs, but this is not always the case. If you know that rough framing done by a carpentry crew is going to prohibit you from installing fixtures with proper placement, you should talk to someone about the impending problem. Whether you talk to the carpenters, a carpentry foreman, a job superintendent, or a general contractor, you should raise the question of what you perceive to be a problem with the framing work. The quicker potential problems can be caught, the easier it will be to correct them.

Who is responsible for the installation of grab bars? It could be the plumbers or the carpenters. This is an issue that must be addressed before a bid is given for a job. Grab bars are not inexpensive, so don't make a mistake by omitting them from a bid price where the person you are bidding the job for expects you to include the bars and their installation in your bid. There's something else to consider on this issue. If you are responsible for the grab bars, you are also responsible for installing proper supports in the framed walls during your rough-in work. Some type of backing, such as a length of framing lumber, must be installed in the wall cavity where a grab bar will be installed. Without the backing, the grab bars will not be solid. Finding out that there is no solid support to secure a grab bar after a job has finished wall coverings is going to be a real problem. I've seen many jobs where backing wasn't installed for wall-hung lavatories and grab bars. This is an expensive and embarrassing mistake.

4094 Atlas Elongated Rim
• 18″ rim height handicapped
• 12″ rough-in
• Antisiphon ballcock
• 3.5 GPF

FIGURE 25.1   Handicap toilet. *(Courtesy of McGraw-Hill.)*

FACILITIES FOR HANDICAP TOILETS

Let's talk about the facilities for handicap toilets (Fig. 25.1). When a handicap toilet is installed in a privacy compartment, the minimum net clear opening for the compartment must be at least 32 inches wide. The door of the compartment must swing out, away from the toilet. The width of such a compartment should be 36 inches, with a depth of 60 inches. Unlike a standard toilet, where the side clearance is 15 inches, handicap toilets require a side distance of 18 inches.

Grab bars must be installed at a height of no less than 33 inches and no more than 36 inches above the finished floor. The bars must have a minimum length of 42 inches. They must be mounted on both sides of the compartment. When the bars are mounted, they must be mounted a maximum of 12 inches from the rear wall and extend a minimum of 54 inches from the rear wall. A rear grab bar of at least 36 inches in length must also be installed. This grab bar must be no more than 6 inches from the closest side wall and extend a minimum of 24 inches beyond the centerline of the toilet away from the closest side wall.

Toilets approved for handicap installations must be higher than a normal toilet. Most of them are 18 inches tall, but the allowable range is anything between 16 and 20 inches above the finished floor. Rules for single-occupant arrangements vary a little from commercial installations. As always, check your local plumbing code for exact regulations in your region.

LAVATORIES

Lavatories installed for handicap use must be of a type that is accessible by a person in a wheelchair (Fig 25.2). The minimum clear space in front of a lavatory must be 30 inches by 30 inches. This is based on a measurement made from the front face of the lavatory, counter, or vanity. Measuring from the finished floor to the top edge of a lavatory or counter should result in a measurement of 35 inches. How much clearance is required under the lavatory? Unobstructed knee clearance with a minimum of 29 inches high by 8 inches deep should be provided. Toe clearance should be a minimum of 9 inches high by 9 inches deep, provided from the lavatory to the wall.

Additional requirements for a handicap lavatory require that all exposed hot-water piping be insulated. Faucets should be installed so that they are no more than 25 inches from the front face of the lavatory, counter, or vanity. The faucet must be able to be turned on and off with a maximum force of 5 pounds. Now, what happens if the lavatory is installed in a privacy compartment of a toilet? When a lavatory is installed in a compartment, the

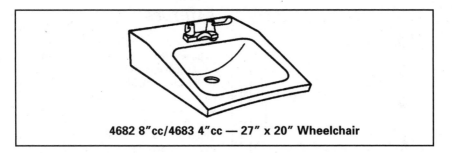

4682 8"cc/4683 4"cc — 27" x 20" Wheelchair

FIGURE 25.2 Handicap lavatory. *(Courtesy of McGraw-Hill.)*

lavatory must be located against the back wall, adjacent to the water closet. The edge of the lavatory must have a minimum of 18 inches of clear space, measured from the center of the toilet.

KITCHEN SINKS

Kitchen sinks require a minimum clear space in front of them that must be 30 inches by 30 inches. This is based on a measurement made from the front face of the kitchen sink, counter, or vanity. Measuring from the finished floor to the top edge of a kitchen sink or counter should result in a maximum measurement of 34 inches. Unobstructed knee clearance with a minimum of 29 inches high by 8 inches deep should be provided. Toe clearance should be a minimum of 9 inches high by 9 inches deep, provided from the sink to the wall.

Additional requirements for a handicap kitchen sink require that all exposed hot-water piping be insulated. Faucets should be installed so that they are no more than 25 inches from the front face of the lavatory, counter, or vanity. The faucet must be able to be turned on and off with a maximum force of 5 pounds.

BATHING UNITS

Bathing units for handicap use are required to be equipped with grab bars. Bathtubs and showers for handicap use are often different in size and equipment from what you would find in a standard fixture (Figs. 25.3 and 25.4). The minimum clear space in front of a bathing unit is 30 inches from the edge of the enclosure away from the unit and 48 inches wide. If a situation exists where a bathing unit is not accessible from the side, the clear space in front of the unit must be increased to a minimum of 48 inches. Faucets for showers and bathtubs must be equipped with a hand-held shower. The hose for these showers must be a minimum of 60 inches in length. The faucets must be able to be opened and closed with a maximum force of 5 pounds.

Grab bars are required in handicap bathing units. Minimum diameters and widths of grab bars must be a minimum of 1¼ inches and a maximum of 1½ inches. The bars must be spaced 1½ inches from the wall. It is not allowable for the bars to rotate. All bars used must be approved for the intended use.

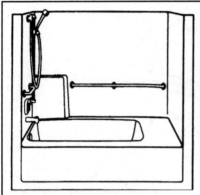

6266-H RHO/6267 LHO Summit 75 TS
• Molded-in seat
• One-piece seamless construction
• 1½″ diameter safety grab bars
• Slip-resistant bottom

FIGURE 25.3   Handicap bathtub. *(Courtesy of McGraw-Hill.)*

BATHTUBS

Bathtubs for handicap use are required to have a seat. The seat may be built in or a detachable model. Grab bars with a minimum length of 24 inches must be mounted against the back wall, in line with each other and parallel to the floor. One of the bars, the top one, must be mounted a minimum of 33 inches and a maximum of 36 inches above the finished floor. The lower bar must be mounted 9 inches above the flood-level rim of the bathtub. A grab bar must be mounted at each end of the bathtub, with the bars being the same height as the top bar on the back wall. The bar used on the faucet end of the tub must be at least 24 inches long. A bar mounted at the other end of the tub must be at least 12 inches long. Faucets must be mounted below the grab bar. If a seat is installed at the end of a bathtub, the grab bar for that end must be omitted.

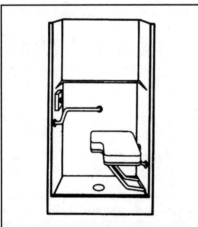

6066-H RHO/6067 LHO Summit 36S
• One-piece seamless construction
• Fold-down bench
• 1½″ diameter safety grab bars
• Meets ANSI standard A117.1-80
• Slip-resistant floor

FIGURE 25.4   Handicap shower. *(Courtesy of McGraw-Hill.)*

SHOWERS

There are two basic types of showers for handicap use. Wide shower enclosures are one type and square shower enclosures are the other. Shower stalls may be made on site or purchased as pre-fab units (Fig. 25.5). When a wide shower enclosure is used, it must have a minimum width of 60 inches. The depth must be no less than 30 inches. Thresholds are prohibited. Showers of this type must be made to allow wheelchairs to enter the enclosure. Shower valves must be mounted on the back wall. The minimum distance for the valve from the shower floor is 38 inches, with a maximum height of 48 inches. A grab bar must be mounted along the entire length of the three walls that form the enclosure. All grab bars are to be set at least 33 inches above the shower floor, but not more than 36 inches above the floor, and they are to be mounted parallel to the shower floor.

A shower enclosure that is square in design has to be at least 36 inches square. Seats for this type of shower may have a seat with a maximum width of 16 inches.

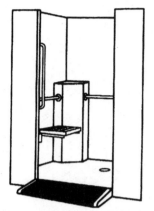

6950 RH Seat/6951 LH Seat Liberte

- Has fold-down seat. Place at 18" height for easy transfer from wheelchair to seat
- Two built-in soap shelves
- Inside diameter of 5' for easy wheelchair turn inside stall
- Entry ramp 36" wide with gentle 8.3% grade
- Lipped door ledge to prevent rolling out of stall
- Anti-skid floor mat included
- White
- Optional dome (6951) available

FIGURE 25.5 Handicap shower with seat and ramp. *(Courtesy of McGraw-Hill.)*

The seat must be mounted along the entire length of the shower. Seat height is established at a minimum of 17 inches above the shower floor, with a maximum height of 19 inches. Grab bars must be installed to extend from the edge of the seat around the side wall opposite the seat. These bars must be at least 33 inches above the shower floor and not more than 36 inches above the floor. A shower valve must be mounted on the side wall opposite the seat. The minimum height of the shower valve is 38 inches above the floor. A maximum height of 48 inches is allowed for the installation of a shower valve.

DRINKING FOUNTAINS

Drinking fountains installed for handicap use shall be installed so that the spout is no more than 36 inches above the finished floor. The spout must be located in the front of the fountain. It is required that the flow of water from the spout shall rise at least 4 inches. Controls for operating the fountain may be mounted on front of the fountain or to the side, so long as the control is side-mounted near the front of the fountain. All handicap fountains require a minimum clear space of 30 inches in front of the fixture. The measurement is made from the front of the unit by 48 inches wide. If a fountain protrudes from a wall, the clear space may be reduced from a width of 48 inches to a width of 30 inches. Handicap fixtures require more attention than standard fixtures. Keeping all the clearances straight in your head can be confusing. Refer to your local code book whenever you need clarification on a measurement.

CHAPTER 26
FACTS AND FIGURES
ABOUT PIPE AND TUBING

When the door to pipe, tubing, and fittings is open, there is a lot to learn. Some of the information is used on a frequent basis, and some of it turns up only in rare situations. We are going to open that door in this chapter and you will learn about various types of pipe and tubing.

I expect that you will find some of the data fascinating and some of it boring. Use what you find valuable. I will present the details in the most user-friendly manner possible. Tables will be used to make the reference material easy to understand. There's much to learn, so let's get started.

THE UNIFIED NUMBERING SYSTEM

Are you aware of the Unified Numbering System (UNS)? This is a system that correlates the many metal-alloy-numbering systems that are used in our country. I could go into a long discussion on this, but I believe that a simple table will give you enough information for now.

Figure 26.1 shows you the various categories of alloys. If you look to the left of the table, you will see letters. The letters are the beginning for understanding types of alloys. For example, if a rating starts with the letter C, it is referring to copper. The letter F at the beginning of a rating indicates cast-iron.

METRIC SIZES

Metric sizes are common in many places of the world. Plumbers in the United States still work, primarily with customary measurements, in terms of inches. However, you may find times when metric equivalents are useful. For this reason, I'm providing figure 26.2 for your use in comparing common measurements from the United States to metric measurements.

26.1

THREADED RODS

Threaded rods often are used to hang various types of pipe. If the size of the threaded pipe is too small in diameter and in its ability to support a proper amount of weight, the use of the rod can be very destructive. If you need to choose threaded rod for hanging pipe, you should find the information in figures 26.3 and 26.4 very helpful.

FIGURING WEIGHT OF A PIPE

Figuring the weight of a pipe and its contents is necessary when you are choosing the strength of a pipe hanger. There is a formula you can use to accomplish this goal. Let's say that you want to know how much a piece of pipe weighs. You will need some information, which can be found in figure 26.5, and you will need the following formula:

$$W = F \times 10.68 \times T \times (O.D. - T)$$

You're probably wondering what all the letters mean. The letter W is the weight of the pipe in pounds per foot. A relative weight factor, which can be found in figure 26.5, is represented by the letter F. Wall thickness of a pipe is known as the letter T. You probably have guessed that O.D. represents the outside diameter of the pipe in inches. I said that you could figure out the weight of pipe and its contents. To determine the weight of water in pipe, refer to figure 26.6.

THERMAL EXPANSION

Thermal expansion can occur in pipes when there are temperature fluctuations. Damage can result from this expansion if the pipe is not installed properly. In order to avoid damage, refer to figures 26.7, 26.8, and 26.9 to learn about the tolerances needed for various types of pipe (Fig. 26.10).

PIPE THREADS

Pipe threads come in different styles. Some are compatible; others are not. You could encounter straight pipe threads, taper pipe threads, or fire-hose coupling straight threads. To understand the types of pipe and hose threads, let me give you some illustrations to consider.

The tables in Figures 26.11, 26.12, and 26.13 show you how many threads per inch to expect with different thread types. Fire-hose threads are not compatible with any other type of thread. The same is true for garden-hose threads. But some threads are compatible with other types. If you have a female NPT thread pattern, it is compatible with male threads of an NPT type. The proper sealant to mate these threads is a thread seal. American Standard Straight Pipe (NPSM) female threads can be mated to either NPSM male threads or NPT male threads. To seal such a connection, a washer seal should be used.

Female threads that are NPSH can be coupled with male threads of NPSH, NPSM, or NPT types. In any of these cases, a washer seal should be used. Threads of a garden hose type are mated with a washer seal. But what happens when you are trying to find compatible matches for a male thread pattern? If you have an NPT male thread, it can be mated to NPT, NPSM, or NPSH threads. When NPT is mated to NPT, a thread sealant should be used. Washer seals are used to mate NPSM or NPSH female threads to male NPT threads. A male NPSM thread can mate with female thread types of NPSM or NPSH. A washer seal should be used for these connections. Garden-hose threads, male or female, can be coupled only to garden-hose threads, and this is done with a washer seal.

HOW MANY TURNS?

How many turns does it take to operate a double-disk valve? It depends on the size of the valve. Refer to figure 26.14 for the answers to how many turns it takes to operate a valve. If you want to know how many turns it takes to operate a metal-seated sewerage valve, look at figure 26.15.

PIPE CAPACITIES

Have you ever wondered about the capacity of a pipe? You could do some heavy math to figure out the answer to your question or you can look at figure 26.16 for quick solutions to your questions.

THE DISCHARGE OF A GIVEN PIPE SIZE UNDER PRESSURE

The pressure and flow are both factors to consider. If you assume that you are dealing with a straight pipe that has no bends or valves, I can give you a reference chart to use for answers to your questions. Further, assume that there will be open flow, with no backpressure, through a pipe with a smoothness rating of C = 100. Refer to figures 26.17, 26.18, and 26.19 for the quick-reference chart.

SOME FACTS ABOUT COPPER PIPE AND TUBING

Would you like some facts about copper pipe and tubing? Well, you're in the right place. Let's go over some data that could serve you well in your plumbing endeavors. Figure 26.20 will show you some size data for copper tubing. Are you interested in size details for copper that is used for drain, waste, and vent (DWV) applications? Refer to figure 26.21 for this information.

Copper is rated in terms of types. For example, Type K copper has a thick wall and is considered a stronger material than Type L or Type M copper. This type of tubing isn't used often in residential work, but it sometimes is used for water services when the copper is supplied in its soft form. Soft copper comes in a roll and allows underground piping, such as that for a water service, to be installed without joints. Type L copper is frequently used for water distribution pipes in homes and can be used in its soft form for water services. A softer type of copper is known as Type M copper. This tubing is used mostly for hot-water baseboard heating systems. It is available only in rigid lengths and is not available in a rolled coil. Many plumbing codes prohibit its use for water distribution systems. Figure 26.22 will show you how different types of copper are available for purchase (Fig. 26.23).

CAST IRON

Cast-iron pipe comes in three basic types. One is known as service-weight pipe and another is called extra-heavy cast iron. These types of pipe may be purchased with either one or two hubs. A third type of cast-iron pipe is no-hub pipe. This type has no hub on either end and is coupled with mechanical joints (Figs. 26.24 and 26.25). Cast iron is still in use and provides years of dependable service.

PLASTIC PIPE FOR DRAINS & VENTS

Plastic pipe for drains and vents are very common in modern plumbing. Polyvinyl Chloride Plastic Pipe (PVC) is probably used more often than any other type of drainage or vent pipe (Fig. 26.26). This type of pipe is strong and resistive to a variety of acids and bases. PVC pipe can be used with water, gas, and drainage systems, but it is not rated for use with hot water. I've found this type of pipe to be sensitive to dirt and water when joints are being made. The areas being joined should be dry, clean, and primed, prior to solvent welding. Also, PVC becomes brittle in cold weather and should not be dropped on hard surfaces.

Acrylonitrile Butadiene Styrene (ABS) pipe is the drainage pipe of preference for me. However, I do use more PVC than ABS at this point in my career. When plastic drainage and vent piping became popular, I cut my teeth on ABS pipe. But PVC pipe is less expensive in most regions and enjoys a less-destructive rating in the case of fires, so most of the industry has moved to PVC. I like ABS because of its durability and its easy working ability. This pipe is so strong that I've seen loaded dump trucks run over a section of ABS on a project and never crush the pipe!

PIPING COLOR CODES

Piping color codes are used when utility companies stake out piping locations. For example, a yellow flag generally indicates one of the following types of pipe:

- Oil
- Steam
- Gas
- Petroleum

When you encounter a blue flag, that indicates the location of piping below ground, the type of piping that you probably are dealing with is one of the following:

- Potable Water
- Irrigation Water
- Slurry Pipes

Green flags tend to mark the locations of sewers and drain lines. You can never count on the colors to be right, and you should always check with the flagging company to know what types of pipes with which you might be dealing. However, the above examples are common choices when color-coded flags are used.

With piping data behind us, let's move on the next chapter and talk about some electrical issues. Plumbers are not usually electricians, but they do often work with electrical connections. With this in mind, please turn to the next chapter.

The first letter (followed by five digits)	Alloy category (assigned to date)
Axxxxx	Aluminum and its alloys
Cxxxxx	Copper and its alloys
Exxxxx	Rare-earth metals, and similar metals and alloys
Fxxxxx	Cast irons
Gxxxxx	AISI and SAE carbon and alloy steels
Hxxxxx	AISI and SAE H-steels
Jxxxxx	Cast steels (except tool steels)
Kxxxxx	Miscellaneous steels and ferrous alloys
Lxxxxx	Low-melting metals and their alloys
Mxxxxx	Miscellaneous nonferrous metals and their alloys
Nxxxxx	Nickel and its alloys
Pxxxxx	Precious metals and their alloys
Rxxxxx	Reactive and refractory metals and their alloys
Sxxxxx	Heat- and corrosion-resistant steels (including stainless), valve steels and iron-based "superalloys"
Txxxxx	Tool steels (wrought and cast)
Wxxxxx	Welding filler metals
Zxxxxx	Zinc and its alloys

FIGURE 26.1 UNS metal family designations. *(Courtesy of McGraw-Hill.)*

Nominal pipe size (NPS), in IP	ASHRAE std. wt. size, mm	AWWA pipe size, mm	NFPA pipe size, mm	ASTM copper tube size, mm	Nominal pipe size DN, mm
⅛	–	–	–	6	6
3/16	–	–	–	8	8
¼	8	–	–	10	10
⅜	10	–	–	12	12
½	15	12.7 & 13	12	15	15
⅝	–	–	–	18	18
¾	20	–	–	22	20
1	25	25	25 & 25.4	28	25
1¼	32	–	33	35	32
1½	40	45	38 & 38.1	42	40
2	50	50 & 50.8	51	54	50
2½	65	63 & 63.5	63.5 & 64	67	65
3	80	75	76 & 80	79	80
3½	–	–	89	–	90
4	100	100	102	105	100
4½	–	114.3			115
5	–	–	127	130	125
6	150	150	152	156	150
8	200	200	203	206	200
10	250	250	–	257	250
12	300	300	305	308	300
14	–	350	–		350
18	–	400	–		400
18	–	–	–		450
20	–	500	–		500
24	–	600	–		600
28					700
30					750
32					800
36					900
40					1000
44					1100
48					1200
52					1300
56					1400
60					1500

FIGURE 26.2 Equivalent metric (SI) pipe sizes. *(Courtesy of McGraw-Hill.)*

Nominal rod diameter, in	Root area of thread, in^2	Maximum safe load at rod temperature of 650°F, lb
¼	0.027	240
⁵⁄₁₆	0.046	410
⅜	0.068	610
½	0.126	1,130
⅝	0.202	1,810
¾	0.302	2,710
⅞	0.419	3,770
1	0.552	4,960
1⅛	0.693	6,230
1¼	0.889	8,000
1⅜	1.053	9,470
1½	1.293	11,630
1⅝	1.515	13,630
1¾	1.744	15,690
1⅞	2.048	18,430
2	2.292	20,690
2¼	3.021	27,200
2½	3.716	33,500
2¾	4.619	41,600
3	5.621	50,600
3¼	6.720	60,500
3½	7.918	71,260

FIGURE 26.3 Load rating of threaded rods. *(Courtesy of McGraw-Hill.)*

Pipe size, in	Rod size, in
2 and smaller	⅜
2½ to 3½	½
4 and 5	⅝
6	¾
8 to 12	⅞
14 and 16	1
18	1⅛
20	1¼
24	1½

FIGURE 26.4 Recommended rod size for individual pipes. *(Courtesy of McGraw-Hill.)*

Pipe	Weight factor*
Aluminum	0.35
Brass	1.12
Cast iron	0.91
Copper	1.14
Stainless steel	1.0
Carbon steel	1.0
Wrought iron	0.98

*Average plastic pipe weights one-fifth as much as carbon steel pipe.

FIGURE 26.5 Relative weight factor for metal pipe. *(Courtesy of McGraw-Hill.)*

IPS, in	Weight per foot, lb	Length in feet containing 1 ft³ of water	Gallons in 1 linear ft
¼	0.42		0.005
⅜	0.57	754	0.0099
½	0.85	473	0.016
¾	1.13	270	0.027
1	1.67	166	0.05
1¼	2.27	96	0.07
1½	2.71	70	0.1
2	3.65	42	0.17
2½	5.8	30	0.24
3	7.5	20	0.38
4	10.8	11	0.66
5	14.6	7	1.03
6	19.0	5	1.5
8	25.5	3	2.6
10	40.5	1.8	4.1
12	53.5	1.2	5.9

FIGURE 26.6 Weight of steel pipe and contained water. *(Courtesy of McGraw-Hill.)*

Pipe material	Coefficient in/in/°F	(°C)
	Metallic pipe	
Carbon steel	0.000005	(14.0)
Stainless steel	0.000115	(69)
Cast iron	0.0000056	(1.0)
Copper	0.000010	(1.8)
Aluminum	0.0000980	(1.7)
Brass (yellow)	0.000001	(1.8)
Brass (red)	0.000009	(1.4)
	Plastic pipe	
ABS	0.00005	(8)
PVC	0.000060	(33)
PB	0.000150	(72)
PE	0.000080	(14.4)
CPVC	0.000035	(6.3)
Styrene	0.000060	(33)
PVDF	0.000085	(14.5)
PP	0.000065	(77)
Saran	0.000038	(6.5)
CAB	0.000080	(14.4)
FRP (average)	0.000011	(1.9)
PVDF	0.000096	(15.1)
CAB	0.000085	(14.5)
HDPE	0.00011	(68)
	Glass	
Borosilicate	0.0000018	(0.33)

FIGURE 26.7 Thermal expansion of piping materials. *(Courtesy of McGraw-Hill.)*

Length (ft)	Temperature Change (°F)						
	40	50	60	70	80	90	100
20	0.278	0.348	0.418	0.487	0.557	0.626	0.696
40	0.557	0.696	0.835	0.974	1.114	1.235	1.392
60	0.835	1.044	1.253	1.462	1.670	1.879	2.088
80	1.134	1.392	1.670	1.879	2.227	2.506	2.784
100	1.192	1.740	2.088	2.436	2.784	3.132	3.480

FIGURE 26.8 Thermal expansion of PVC-DWV. *(Courtesy of McGraw-Hill.)*

Length (ft)	Temperature Change (°F)						
	40	50	60	70	80	90	100
20	0.536	0.670	0.804	0.938	1.072	1.206	1.340
40	1.070	1.340	1.610	1.880	2.050	2.420	2.690
60	1.609	2.010	2.410	2.820	3.220	3.620	4.020
80	2.143	2.680	3.220	3.760	4.290	4.830	5.360
100	2.680	3.350	4.020	4.700	5.360	6.030	6.700

FIGURE 26.9 Thermal expansion of all pipes (except PVC-DWV). *(Courtesy of McGraw-Hill.)*

A hole in a pipe that is not more than .63 centimeters in diameter can result in a loss of 14,952 gallons of water a day! Even a pinhole leak can amount to a loss of over 18,000 gallons of water in a three-month period.

FIGURE 26.10 Tech tips. *(Courtesy of McGraw-Hill.)*

Pipe size (in inches)	Maximum outside diameter	Threads per inch
¼	1.375	8
1	1.375	8
1¼	1.6718	9
1½	1.990	9
2	2.5156	8
3	3.6239	6
4	5.0109	4
5	6.260	4
6	7.025	4

FIGURE 26.11 Threads per inch for national standard threads. *(Courtesy of McGraw-Hill.)*

Pipe size (in inches)	Maximum outside diameter	Threads per inch
¼	1.0353	14
1	1.295	11.5
1¼	1.6399	11.5
1½	1.8788	11.5
2	2.5156	8
3	3.470	8
4	4.470	8

FIGURE 26.12 Threads per inch for American Standard Straight Pipe. *(Courtesy of McGraw-Hill.)*

Hose size (in inches)	Maximum outside diameter	Threads per inch
¼	1.0625	11.5

FIGURE 26.13 Threads per inch for garden hose. *(Courtesy of McGraw-Hill.)*

Valve size (in inches)	Number of turns required to operate valve
3	7.5
4	14.5
6	20.5
8	27
10	33.5

FIGURE 26.14 The number of turns required to operate a double-disk valve. *(Courtesy of McGraw-Hill.)*

Valve size (in inches)	Number of turns required to operate valve
3	11
4	14
6	20
8	27
10	33

FIGURE 26.15 Number of turns required to operate a metal-seated sewerage valve. *(Courtesy of McGraw-Hill.)*

Pipe diameter (in inches)	Approximate capacity (in U.S. gallons) per foot of pipe
¾	.0230
1	.0408
1¼	.0638
1½	.0918
2	.1632
3	.3673
4	.6528
6	1.469
8	2.611
10	4.018

FIGURE 26.16 Pipe capacities. *(Courtesy of McGraw-Hill.)*

Pipe size (in inches)	PSI	Length of pipe is 50 feet
¾	20	16
¾	40	24
¾	60	29
¾	80	34
1	20	31
1	40	44
1	60	55
1	80	65
1¼	20	84
1¼	40	121
1¼	60	151
1¼	80	177
1½	20	94
1½	40	137
1½	60	170
1½	80	200

FIGURE 26.17 Discharge of pipes in gallons per minute. *(Courtesy of McGraw-Hill.)*

Pipe size (in inches)	PSI	Length of pipe is 100 feet
¾	20	11
¾	40	16
¾	60	20
¾	80	24
1	20	21
1	40	31
1	60	38
1	80	44
1¼	20	58
1¼	40	84
1¼	60	104
1¼	80	121
1½	20	65
1½	40	94
1½	60	117
1½	80	137

FIGURE 26.18 Discharge of pipes in gallons per minute. *(Courtesy of McGraw-Hill.)*

Pipe size (in inches)	PSI	Length of pipe is 200 feet
¾	20	8
¾	40	11
¾	60	14
¾	80	16
1	20	14
1	40	21
1	60	26
1	80	31
1¼	20	39
1¼	40	58
1¼	60	72
1¼	80	84
1½	20	45
1½	40	65
1½	60	81
1½	80	94

FIGURE 26.19 Discharge of pipes in gallons per minute. *(Courtesy of McGraw-Hill.)*

Nominal pipe size (inches)	Outside diameter (inches)	Inside diameter (inches)
Type K		
¼	0.375	0.305
⅜	0.500	0.402
½	0.625	0.527
⅝	0.750	0.652
¾	0.875	0.745
1	1.125	0.995
1¼	1.375	1.245
1½	1.625	1.481
2	2.125	1.959
2½	2.625	2.435
3	3.125	2.907
3½	3.625	3.385
4	4.125	3.857
5	5.125	4.805
6	6.125	5.741
8	8.125	7.583
10	10.125	9.449
12	12.125	11.315
Type L		
¼	0.375	0.315
⅜	0.500	0.430
½	0.625	0.545
⅝	0.750	0.666
¾	0.875	0.785
1	1.125	1.025
1¼	1.375	1.265
1½	1.625	1.505
2	2.125	1.985
2½	2.625	2.465
3	3.125	2.945
3½	3.625	3.425
4	4.125	3.905
5	5.125	4.875
6	6.125	5.845
8	8.125	7.725
10	10.125	9.625
12	12.125	11.565

FIGURE 26.20 Sizing data for copper tubing. *(Courtesy of McGraw-Hill.)*

Inside diameter (inches)	Nominal size (inches)	Outside diameter (inches)
Type DWV		
N/A	¼	0.375
N/A	⅜	0.500
N/A	½	0.625
N/A	⅝	0.750
N/A	¾	0.875
N/A	1	1.125
1.295	1¼	1.375
1.511	1½	1.625
2.041	2	2.125
3.035	2½	2.625
N/A	3	3.125
N/A	3½	3.625
4.009	4	4.125
4.981	5	5.125
5.959	6	6.125
N/A	8	8.125
N/A	10	10.125
N/A	12	12.125

FIGURE 26.21 Copper tube-DWV. *(Courtesy of McGraw-Hill.)*

Drawn (hard copper) (feet)		Annealed (soft copper) (feet)	
Type K Tube			
Straight Lengths:		Straight Lengths:	
Up to 8-in. diameter	20	Up to 8-in. diameter	20
10-in. diameter	18	10-in. diameter	18
112-in. diameter	12	12-in. diameter	12
		Coils:	
		Up to 1-in. diameter	60
		1¼-in. diameter	60
			100
		2-in. diameter	40
			45
Type L Tube			
Straight Lengths:		Straight Lengths:	
Up to 10-in. diameter	20	Up to 10-in. diameter	20
12-in. diameter	18	12-in. diameter	18
		Coils:	
		Up to 1-in. diameter	60
			100
		1¼- and 1½-in. diameter	60
			100
		2-in. diameter	40
			45
DWV Tube			
Straight Lengths:		Not available	
All diameters	20		
Type M Tube			
Straight Lengths:		Not available	
All diameters	20		

FIGURE 26.22 Available lengths of copper plumbing tube. *(Courtesy of McGraw-Hill.)*

Hard copper is also known as drawn copper, while soft copper tubing is known as annealed copper.

FIGURE 26.23 Tech tips. *(Courtesy of McGraw-Hill.)*

Size (inches)	Service weight per linear foot (pounds)	Extra heavy size (inches)	Per linear foot (pounds)
2	4	2	5
3	6	3	9
4	9	4	12
5	12	5	15
6	15	6	19
7	20	8	30
8	25	10	43
		12	54
		15	75

FIGURE 26.24 Weight of cast-iron soil pipe. *(Courtesy of McGraw-Hill.)*

	Diameter (inches)	Service weight (lb)	Extra heavy weight (lb)
Double hub, 5-ft lengths	2	21	26
	3	31	47
	4	42	63
	5	54	78
	6	68	100
	8	105	157
	10	150	225
Double hub, 30-ft length	2	11	14
	3	17	26
	4	23	33
Single hub, 5-ft lengths	2	20	25
	3	30	45
	4	40	60
	5	52	75
	6	65	95
	8	100	150
	10	145	215
Single hub, 10-ft lengths	2	38	43
	3	56	83
	4	75	108
	5	98	133
	6	124	160
	8	185	265
	10	270	400
No-hub pipe, 10-ft lengths	1½	27	
	2	38	
	3	54	
	4	74	
	5	95	
	6	118	
	8	180	

FIGURE 26.25 Weight of cast-iron pipe. *(Courtesy of McGraw-Hill.)*

CHAPTER 27
PIPE-FITTING CALCULATIONS

Plumbing and pipe fitting are similar, but not always the same. Modern plumbers generally work with copper tubing and various forms of plastic piping. Cast-iron pipe still is encountered, and steel pipe is used for gas work. Finding a plumber who works with threaded joints is not nearly as common as it once was. But threaded pipe still is used in plumbing, and it is used frequently in pipe fitting. Figuring the fit for a pipe when threads are to be inserted into a fitting is a little different from sliding copper or plastic pipe into a hub fitting. However, many of the calculations used with threaded pipe apply to other types of pipe.

Many plumbers don't spend a lot of time using mathematical functions to figure offsets. Heck, I'm one of them. How often have you taken a 45 and held it out to guestimate a length for a piece of pipe? If you have a lot of experience, your trained eye and skill probably gave you a measurement that was close enough for plastic pipe or copper tubing. I assume this, because I do it all the time. But there are times when it helps to know how to use a precise formula to get an accurate measurement. The need for accuracy is more important when installing threaded pipe. For example, you can't afford to guess at a piece of gas pipe and find out the hard way that the threads did not go far enough into the receiving fitting.

In the old days, when I was first learning the trade, plumbers taught their helpers and apprentices. Those were the good old days. In today's competitive market, plumbing companies don't spend nearly as much time or money training their up-and-coming plumbers. As the owner of a plumbing company, I understand why this occurs, but I don't agree with it. And the net result is a crop of plumbers who are not well prepared for what their trade requires. Sure, they can do the basics of gluing, soldering, and simple layouts, but many of the new breed don't possess the knowledge needed to be true master plumbers. Don't get me wrong, it's not really the fault of the new plumbers. Responsibility for becoming an excellent plumber rests on many shoulders.

Ideally, plumbing apprentices and helpers should have classroom training. Company supervisors should authorize field plumbers some additional time for in-the-field training for apprentices. Working apprentices should go the extra mile to do research and study on their own. When I was a helper, I used to spend my lunch break reading the code book. There is no single individual to blame for the quality of education that some new plumbers are, or are not, receiving. Money is probably the root of the problem. Customers are looking for low bids. Contractors must be competitive, and this eliminates the ability to have a solid on-the-job training program. Many helpers today seem to be more interested in getting their check than getting an education.So here we are, with a lot of plumbers who don't know the inner workings of the finer points of plumbing.

I was fortunate enough to be in what might have been the last generation of plumbers to get company support in learning the trade. Plenty of time was spent running jack-hammers and using shovels, but my field plumber took the time to explain procedures to me. I learned quickly how to plumb a basic house. Then I learned how to run gas pipes and to do commercial buildings. As a part of my learning process, I read voraciously. Later, I became a supervisor, then the owner of my own company, and eventually an educator for other plumbers and for apprentices. I could have stopped anywhere along the way, but I've taken my interest in the trade to the limits, and I continue to push ahead. No, I don't know all there is to know, but I've worked hard to gain the knowledge I have. Now is the time for me to share my knowledge of pipe-fitting math with you.

45° OFFSETS

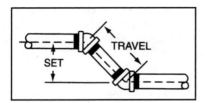

FIGURE 27.1 Calculated 45° offsets.

Offsets for 45° bends are common needs in both plumbing and pipe fitting. In fact, this degree of offset is one of the most common in the trade. I mentioned earlier that many plumbers eyeball such offset measurements. The method works for a lot of plumbers, but let's take a little time to see how the math of such offsets can help you in your career.

To start our tutorial, let's discuss terms that apply to offsets. Envision a horizontal pipe in which you want to install a 45° offset. For ease of vision, think of the horizontal pipe resting in a pipe hanger. You have to offset the pipe over a piece of ductwork. This will have a 45° fitting looking up from your horizontal pipe. There will be a piece of pipe in the upturned end of the fitting that will come into the bottom of the second 45° fitting.

As we talk about measurements here, they will all be measured from the center of the pipe. There are two terms you need to know for this calculation. Travel, the first term, is the length of the pipe between the two 45° fittings. The length of travel begins and ends at the center of each fitting. The distance from the center of the lower horizontal pipe to the center of the upper horizontal pipe is called the set. Now that you know the terms, we can do the math. To make doing the math easier, I am including tables for you to work from (Fig. 27.2). Let's say that the set is 7¼ inches. Find this measurement in the table in figure 27.2. This will show you that the travel is 10.251 inches. Now you can use the table for converting decimal equivalents of fractions of an inch (Fig. 27.3) to convert the 10.251 inches. Finding the decimal equivalent of a fraction is a matter of dividing the numerator by the denominator. The chart in figure 27.3 proves the measurement to be 10¼ inches. You can find the set if you know the travel by reversing the procedure.

If the travel is known to be 10¼ inches, what is the set? We both know that it is 7¼ inches, but how would you find it? Use the table in figure 27.2 and look under the heading of travel. Find the 10.251 listing that represents 10¼ inches. Refer to the set heading. What does it say? Of course, it says 7¼. It's that easy. All you have to do is use the tables that I've provided to make your life easier in calculating 45° offsets.

Set	Travel	Set	Travel	Set	Travel
2	2.828	¼	15.907	½	28.987
¼	3.181	½	16.261	¾	29.340
½	3.531	¾	16.614	21	29.694
¾	3.888	12	16.968	¼	30.047
3	4.242	¼	17.321	½	30.401
¼	4.575	½	17.675	¾	30.754
½	4.949	¾	18.028	22	31.108
¾	5.302	13	18.382	¼	31.461
4	5.656	¼	18.735	½	31.815
¼	6.009	½	19.089	¾	32.168
½	6.363	¾	19.442	23	32.522
¾	6.716	14	19.796	¼	32.875
5	7.070	¼	20.149	½	33.229
¼	7.423	½	20.503	¾	33.582
½	7.777	¾	20.856	24	33.936
¾	8.130	15	21.210	¼	34.289
6	8.484	¼	21.563	½	34.643
¼	8.837	½	21.917	¾	34.996
½	9.191	¾	22.270	25	35.350
¾	9.544	16	22.624	¼	35.703
7	9.898	¼	22.977	½	36.057
¼	10.251	½	23.331	¾	36.410
½	10.605	¾	23.684	26	36.764
¾	10.958	17	24.038	¼	37.117
8	11.312	¼	24.391	½	37.471
¼	11.665	½	24.745	¾	37.824
½	12.019	¾	25.098	27	38.178
¾	12.372	18	25.452	¼	38.531
9	12.726	¼	25.805	½	38.885
¼	13.079	½	26.159	¾	39.238
½	13.433	¾	26.512	28	39.592
¾	13.786	19	26.866	¼	39.945
10	14.140	¼	27.219	½	40.299
¼	14.493	½	27.573	¾	40.652
½	14.847	¾	27.926	29	41.006
¾	15.200	20	28.280	¼	41.359
11	15.554	¼	28.635	½	41.713

FIGURE 27.2 Set and travel relationships in inches for 45° offsets.

BASIC OFFSETS

Basic offsets are all based on the use of right triangles. You now know about set and travel. It is time that you learned about a term known as run. Travel, as I said earlier, is the distance between center of two offset fittings that creates the length of a piece of pipe. This pipe's length is determined as it develops from fitting to fitting, traveling along the angle of the offset. When you want to know the run, you are interested in the distance measured along a straight line from the bottom horizontal pipe. Refer to figure

Inches	Decimal of an inch	Inches	Decimal of an inch
1/64	.015625	33/64	.515625
1/32	.03125	17/32	.53125
3/64	.046875	35/64	.546875
1/16	.0625	9/16	.5625
5/64	.078125	37/64	.578125
3/32	.09375	19/32	.59375
7/64	.109375	39/64	.609375
1/8	.125	5/8	.625
9/64	.140625	41/64	.640625
5/32	.15625	21/32	.65625
11/64	.171875	43/64	.671875
3/16	.1875	11/16	.6875
13/64	.203125	45/64	.703125
7/32	.21875	23/32	.71875
15/64	.234375	47/64	.734375
1/4	.25	3/4	.75
17/64	.265625	49/64	.765625
9/32	.28125	25/32	.78125
19/64	.296875	51/64	.796875
5/16	.3125	13/16	.8125
21/64	.328125	53/64	.828125
11/32	.34375	27/32	.84375
23/64	.359375	55/64	.859375
3/8	.375	7/8	.875
35/64	.390625	57/64	.890625
13/32	.40625	29/32	.90625
27/64	.421875	59/64	.921875
7/16	.4375	15/16	.9375
29/64	.453125	61/64	.953125
15/32	.46875	31/32	.96875
31/64	.484375	63/64	.984375
1/2	.5	1	1

FIGURE 27.3 Decimal equivalents of fractions of an inch

27.4 for an example of what I'm talking about. Run is a term applied to the horizontal measurement from the center of one offset fitting to the center of the other offset fitting.

Most charts and tables assign letters to terms used in formulas. For our purposes, let's establish our own symbols. We will call the letter S set, the letter R run, and the letter T travel. What are common offsets in the plumbing and pipe-fitting trade? A 45° offset is the most common. Two other offsets sometimes used are 60° bends and 22½° bends. These are the three most frequently used offsets and the ones on which we will concentrate our efforts.

The use of the right triangle is important when dealing with piping offsets. The combination of set, travel, and run form the triangle. I can provide you with a table that will

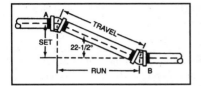

FIGURE 27.4 Simple offsets.

make calculating offsets easier (Fig. 27.5), but you must still do some of the math yourself, or at least know some of the existing figures. This may seem a bit intimidating, but it is not as bad as you might think. Let me explain.

As a working plumber or pipe fitter, you know where your first pipe is. In our example earlier, where there was ductwork that needed to be cleared, you can easily determine what the measurement of the higher pipe must be. This might be determined by measuring the distance from a floor or ceiling. Either way, you will know the center measurement of your existing pipe and the center measurement for the higher pipe's location. Knowing these two numbers will give you the Set figure. Remember, Set is measured as the vertical distance between the centers of two pipes. Refer back to figure 27.1 if you need a reminder on this concept.

Let's assume that you know what your Set distance is. You want to know what the Travel is. To do this, use the table in figure 27.5. For example, if you were looking for the Travel of a 45° offset when the Set is known, you would multiply the Set measurement by a factor of 1.414. Now let's assume that you know the Travel and want to know the Set. For the same 45° offset, you would multiply the Travel measurement by .707. It's really simple, as long you have the chart to use. The procedure is the same for different degrees of offset. Just refer to the chart and you will find your answers quickly and easily.

Finding Run measurements is no more difficult than Set or Travel. Say you have the Set measurement and want to know the Run figure for a 45° offset. Multiply the Set figure by 1.000 to get the run number. If you are working with the Travel number, multiply that number by .707 to get the run number for a 45° offset.

SPREADING OFFSETS EQUALLY

If you take a lot of pride in your work or are working to detailed piping diagrams, you may find that the spacing of your offsets must be equal. Equally spaced offsets are not only more attractive and more professional looking, they might be required. You can guess and eyeball measurements to get them close, but you will need a formula to work

To find side*	When known side is	Multi- ply Side	For 60° ells by	For 45° ells by	For 30° ells by	For 22½° ells by	For 11¼° ells by	For 5⅝° ells by
T	S	S	1.155	1.414	2.000	2.613	5.125	10.187
S	T	T	.866	.707	.500	.383	.195	.098
R	S	S	.577	1.000	1.732	2.414	5.027	10.158
S	R	R	1.732	1.000	.577	.414	.198	.098
T	R	R	2.000	1.414	1.155	1.082	1.019	1.004
R	T	T	.500	.707	.866	.924	.980	.995

*S = set, R = run, T = travel.

FIGURE 27.5 Multipliers for calculating simple offsets.

with if you want the offsets to be accurate. Fortunately, I can provide you with such a formula, and I will.

Again, we will concentrate on 45°, 60°, and 22½° bends, since these are the three most often used in plumbing and pipe fitting. We will start with the 45° turns. In our example, you should envision two pipes rising vertically. Each pipe will be offset to the left and then the pipes will continue to rise vertically. For a visual example, refer to figure 27.6. It is necessary for us to determine uniform symbols for what we are doing, so let's get that out of the way right now.

In our measurement examples, we will refer to Spread, the distance between the two offsetting pipes from center to center, as A. Set will remain with the symbol of S. Travel will be T and it will be the same as Distance D. Run will be noted by the letter R. The letter F will be the length of pipe threads.

Now for the deal. Travel is determined in an equally offset pipe run at a 45° angle by multiplying the Set by 1.414. Run is found by multiplying Set by 1.000. The F measurement is found by multiplying the spread (A) by .4142. Remember that T and D are the same.

Want to do the same exercise with a 60° setup? Why not? To run a similar deal on a 60° angles of equally-spaced offset pipes you follow the same basic principles used in the previous example. Multiply the Set by 1.155 to find the Travel. Run is found by multiplying Set by .5773. The F measurement is found by multiplying the spread (A) by .5773. Remember that T and D are the same.

Need to find numbers for 22½° bends? Well, it's not difficult. To find figures for equally spaced pipes with 22½° bends, multiply the Set by 2.613 to find the Travel. Run is found by multiplying Set by 2.414. The F measurement is found by multiplying the spread (A) by .1989. Remember that T and D are the same.

GETTING AROUND PROBLEMS

Getting around problems and obstacles is part of the plumbing and pipe fitting trades. Few jobs run without problems or obstacles. As any experienced piping contractor knows, there are always some obstructions in the preferred path of piping. Many times the obstacle is ductwork, but the obstacles can involve electrical work, beams, walls, and other objects that are not easily relocated. This means that the pipes must be rerouted. This section is going to deal with the mathematics required to compensate for immovable objects.

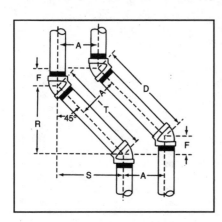

Let me set the stage for a graphic example of getting around an overhead obstruction. Assume that you are bringing a pipe up out of a concrete floor in a basement. There is a window directly above the pipe that you must offset around. The window was an afterthought. Having the pipe under the window was not a mistake in the groundworks rough-in. However, it is your job to move the pipe, without breaking up the floor, to get around the window.

FIGURE 27.6 Two-pipe 45° equal-spread offset.

In many cases, you might just cut the pipe off, close to the floor, stick a 45° fitting on it, and bring a piece of pipe over to another 45° fitting. This is usually enough, but suppose you have a very tight space to work with and must make an extremely accurate measurement. Do you know how to do it? Imagine a situation where an engineer has indicated an exact location for the relocated riser. Can you hit the spot accurately? Do you know what type of formula to use in order to comply with the job requirements? If not, consider the following information as your ticket to success.

Our formula will involve three symbols. The first symbol will be an A, and it will be representative of the distance from the center of your 45° fitting to the bottom edge of the window. The distance from the center of the rising pipe to the outside edge of the window will be known by the letter B. We will use the letter C to indicate the distance from the center of the travel piece of pipe from the edge of the window. Using E to indicate the distance of the center of the rising pipe from the right edge of the window and D to indicate the center of the offset rising pipe from the right edge of the window, we can use the formula. Let's see how it works.

To find the distance from the bottom of the window to the starting point of the offset, you would take the distance from the center of the riser to the left edge of the window (B) and add the distance from the corner of the left window edge to the center of the pipe (C) times 1.414. The formula would look like this: $A = B + C \times 1.414$. Refer to figure 27.7 for an example. Now let's put measurement numbers into the formula.

Assume that you want to find A. Further assume that B is equal to 1 foot and that C is equal to 6 inches. The numerical formula would be like this: $A = 12$ inches (B) + 6 inches (C) x 1.414 = 12 inches + $8\frac{1}{2}$ inches = $20\frac{1}{2}$ inches. This would prove that the upper 45° fitting would be $20\frac{1}{2}$ inches from the edge of the right edge of the window. As you can see, the actual procedure is not as difficult as the intimidation of using formulas might imply.

Round Obstacles

You've just seen how to get around what many would to as a typical problem. Most offsets are used to get around square or rectangular objects. But what happens when you have to bypass a round object, such as a pressure tank? Don't worry, there is a simple way to get around most any problem, so let's talk about going around circular objects.

Okay, we have a pipe that has to rise vertically, but there is a horizontal expansion tank hanging in the ceiling that is blocking the path of our pipe. We have a very limited amount of space to either side of the tank to work within, so our measurements have to be precise. Assume that an eyeball measurement will not work in this case. So let's set up the symbols that we will use in this formula.

Let's use the letter A to indicate the center of the offset rising pipe from the center of the expansion tank. The letter B will represent the center of the offset rising pipe from the edge of the tank. One-half the diameter of the tank will be identified by the letter C. We will use the letter D to indicate the distance from the center line of the tank to the starting point of the offset. Additional information needed is that $A = B + C$ and $D = A \times .4142$. See figure 27.8 for a drawing to help you visualize the setup.

To put the letters into numbers, let's plug in some hypothetical numbers. Assign a number of 18 inches to C and 8 inches to B. What is D? Here's how it works. $A = B + C = 8 + 18 = 26$ inches. D will equal A x .4142 = 26 x .4142 = $10\frac{3}{4}$ inches. This makes the center of the fitting $10\frac{3}{4}$ inches from the center of the the the tank.

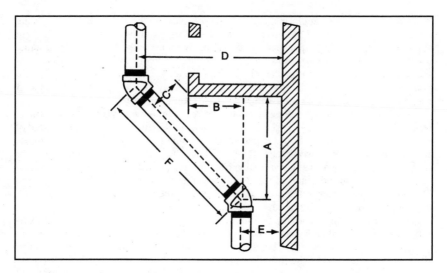

FIGURE 27.7 Pipe offsets around obstructions.

Rolling Offsets

Rolling offsets can be figured with a complex method or with a simplified method. Since I assume that you are most interested in the most accurate information that you can get in the shortest amount of time, I will give you the simplified version. The results will be the same as the more complicated method, but you will not pull out as much hair or lose as much time as you would with the other exercise, and you will arrive at the same solution.

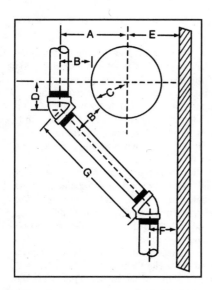

FIGURE 27.8 Starting point of a 45° offset around a tank.

To figure rolling offsets simply, you will need a framing square – just a typical, steel, framing square. The corner of any flat surface is also needed, so that you can form a right angle. You will also need a simple ruler. The last tool needed is the table that I am providing in figure 27.9. This is going to be really easy, so don't run away. Let me explain how you will use these simple elements to figure rolling offsets.

Stand your framing square up on a flat surface. The long edge should be vertical and the short edge should be horizontal. The long, vertical section will be the Set, and the short, horizontal section will be the Roll. Your ruler will be used to tie the Set together with the Roll (Fig. 27.10 square and ruler). A constant will be needed to arrive at a solution, and you will find constants in the table I've provided in figure 27.9. Once again, the three main angles are addressed.

Angle	Constant
5⅝°	10.207
11¼°	5.125
22½°	2.613
30°	2.000
45°	1.414
60°	1.154

FIGURE 27.9 Simplified method of figuring a rolling offset.

When you refer to figure 27.9, you will find that the constant for a 45° bend is 1.414. The number for a 60° bend is 1.154, and the constant for a 22½° bend is 2.613. If you were working with a 45° angle that had a Set of 15 inches and a Roll of 8 inches, you would use your ruler to measure the distances between the two marks on the framing square. In this case, the measurement from the ruler would be 17 inches. You would multiply the 17-inch number by the 45° constant of 1.414 (found in Fig. 27.10) and arrive at a figure of 24³⁄₃₂ of an inch. This would be the length of the pipe, from center to center, needed to make your rolling offset. Could it get any easier?

RUNNING THE NUMBERS

Running the numbers of pipe fitting is not always necessary to complete a job. If you have the experience and the eye to get the job done, without going through mathematical functions, that's great. I admit, I rarely have to use sophisticated math to figure out my piping layouts. But I do know how to hit the mark right on the spot when I need to, and so should you. Accuracy can be critical. If you don't invest the time to learn the proper methods for figuring offsets, you might cut your career opportunities short. Believe me, you owe it to yourself to expand your knowledge. Sitting still can cost you. Reach out, as you are doing by reading this book, and expand your knowledge.

Some people see plumbers and pipe fitters as blue-collar workers. This may true. If it is, I'm proud to wear a blue collar. Yet, if you proceed in your career, you may own your own business, and this will, by society standards, graduate you to a white collar. As far as I am concerned, the color of a person's collar has no bearing on the person's worth. Blue collar or white collar, individuals are what they are. We all bring something to the table. Yes, some people do prosper more than others, and education does play a role in most career advancements.

You might or might not need what you've learned in this chapter. However, knowing some simple math and having access to the tables in this chapter probably will give you an edge on many of the people with whom you work or compete. Like it or not, making a living in today's world is competitive. So why not be as well-prepared as possible?

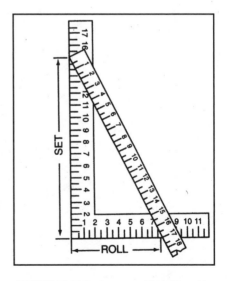

FIGURE 27.10 Laying out a rolling offset with a steel square.

CHAPTER 28
WATER PUMPS

Some plumbers work their entire careers without ever having to know anything about water pumps. Other plumbers deal with pumps on a frequent basis. The difference is where the plumbers work. I've never worked in New York City, but I suppose that are not many water pumps to be installed or serviced there. But where I live, in Maine, there are more homes served by private water wells than you can shake a stick at. When I lived in Virginia, there were plenty of water pumps, too. Some of the pumps are jet pumps and others are submersible pumps. The two are very different, even though they do the same job.

Jet pumps are at their best when used in conjunction with shallow wells, with depths of approximately 25 feet or less. Two-pipe jet pumps can be used with deep wells, but a submersible pump is usually a better option for deep wells. Sizing water pumps and pressure tanks is routine for some plumbers and foreign to others. This chapter is going to give you plenty of data to use when working with pump systems.

The illustrations I have to offer you in this chapter are detailed and self-explanatory. I believe that you will be able to use this chapter as a quick-reference guide to answer most of your pump questions. Look over the following illustrations and you will find data on jet pumps, submersible pumps, and pressure tanks. The data will prove very helpful if you become involved with the installation, sizing, or repair of water pumps.

Average water requirements for general service around the home and farm

Each person per day, for all purposes	75 gal.
Each horse, dry cow, or beef animal	12 gal.
Each milking cow	35 gal.
Each hog per day	4 gal.
Each sheep per day	2 gal.
Each 100 chickens per day	4 gal.

Average amount of water required by various home and yard fixtures

Drinking fountain, continuously flowing	50 to 100 gal. per day
Each shower bath	Up to 30 gal. @ 3–5 gpm
To fill bathtub	30 gal.
To flush toilet	6 gal.
To fill lavatory	2 gal.
To sprinkle 1/4" of water on each 1000 square feet of lawn	160 gal.
Dishwashing machine — per load	7 gal. @ 4 gpm
Automatic washer —per load	Up to 50 gal. @ 4–6 gpm
Regeneration of domestic water softener	50–100 gal.

Average flow rate requirements by various fixtures
(gpm = gal. per minute; gph = gal. per hour)

Shower	3–5 gpm
Bathtub	3–5 gpm
Toilet	3 gpm
Lavatory	3 gpm
Kitchen sink	2–3 gpm
1/2" hose and nozzle	200 gph
3/4" hose and nozzle	300 gph
Lawn sprinkler	120 gph

FIGURE 28.1 Average water requirements for general service. *(Courtesy of McGraw-Hill.)*

	Approx. Gallons Per Day
Each horse	12
Each producing cow	15
Each nonproducing cow	12
Each producing cow with drinking cups	30–40
Each nonproducing cow with drinking cups	20
Each hog	4
Each sheep	2
Each 100 chickens	4–10
Yard fixtures:	
½-inch hose with nozzle	200
¾-inch hose with nozzle	300
Bath houses	10
Camp	
Construction, semipermanent	50
Day (with no meals served)	15
Luxury	100–150
Resorts (day and night, with limited plumbing)	50
Tourists with central bath and toilet facilities	35
Cottages with seasonal occupancy	50
Courts, tourists with individual bath units	50
Clubs	
Country (per resident member)	100
Country (per nonresident member present)	25

FIGURE 28.2 Daily water requirements. *(Courtesy of McGraw-Hill.)*

Dwellings
Luxury	75
Multiple family, apartments (per resident)	60
Rooming houses (per resident)	50
Single family	75
Estates	100–150
Factories (gallons/person/shift)	15–35
Institutions other than hospitals	75–125
Hospitals (per bed)	250–400
Laundries, self-serviced (gallons per washing, i.e., per customer)	50
Motels	
With bath and toilet (per bed space)	100
Parks	
Overnight with flush toilets	25
Trailers with individual bath units	50
Picnic	
With bath houses, showers, and flush toilets	20
With only toilet facilities (gal./picnicker)	10
Restaurants with toilet facilities (per patron)	10
Without toilet facilities (per patron)	3
With bars and cocktail lounge (additional quantity)	2
Schools	
Boarding	50–70
Day with cafeteria, gymnasiums and showers	25
Day with cafeteria but no gymnasiums or showers	20
Service stations (per vehicle)	10
Stores (per toilet room)	400
Swimming pools	10
Theaters	
Drive-in (per car space)	5
Movie (per auditorium seat)	5
Workers	
Construction (semipermanent)	50
Day (school or offices per shift)	15

Providing an adequate water supply provides for a healthy family and higher production from livestock. Assuming the total daily requirement is calculated to be 1200 gpd (gallons per day), a pump would be selected for a capacity of 10 gpm (gallons per minute) based on the following formula:

$$1200 \text{ gph} \div 2 \text{ equals } 600 \text{ gph (gal. per hr.)}$$

Example: 5 in family @ 75 gpd each person	375
1¾" hose with nozzle @ 300	300
10 non-producing cows with cups @ 20	200
Total 24 hr. req.	875

$$875 \div 2 = 438 \text{ gph or } 7.3 \text{ gpm pump selection}$$

FIGURE 28.2 Daily water requirements. *(Courtesy of McGraw-Hill.) (Continued)*

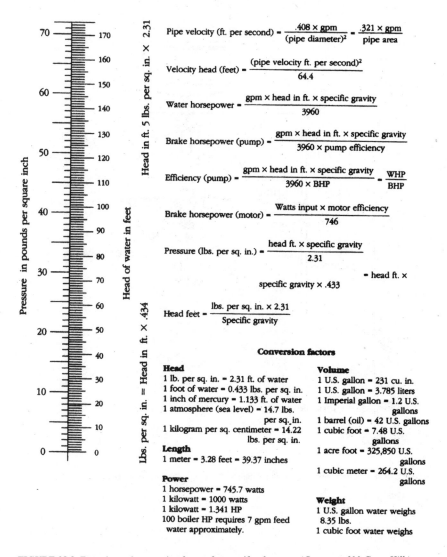

Engineering data
Formulas and conversion factors
Centrifugal pumps

Pipe velocity (ft. per second) = $\dfrac{.408 \times \text{gpm}}{(\text{pipe diameter})^2} = \dfrac{.321 \times \text{gpm}}{\text{pipe area}}$

Velocity head (feet) = $\dfrac{(\text{pipe velocity ft. per second})^2}{64.4}$

Water horsepower = $\dfrac{\text{gpm} \times \text{head in ft.} \times \text{specific gravity}}{3960}$

Brake horsepower (pump) = $\dfrac{\text{gpm} \times \text{head in ft.} \times \text{specific gravity}}{3960 \times \text{pump efficiency}}$

Efficiency (pump) = $\dfrac{\text{gpm} \times \text{head in ft.} \times \text{specific gravity}}{3960 \times \text{BHP}} = \dfrac{\text{WHP}}{\text{BHP}}$

Brake horsepower (motor) = $\dfrac{\text{Watts input} \times \text{motor efficiency}}{746}$

Pressure (lbs. per sq. in.) = $\dfrac{\text{head ft.} \times \text{specific gravity}}{2.31}$

= head ft. × specific gravity × .433

Head feet = $\dfrac{\text{lbs. per sq. in.} \times 2.31}{\text{Specific gravity}}$

Conversion factors

Head
1 lb. per sq. in. = 2.31 ft. of water
1 foot of water = 0.433 lbs. per sq. in.
1 inch of mercury = 1.133 ft. of water
1 atmosphere (sea level) = 14.7 lbs. per sq. in.
1 kilogram per sq. centimeter = 14.22 lbs. per sq. in.

Length
1 meter = 3.28 feet = 39.37 inches

Power
1 horsepower = 745.7 watts
1 kilowatt = 1000 watts
1 kilowatt = 1.341 HP
100 boiler HP requires 7 gpm feed water approximately.

Volume
1 U.S. gallon = 231 cu. in.
1 U.S. gallon = 3.785 liters
1 Imperial gallon = 1.2 U.S. gallons
1 barrel (oil) = 42 U.S. gallons
1 cubic foot = 7.48 U.S. gallons
1 acre foot = 325,850 U.S. gallons
1 cubic meter = 264.2 U.S. gallons

Weight
1 U.S. gallon water weighs 8.35 lbs.
1 cubic foot water weighs

FIGURE 28.3 Formulas and conversion factors for centrifugal pumps. *(Courtesy of McGraw-Hill.)*

28.5

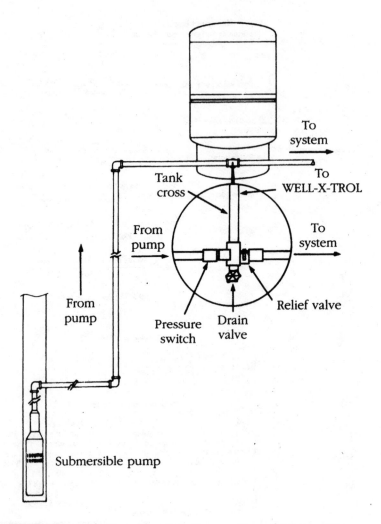

FIGURE 28.4 Pressure tank in use with a submersible pump. *(Courtesy of McGraw-Hill.)*

Model no.	HP	Volts	Impeller material	Pres. switch setting	Suction pipe size	Discharge size	Shipping weight
8130	⅓	115	Plastic	20-40	1¼"	¾"	46 lbs.
8131	⅓	115	Brass	20-40	1¼"	¾"	48 lbs.
8150	½	115/230	Plastic	20-40	1¼"	¾"	48 lbs.
8151	½	115/230	Brass	20-40	1¼"	¾"	50 lbs.
8170	¾	115/230	Plastic	30-50	1¼"	¾"	50 lbs.
8171	¾	115/230	Brass	30-50	1¼"	¾"	52 lbs.
8110	1	115/230	Plastic	30-50	1¼"	¾"	52 lbs.
8111	1	115/230	Brass	30-50	1¼"	¾"	53 lbs.

FIGURE 28.5 Performance ratings for jet pumps. (*Courtesy of McGraw-Hill.*)

Model	Model	HP	Volts	Impeller material	Pres. switch setting	Suction pipe size	Twin type drop pipe	Shipping weight
1550	1050	½	115/230	Brass	30-50	1¼"	1" × 1¼"	65 lbs.
1575	1075	¾	115/230	Brass	30-50	1¼"	1" × 1¼"	71 lbs.
1575SW	1075SW	¾	115/230	Brass	30-50	1¼"	1" × 1¼"	66 lbs.
1510	1010	1	115/230	Brass	30-50	1¼"	1" × 1¼"	74 lbs.
1510SW	1010SW	1	115/230	Brass	30-50	1¼"	1" × 1¼"	67 lbs.
1515SW	1015SW	1½	115/230	Brass	30-50	1¼"	1" × 1¼"	72 lbs.

FIGURE 28.6 Performance ratings for multi-stage pumps. *(Courtesy of McGraw-Hill.)*

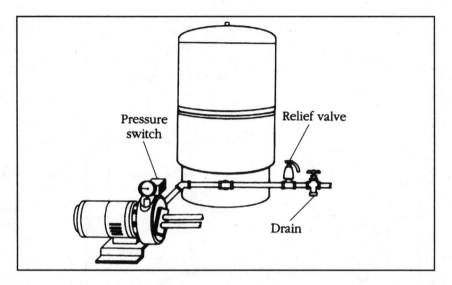

FIGURE 28.7 A typical jet-pump setup. *(Courtesy of McGraw-Hill.)*

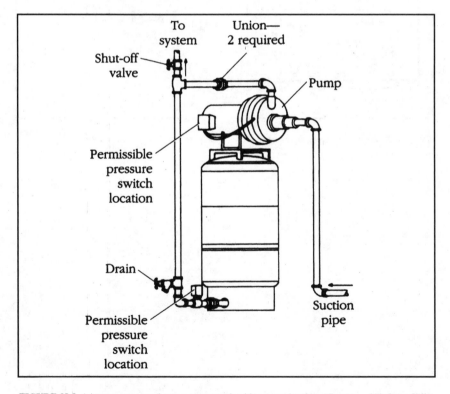

FIGURE 28.8 A jet pump mounted on a pressure tank with a pump bracket. *(Courtesy of McGraw-Hill.)*

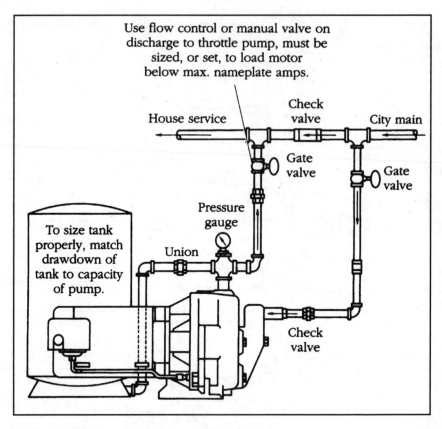

FIGURE 28.9 A typical piping arrangement for a jet pump. *(Courtesy of McGraw-Hill.)*

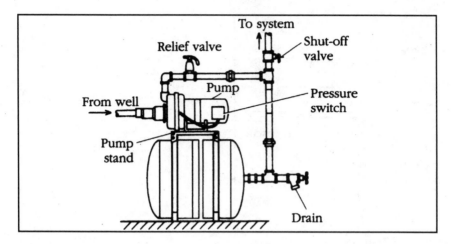

FIGURE 28.10 Bracket-mounted jet pump on a horizontal pressure tank. *(Courtesy of McGraw-Hill.)*

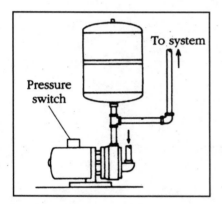

FIGURE 28.11 Small vertical pressure tank installed above pump. *(Courtesy of McGraw-Hill.)*

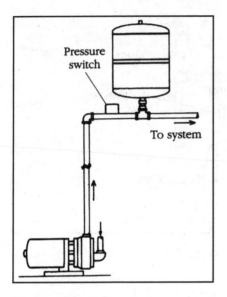

FIGURE 28.12 Small vertical pressure tank installed above pump. *(Courtesy of McGraw-Hill.)*

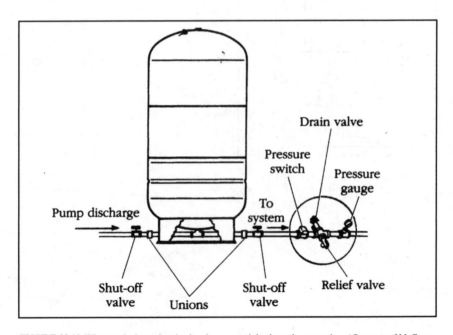

FIGURE 28.13 When replacing galvanized tanks, use straight-through connection. *(Courtesy of McGraw-Hill.)*

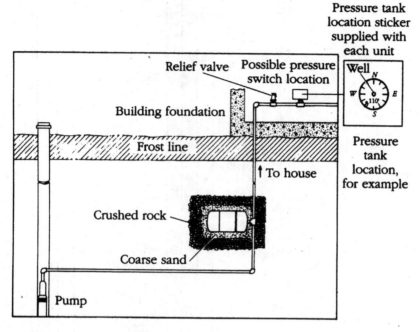

FIGURE 28.14 An underground installation of a pressure tank. *(Courtesy of McGraw-Hill.)*

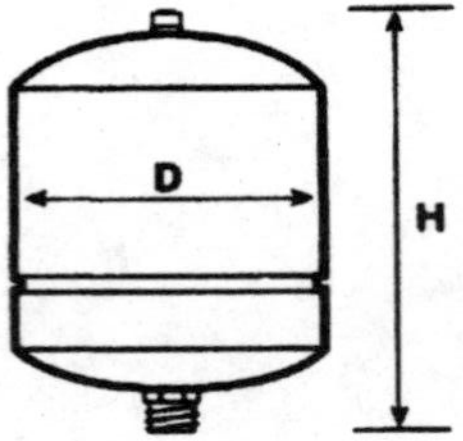

Model No.	Dimensions		Total Volume	Drawdown			System Connection	Precharge Pressure	Shipping Wt./Vol.
	Diameter	Height		1,5/3,0 bar	2,0/3,5 bar	2,5/4,0 bar			
	mm	mm	Ltr	Liter	Liter	Liter	R"	bar	KG / m3
WX 2,6	156	228	2,6	1,0	0,9	0,8	3/4	1,5	1,0 / ,005
WX 4	156	302	4,1	1,5	1,4	1,2	3/4	1,5	1,5 / ,007
WX 8	200	320	8	3,0	2,6	2,4	3/4	1,5	2,3 / ,02
WX 18	280	380	18	6,7	6,0	5,4	3/4	1,5	4,1 / ,03
WX 33	280	630	33	12,4	10,9	9,9	3/4	1,5	6,8 / ,05

FIGURE 28.15 Specifications for in-line pressure tanks. (*Courtesy of McGraw-Hill.*)

10 bar series

Model No.	Dimensions		Total Volume
	Diameter	Height	
	mm	mm	Ltr
WL 1855	560	805	80
WL 1856	560	1240	180
WL 1858	560	1700	300
WL 1859	750	1880	600
WL 1860	750	2340	800
WL 1861	1000	1960	1000
WL 1862	1000	2740	1600
WL 1863	1200	2493	2000

Design with stainless steel system connection (V4A)

WL 1935	560	805	80
WL 1936	560	1240	180
WL 1938	560	1700	300
WL 1939	750	1880	600
WL 1940	750	2340	800
WL 1941	1000	1960	1000
WL 1942	1000	2740	1600
WL 1943	1200	2493	2000

16 bar series

WL 1955	560	805	80
WL 1956	560	1240	180
WL 1958	560	1700	300
WL 1959	750	1880	600
WL 1960	750	2340	800
WL 1961	1000	1960	1000
WL 1962	1000	2740	1600
WL 1963	1200	2493	2000

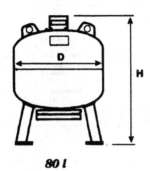

80 l

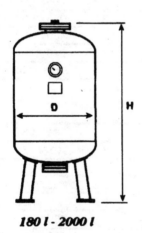

180 l - 2000 l

Maximum Operating Temperature = 90°C.
Horizontal designs and tanks for Operating Pressures of 25 bar are available on request.
Note: Drawdown can be affected by ambient and system conditions, including temperature and pressure.

FIGURE 28.16 Specifications for pressure tanks with replaceable bladder designs. *(Courtesy of McGraw-Hill.)*

| 1,5/3,0 bar | 2,0/3,5 bar | 2,5/4,0 bar | System | Precharge | Shipping |
| | Drawdown | | Connection | Pressure | Wt./Vol. |
Liter	Liter	Liter	R"	bar	KG / m3
30	27	24	2	3,5	59 / ,25
68	60	54	2	3,5	83 / ,39
113	99	90	2	3,5	155 / ,53
225	198	180	2	3,5	285 / 1,06
300	264	240	2	3,5	360 / 1,32
375	330	300	3	3,5	400 / 1,96
600	528	480	3	3,5	540 / 2,74
750	660	600	3	3,5	780 / 3,59

30	27	24	2	3,5	59 / ,25
68	60	54	2	3,5	83 / ,39
113	99	90	2	3,5	155 / ,53
225	198	180	2	3,5	285 / 1,06
300	264	240	2	3,5	360 / 1,32
375	330	300	3	3,5	400 / 1,96
600	528	480	3	3,5	540 / 2,74
750	660	600	3	3,5	7890 / 3,59

30	27	24	2	3,5	64 / ,25
68	60	54	2	3,5	102 / ,39
113	99	90	2	3,5	220 / ,53
225	198	180	2	3,5	400 / 1,06
300	264	240	2	3,5	505 / 1,32
375	330	300	3	3,5	560 / 1,96
600	528	480	3	3,5	756 / 2,74
750	660	600	3	3,5	1330 / 3,5

FIGURE 28.17 Specifications for pressure tanks with replaceable bladder designs. *(Courtesy of McGraw-Hill.) (Continued)*

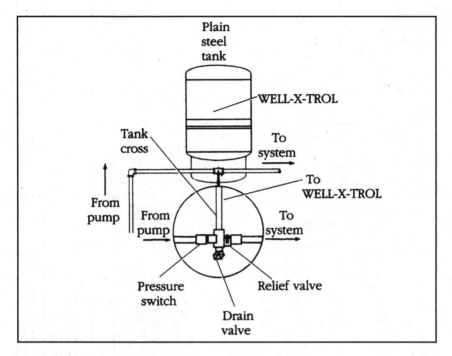

FIGURE 28.18 Detail for a tank-tee setup. *(Courtesy of McGraw-Hill.)*

		Dimensions		Total	Max.		Drawdown		Shipping
Model No.	Diameter (ins)	Height (ins)	Volume (gals)	Accept. Factor	20/40 (gals)	30/50 (gals)	40/60 (gals)	Wt. (Vol.) lbs (cu ft)	
WX-103-PS	12 1/2	11 1/4	8.6	0.28	3.1	2.7	2.2	20 (2.4)
WX-200-PS	17 1/4	15 5/8	14.0	0.81	5.2	4.3	3.8	29 (4.0)

Precharge Pressure is 30 PSIG and Sys. Conn. is 3/4" NPTM Fitting for 103-PS and 1" NPTF Coupling for 200-PS.
Maximum Working Pressure is 100 PSIG and Maximum Working Temperature is 200° F.

FIGURE 28.19 Pump-stand type of pressure tank. *(Courtesy of McGraw-Hill.)*

		Dimensions		Total	Max.		Drawdown		Shipping
Model No.	Diameter (ins)	Height (ins)	Volume (gals)	Accept. Factor	20/40 (gals)	30/50 (gals)	40/60 (gals)	Wt. (Vol.) lbs (cu ft)	
WX-200-UG	15 3/8	22	14.0	0.81	5.2	4.3	3.8	25 (3.8)
WX-202-UG	15 3/8	29 3/4	20.0	0.57	7.4	6.2	5.4	33 (4.9)
WX-250-UG	22	33 3/8	44.0	0.77	16.3	13.6	11.9	63 (11.0)
WX-251-UG	22	44 1/2	62.0	0.55	22.9	19.2	16.7	83 (13.9)

Precharge Pressure for Models 200-UG and 202-UG is 30 PSIG and Sys. Conn. is 1" NPTF Coupling.
Precharge Pressure for Models 250-UG and 251-UG is 38 PSIG; Sys. Conn. is 1 1/4" NPTF Coupling.
Maximum Working Pressure is 100 PSIG and Maximum Working Temperature is 200° F.

FIGURE 28.20 Underground pressure tank specifications. *(Courtesy of McGraw-Hill.)*

1.
WELL-X-TROL has a sealed-in air chamber that is pre-pressurized before it leaves our factory. Air and water do not mix eliminating any chance of "waterlogging" through loss of air to system water.

2.
When the pump starts, water enters the WELL-X-TROL as system pressure passes the minimum pressure precharge. Only usable water is stored.

3.
When the pressure in the chamber reaches maximum system pressure, the pump stops. The WELL-X-TROL is filled to maximum capacity.

4.
When water is demanded, pressure in the air chamber forces water into the system. Since WELL-X-TROL does not waterlog and consistently delivers the maximum usable water, minimum pump starts are assured.

FIGURE 28.21 How diaphragm pressure tanks work. *(Courtesy of McGraw-Hill.)*

Motor rating	Maximum starts per 24 hr. day	
	Single phase	Three phase
Up to ¾ hp	300	300
1 hp thru 5 hp	100	300
7½ hp thru 30 hp	50	100
40 hp and over		100

FIGURE 28.22 Recommended maximum number of times a pump should start in a 24-hour period. *(Courtesy of McGraw-Hill.)*

Pump discharge rate gpm (approx.)	Operating pressure—psig					
	20/40		30/50		40/60	
	ESP I	ESP II	ESP I	ESP II	ESP I	ESP II
2.5	WX-104	WX-201	WX-104	WX-202	WX-104	WX-202
5	WX-201	WX-205	WX-202	WX-205	WX-202	WX-250
7	WX-202	WX-250	WX-203	WX-251	WX-205	WX-251
10	WX-203	WX-251	WX-205	WX-302	WX-250	WX-302
12	WX-205	WX-302	WX-250	WX-302	WX-251	WX-350
15	WX-250	WX-302	WX-251	WX-350	WX-251	WX-350
20	WX-251	WX-350	WX-302	(2)WX-251	WX-302	(2)WX-302
25	WX-302	(2)WX-302	WX-302	(2)WX-302	WX-350	(3)WX-251
30	WX-302	(2)WX-302	WX-350	(1)WX-302 (1)WX-350	WX-350	(2)WX-350
35	WX-350	(1)WX-302 (1)WX-350	WX-350	(2)WX-350	(2)WX-251	(3)WS-302
40	WX-350	(2)WX-350	(2)WX-251	(3)WX-302	(2)WX-302	(1)WX-302 (2)WX-350

FIGURE 28.23 Sizing and selection information for pressure tanks. (*Courtesy of McGraw-Hill.*)

WELL-X-TROL QUICK SIZING FORM

(We suggest you make an office copy of this page when ready to calculate.)

For selecting WELL-X-TROLs for a different running time than ESP I or ESP II, and/or at pressure ranges the same or different than 20/40, 30/50, 40/60:

THINGS YOU MUST KNOW

1. System flow rate (pump capacity or discharge) _____ GPM

2. Desired running time, in minutes and fractions
of minutes (1.5 min. = 1 min. 30 sec.) _____ Min.

3. Pump cut-in, in gauge pressure _____ Psig

4. Pump cut-out, in gauge pressure _____ Psig

CALCULATING TANK SIZE

5. Multiply Line 1 by Line 2 and enter ESP Volume _____ ESP Vol.

6. Refer to Table 1. Find pressure factor for
Line 3 and Line 4 and enter _____ P.F.

7. Divide Line 5 by Line 6 and enter minimum
total WELL-X-TROL Volume _____ Gals.

8. Refer to Table 2 and select WELL-X-TROL model
that is greater than Line 7 for "Total Volume"
and Line 5 is less than "Maximum ESP Volume" _____ WX No.

9. Select precharge pressure _____ Psig

> NOTE: The precharge pressure *must* be adjusted to the pump cut-in pressure. For the following example reduce from 40 to 25 psig.

> *EXAMPLE:* A system flow will be delivered by a pump at a rate of 12.5 GPM. The pump switch is to be installed at the WELL-X-TROLL and has been determined to cut-in the pump at 25 psig. Its differential, or operating range, is 20 psi. It is desired to have the pump run *at least* one minute and 30 seconds every time it starts. Which WELL-X-TROL will provide "ESP"?

THINGS YOU MUST KNOW

1. System flow rate (pump delivery) 12.5 GPM

2. Desired running time, in minutes and
fractions of minutes (1.5 min. = 1 min. 30 sec.) 1.5 Min.

3. Pump cut-in, in gauge pressure 25 Psig

4. Pump cut-out, in gauge pressure 45 Psig

CALCULATING TANK SIZE

5. Multiply Line 1 by Line 2 and enter ESP Volume 18.8 ESP Vol.

6. Find Pressure factor for Line 3 and Line 4 in Table 1, and enter 34 P.F.

7. Divide Line 5 by Line 6 and enter minimum total
WELL-X-TROL volume 55.2 Gals.

8. Refer to Table 2 and select WELL-X-TROL model
that is greater than Line 7 for "Total Volume"
and Line 5 is less than "Maximum ESP Volume" WX-251

**A WX-251 has a total volume of 62 gallons
and a maximum ESP volume of 34 gallons.**

FIGURE 28.24 Pressure tank sizing form. *(Courtesy of McGraw-Hill.)*

CHAPTER 29
PROFESSIONAL PRINCIPLES

What are professional principles? Professional principles are the qualities that separate professional plumbers from renegades. These principles are what make the plumbing profession a trade to be proud of. It is no longer enough to pick up a torch or a pipe wrench and call yourself a plumber. Today, in most areas, a professional plumber must be educated and tested about that education before a plumber's license will be issued. The plumbing codes, the plumbing inspectors, and the plumbers are all a part of what has made plumbing a respectable and profitable career.

For anyone in the plumbing trade, ethics and professional principles are paramount to maintaining the good image plumbers have come to enjoy. Before you have a plumber's license, you might wish licenses were not required. After you hold a license, however, you will be glad the profession is regulated. It is the regulation of the trade that keeps it high on the list of quality incomes. Licensing requirements prohibit just anyone from being a plumber. By requiring all plumbers to be licensed properly, the field of competition is narrowed, and demand for good plumbers remains high. It is this high demand that opens employment opportunities for licensed plumbers and affords them a better-than-average income.

The plumbing trade has changed a lot since I entered it. When I started plumbing, I was required to spend my first six months in the trade digging ditches and running a jackhammer. Being young, and not well paid, I considered quitting the trade. It was not until after I buckled down and applied myself that I learned the reason for the hard labor in those first six months.

My employer didn't want to spend the time and money training me to be a plumber, unless I proved I wanted it badly enough. By sticking with it, I was given the opportunity to work with one of the best plumbers I have ever seen, even to this day. That man, Jerry, trained me to be a real plumber. Jerry didn't cut corners, and he showed me every aspect of the plumbing trade. I learned how to produce quality work in a reasonable time. I learned how to work with inspectors, not against them. In general, Jerry made a very good plumber out of a boy who was not too mechanically inclined, at the time.

Times changed and I moved from the foothills of the Blue Ridge Mountains to the big city. When I moved, I was given the position of working foreman. I learned quickly that plumbing wasn't done with the same care and quality in the city that I had learned in the country. I had to adjust my work habits to allow for faster production. Although I increased my speed and didn't spend as much time wiping pipe joints and making my work perfect, I refused to do shoddy work.

My personal requirements for professional work cost me a few jobs. Employers were not interested in the quality of the work—only the speed and profit from the job. Being spurred by bad feelings, I tested for my master's license and got it. I was as proud of that license as any other accomplishment I had ever made. I continued to work for other people for some time, but it wasn't long before I opened my own business.

The rest is history. I have been a successful plumbing contractor, along with other ventures, for many years. I attribute much of my personal and financial success to my plumbing endeavors. There have been times when it seemed that if I wasn't a master plumber, I would have gone hungry. But my plumber's license has never let me down; it always has enabled me to survive and prosper, even through two recessions.

It might seem that I have been rambling about myself, and I have, but for a good reason. I've been involved with plumbing for more than twenty years, and I know what the trade is all about. For you, the trade might only seem like a job, but it is much more, if you want it to be.

Holding a master plumber's license gives you the opportunity to own your own business. One that is known for its high profits. This opportunity doesn't come without its price. You must dedicate yourself to the trade. You must be willing to pay your dues before you can reap your rewards. If you are looking for a fast track to wealth, plumbing isn't it. But if you are looking for a good career, with unlimited opportunity, plumbing fits the bill.

What does all of this have to do with professional principles? Everything. Professional principles and ethics are the core of any honorable profession. When you assume the role of a plumber, you must bear the responsibilities that go with the position. This book already has taught you many of the rules, regulations, and laws you must follow. This chapter is going to go a little deeper, and teach you how to be a better plumber.

LICENSING

The first step toward becoming a professional plumber is earning a license to practice the trade. For many years a license was not a big deal. Today, however, most jurisdictions require that anyone earning a living with plumbing have some type of license. In the old days, there were plumber's helpers; today, there are apprentices. By current standards, if you are going to solder joints or glue pipe, you generally are required to have at least an apprentice license.

Digging ditches and carrying pipe can be done without a license, but to become involved enough to learn the trade, you usually need an apprentice license. These licenses are easy to obtain, and they are the first step toward becoming a full-fledged plumber. Generally, all that is required to obtain an apprentice license is the filling out and filing of an application. Apprentice plumbers are not allowed to perform plumbing duties without the direct supervision of either a journeyman or master plumber.

The next license to work toward is a journeyman's license. These licenses are not as easy to come by. Requirements vary, but they might include educational requirements, as well as on-the-job training. The educational requirements can mean spending years in vocational school. The on-the-job training period is usually 4,000 hours of work. This works out to be about two years of full-time work. In most cases, a written test will be administered to determine your knowledge of proper plumbing procedures.

After educational requirements and on-the-job training are completed, an individual is qualified to take the licensing test. These licensing tests are not easy. In fact, they are often somewhat tricky. The tests usually consist of multiple-choice questions. The degree of difficulty encountered with the test for a journeyman license is much less than that for a master's license, but it is not a task to be taken lightly.

Some areas might waive the requirement for on-the-job training if enough classroom education is received. Likewise, some areas might waive the educational requirements, as long as the amount of on-the-job training is adequate. Students who gain all of their experience in the classroom might not have to take an outside licensing test. Their tests might be administered in the classroom apprenticeship program.

After a person becomes a journeyman plumber, that individual is allowed to perform plumbing, without direct supervision. Journeyman plumbers generally are not allowed to go into business for themselves, unless they employ a master plumber. Most areas require journeyman plumbers to work under the supervision of a master plumber, but not direct supervision. This basically means that a journeyman plumber can go out on a job and work alone, as long as a master plumber reviews the work from time to time.

After becoming licensed as a journeyman plumber, an individual usually is required to work for a period of time before sitting for the master plumber's test. This period of time is usually 2,000 hours, or about one year of full-time work. After the time requirements are met, an individual can take the test for a master's license. These tests are the most difficult of all.

Like the journeyman's test, the master's test also contains numerous multiple-choice questions. The type and complexity of these questions, however, are more demanding than those for a journeyman's license. A master's applicant generally will have to perform some type of pipe sizing for plumbing and storm drainage. Master licenses are awarded only to those with a thorough knowledge of the plumbing code. Many first-time applicants are unsuccessful in their quest for a passing grade.

To improve the odds of passing either the journeyman or master test, individuals can attend test-preparation courses. These courses, like the one I taught, are designed to hone the knowledge of students and to prepare them for taking their licensing exams. These courses usually are offered in the evenings. The length of the course varies, but they usually are held one or two nights a week, for several weeks.

If students pay attention, these classes offer a big advantage. The instructor usually has a good feel for the type of material the students will be tested on for their licensing requirements. It is standard procedure in these exam preparation courses for the instructor to give many tests during the course. The tests often are similar to the ones encountered when taking a licensing exam. By taking these sample tests, the students learn many aspects of passing the real test.

The students learn what areas they are deficient in. They learn how to study, and this is a big factor in passing any test. Instructors help clear confusion students might have in interpreting the local plumbing code. By the end of a code-awareness class, students have a significant edge on the day of their licensing exams.

One of the most important aspects of passing a licensing exam is understanding the plumbing code. The way licensing tests often are worded, simply memorizing the code book is not enough to pass the tests. To achieve good scores, applicants must be able to interpret and apply the code. This is often much more difficult than simply being able to quote the code in chapter and verse.

After the licensing tests are taken and passed, a license will be forthcoming. It is very important to pay particular detail to the answers given on the application for a license. By not providing a truthful account of your past, you might be rejected from consideration for licensing. If a license is issued based on false or incomplete application information, the license might be revoked. Something as simple as forgetting traffic violations can cause trouble with your application. Answer all questions, and answer them honestly.

After the license is issued, it usually is required to be displayed in the place of business where you will work. This generally holds true for apprentices, journeymen, and masters. Most areas also issue a small, pocket-sized license, along with the wall certificate. These pocket cards generally are required to be carried at all times, when plumbing is being performed. The cards allow a means for a code enforcement officer to verify the license status of all people working on a job.

After you have your license, don't do anything to risk losing it. If a permit is required for the work you will be performing, obtain a permit before starting the work. Do all work in compliance with the local code requirements. Never try to fool an inspector with tricks. Most inspectors know the tricks and will catch you. Even if you don't lose your license, you will lose the respect of the code officer. Consequently, all of your future work will be inspected very closely.

Abide by all code requirements. Some jurisdictions require a plumber's license number to be listed in all advertisements and on all work vehicles. If this is a requirement in your area, comply with it. Some jurisdictions require continuing education for licensed plumbers. If you want to keep your license, you must attend these seminars. You will be doing yourself a favor to learn from the seminars, rather than simply attending them.

For your own benefit, join plumbing organizations and read; read a lot. The more you learn, the better prepared you will be to make a better living. Don't argue with code officers. If you believe they are wrong, approach the subject, but do so diplomatically. Remember, much of the plumbing code is left to the interpretation of the local code officer.

Avoid doing sloppy work. Not only will inspectors check your work more closely when it is shoddy, you will not be in as much demand as a plumber. Quality work pays off in many ways. When a job is done right, you eliminate callbacks. Having to go back and fix improper work gets expensive. If you have your own business, it will not take long to see the advantages of doing the job right the first time. If you are an employee who is frequently creating call-backs, you might not be employed for long.

Good work will result in referral business. There is no better source of new business than referrals. By giving customers their money's worth, you are building business for yourself and your employer. Even if you don't own the company, if the company doesn't have ongoing work, neither do you.

Take pride in your chosen field, and don't let the new image of plumbers be tarnished. It takes a multitude of professional plumbers to build a strong image, but it only takes a few bad plumbers to ruin the image.

Lastly, take the time to train new plumbers. Remember, you wouldn't have your present opportunities if someone had not taken the time to train you properly. With rising costs and the desire to cut expenses, too many plumbers don't spend the time to train new help. This creates a void in the trade. It is also a fact that today's worker is not as interested in learning the trade as in making money. Many young people don't understand that they must learn, before they can earn.

In closing, remember this, plumbers protect the health of the nation. Without plumbers, our world would be a much worse place to live, work, and play. Today, more than ever, water conservation, pollution control, and sanitary conditions are primary concerns. It is up to professional plumbers to see that these needs are met and improved.

APPENDIX A
TRADE ASSOCIATIONS

Acoustical Society of America (ASA)
335 East 45th Street
New York, NY 10017
(212) 349-7800

Adhesive and Sealant Council (ASC)
1500 Wilson Blvd., Suite 515
Arlington, VA 22209
(703) 841-1112

Air and Waste Management Association
(AWMA)
P.O. Box 2861
Pittsburgh, PA 15230
(412) 232-3444

Air Conditioning and Refrigeration Institute
(ARI)
1501 Wilson Boulevard
Arlington, VA 22209
(703) 524-8800

Air Conditioning Contractors of America
(ACCA)
1513 16th Street NW
Washington, DC 20036
(202) 483-9370

Air Diffusion Council (ADC)
230 North Michigan Avenue, Suite 1200
Chicago, IL, 60601
(312) 372-9800

Air Distribution Institute (ADI)
4415 West Harrison Street, Suite 242-C
Hillside, IL 60162
(708) 449-2933

Air Movement and Control Association
(AMCA)
30 West University Drive
Arlington Heights, IL 60004
(708) 449-2933

Alliance for Engineering in Medicine and
Biology
1101 Connecticut Avenue NW, Suite 700
Washington, DC 20038
(202) 857-0666

Alliance of American Insurers (AAI)
1501 Woodfield Avenue
Schaumburg, IL 60173
(708) 330-8500

Alliance to Save Energy
1725 K Street NW, Suite 914
Washington, DC 20006
(202) 857-0666

Aluminum Association (AA)
900 19th Street NW, Suite 300
Washington, DC 20006
(202) 862-5100

Amateur Athletic Union (AAU)
3400 West 86th Street
Indianapolis, IN 46468
(317) 872-2900

American Architectural Manufacturers
   Association (AAMA)
2700 River Road, Suite 118
Des Plaines, IL 60018
(708) 699-7310

American Association for Accreditation of
   Laboratory Animal Care (AAALAC)
208A North Cedar Road
New Lenox, IL 60451
(815) 485-7101

American Association for Laboratory
   Accreditation
656 Quince Orchard Road
Gaithersburg, MD 20878
(301) 670-1377

American Association of Cost Engineers
   (AACE)
P.O. Box 1557
Morgantown, WV 26507
(304) 296-8444

American Association of Engineering Societies
   (AAEC)
415 Second Street NE, Suite 200
Washington, DC 20002
(202) 546-2237

American Association of Nurserymen (AAN)
1250 I Street NW, Suite 500
Washington, DC 20005
(202) 789-2900

American Association of State Highway and
   Transportation Officials (AASHTO)
444 North Capitol Street NW, Suite 225
Washington, DC 20001 (202) 624-5811

American Backflow Prevention Association
   (ABPA)
P.O. Box 1563
Akron, OH 44309-1563
(216) 375-2637

American Boiler Manufacturers Association
   (ABMA)
950 North Globe Road, Suite 160
Arlington, VA 22203
(703) 522-7350

American Builders and Contractors (ABC)
729 15th Street, NW
Washington, DC 20005
(202) 637-8800

American Chemical Society (ACS)
1155 16th Street, NW
Washington, DC 20038
(202) 872-4600

American Concrete Institute International
   (ACI)
P.O. Box 19151
Detroit, MI 48219
(313) 532-2600

American Concrete Pipe Association (ACPA)
8300 Boone Boulevard, Suite 400
Vienna, VA 22182
(703) 821-1990

American Concrete Pressure Pipe Association
8300 Boone Boulevard, Suite 400
Vienna, VA 22182
(703) 821-3054

American Conference of Government Industrial
   Hygienists
6500 Glenway Avenue
Bridgetown, OH
(513) 661-7881

American Consulting Engineers Council
   (ACEC)
1015 15th Street NW, Suite 802
Washington, DC 20005
(202) 347-7474

American Contractors Association (ACA)
1004 Duke Road
Alexandria, VA 22314
(703) 684-3450

American Council of Independent Laboratories
   (ACIL)
1725 K Street NW, Suite 412
Washington, DC 20006
(202) 887-5872

American Design Drafting Association
   (ADDA)
966 Hungerford Drive, Suite 10-B
Rockville, MD 20850
(301) 294-8712

American Engineering Model Society (AEMS)
P.O. Box 2066
Aiken, SC 29802
(803) 649-6710

American Filtration Society
2360 Highway 59
Kingswood, TX
(713) 540-2116

American Fire Sprinkler Association (AESA)
11325 Pegasus, Suite S-20
Dallas, TX 75328
(214) 349-5965

American Galvanizers Association
315 South Patrick Street
Alexandria, VA 22314
(800) 468-7723
(703) 549-0000

American Gas Association (AGA)
1515 Wilson Blvd.
Arlington, VA 22209
(800) 336-4795
(703) 841-8400

American Hardboard Association (AHA)
520 North Hicks Road
Palatine, IL 60067
(708) 934-8800

American Health Care Association (AHCA)
1201 L Street NW
Washington, DC 20005
(202) 842-4444

American Hospital Association
840 North Lake Shore Drive
Chicago, IL 60611
(800) 242-2626
(312) 180-5223

American Industrial Hygiene Association
475 Wolf Ledges Parkway
Akron, OH 44311
(216) 762-7294

American Institute of Architects (AIA)
1735 New York Avenue NW
Washington, DC 20006
(202) 626-7300

American Institute of Chemical Engineers
345 East 47th St.
New York, NY 10017
(212) 705-7338

American Institute of Plant Engineers (AIPE)
3975 Erie Avenue
Cincinnati, OH 45208
(513) 561-6000

American Institute of Timber Construction
11818 Southeast Mill Plains Blvd., Suite 415
Vancouver, WA 98484
(206) 254-9132

American Institute of Steel Construction
(AISC)
400 North Michigan Avenue, 8th Floor
Chicago, IL 60611
(312) 670-2400

American Insurance Association (AIA)
1130 Connecticut Avenue NW, Suite 1000
Washington, DC 20036
(202) 828-7100

American Iron and Steel Institute (AISI)
1000 16th Street NW
Washington, DC 20036
(216) 835-3040

American Lumber Standards Committee
(ALSC)
P.O. Box 210
Germantown, MD 20874
(301) 972-1700

American National Metric Council (ANMC)
1010 Vermont Avenue NW, Suite 320
Washington, DC 20005
(202) 628-5757

American National Standards Institute (ANSI)
1430 Broadway
New York, NY 10018
(212) 642-4900

American Nuclear Insurers (ANI)
The Exchange, Suite 245
Farmington, CT 06032
(203) 677-7305

American Nuclear Society (ANS)
555 North Kensington Avenue
La Grange Park, IL 60525
(708) 352-6611

American Petroleum Institute (API)
1220 L Street NW
Washington, DC 20005
(202) 682-8000

American Pharmaceutical Association (APA)
2215 Constitution Avenue NW
Washington, DC 20037
(202) 628-4410

American Pipe Fittings Association (APFA)
203 Old Keene Mill Court
Springfield, VA 22152
(703) 644-0001

American Plywood Association (APA)
7011 South 19th Street, P.O. Box 11700
Tacoma, WA 98411
(206) 565-6600

American Production & Inventory Control
Society
500 West Annandale Road
Falls Church, VA 22046
(703) 237-8344

American Public Gas Association (APGA)
P.O. Box 1426
Vienna, VA 22183
(703) 281-2910

American Public Health Association
1015 15th Street NW
Washington, DC 20005
(202) 789-5600

American Public Power Association (APPA)
2301 M Street NW
Washington, DC 20037
(202) 775-8300

American Public Works Association (APWA)
1313 East 60th Street
Chicago, IL 60637
(312) 667-2200

American Railway Engineering Association
50 F Street NW
Washington, DC 20001
(202) 639-2190

American Rental Association (ARA)
1900 19th Street
Moline, IL 61265
(309) 764-2475

American Road and Transportation Builders
Association (ARTBA)
525 School Street, ARTBA Building
Washington, DC 20024
(202) 488-2722

American Society for Engineering Education
(ASEE)
11 DuPont Circle
Washington, DC 20036
(202) 293-7080

American Society for NonDestructive Testing
(ASNT)
4153 Arlington Plaza
Columbus, OH 43228
(614) 274-6003

American Society for Quality Control (ASQC)
P.O. Box 3005
Milwaukee, WI 53201-3005
(414) 272-8575

American Society for Testing and Materials
(ASTM)
1916 Race Street
Philadelphia, PA 19103
(215) 299-5400

American Society for Training and
Development
1630 Duke Street, P.O. Box 1443
Alexandria, VA 22313
(703) 683-8100

American Society of Agricultural Engineers
(ASEA)
2950 Niles Road
St. Joseph, MI 49085
(616) 429-0300

American Society of Anesthesiologists (ASA)
520 North Northwest Highway
Park Ridge, IL 60068
(708) 825-5586

American Society of Association Executives
(ASAE)
1575 I Street NW
Washington, DC 20005-1168
(202) 626-2723

American Society of Gas Engineers (ASGE)
P.O. Box 936
Tinley Park, IL 60477
(708) 532-5707

American Society of Heating,Refrigeration
and Air Conditioning Engineers (ASHRAE)
1791 Tullie Circle NE
Atlanta, GA 30329
(404) 636-8400

American Society of Horticultural Sciences
(ASHS)
113 South West Street
Alexandria, VA 22307
(703) 836-4606

American Society of Hospital Engineers
840 North Lake Shore Drive
Chicago, IL 60611
(312) 280-6144

American Society of Interior Designers (ASID)
1430 Broadway
New York, NY 10018
(212) 944-9220

American Society of Landscape Architects
(ASLA)
4401 Connecticut Avenue NW,, 5th Floor
Washington, DC 20008 (202) 686-2752

American Society of Mechanical Engineers
(ASME)
345 East 47th Street
New York, NY 10017
(212) 705-7722

American Society of Microbiology (ASM)
1913 I Street NW
Washington, DC 20006
(202) 737-3600

American Society of Plumbing Engineers
(ASPE)
3617 Thousand Oaks Blvd., Suite 210
Westlake, CA 91362
(805) 495-7120

American Society of Professional Estimators
(ASPE)
6911 Richmond Highway, Suite 230
Alexandria, VA 22306
(703) 765-2700

American Society of Safety Engineers
1800 East Oakton Street
Des Plaines IL 60018
(708) 692-4121

American Society of Sanitary Engineering
P.O. Box 40362
Bay Village, OH 44140
(216) 835-3040

American Solar Energy Society (ASES)
2400 Central Avenue
Boulder, CO 80301
(303) 443-3130

American Subcontractors Association (ASA)
1004 Duke Street
Alexandria, VA 22314
(703) 684-3450

American Supply Association
20 North Wacker Drive, Suite 2260
Chicago, IL 60606
(312) 236-4082

American Water Works Association (AWWA)
6666 West Quincy Drive
Denver, CO 80235
(303) 794-7711

American Welding Society (AWS)
550 Le June Road NW, P.O. Box 351040
Miami, FL 33135
(305) 443-9353

American Wood Preservers Association
(AWPA)
P.O. Box 5283
Springfield, VA 21666
(703) 339-6660

American Wood Preservers Bureau (AWPB)
P.O. Box 6058
2772 South Randolph Street
Arlington, VA 22206
(703) 931-8180

Anthracite Industry Association
1110 Penn Plaza, Suite 1000
New York, NY 10001
(212) 279-3580

Arbitration Forums
200 White Plains Road
Tarrytown, NY 10591-0066
(914) 332-4770

Architectural Woodwork Institute (AWI)
2310 South Walter Reed Drive
Arlington, VA 22206
(703) 671-9100

Asbestos Abatement Council (AAC)
1600 Cameron Street
Alexandria, VA 22314
(703) 684-2924

Asbestos Abatement Equipment Distributors
Association
5875 Peachtree Industrial Blvd., Suite 370
Norcross, GA 30092
(800) 222-5252

Asbestos Information Association
1745 Jefferson Davis Highway
Arlington, VA 22202
(703) 979-1150

Asbestos Litigation Group
151 Meeting Street, Suite 600
Charleston, SC 29402
(803) 577-6747

Asphalt Institute (AI)
Asphalt Institute Building
College Park, MD 20740
(301) 277-4258

Associated Air Balance Council (AABC)
1518 K Street NW, Suite 503
Washington, DC 20005
(202) 737-0202

Associated Equipment Distributors (AED)
615 West 22nd Street
Oak Brook, IL 60521
(708) 574-0650

Associated General Contractors of America
  (AGC)
1957 East Street NW
Washington, DC 20006
(202) 393-2040

Associated Laboratories (ALI)
8 Brush Street
Pontiac, MI 48053
(312) 358-7400

Associated Specialty Contractors (ASC)
7315 Wisconsin Avenue, 13th Floor
Bethesda, MD 20814
(301) 657-3110

Association for the Advancement of Medical
  Instrumentation (AAMI)
3330 Washington Blvd., Suite 400
Arlington, VA 22201
(703) 525-4890

Association of American Railroads (AAR)
50 F Street
Washington, DC 20001
(202) 639-2190

Association of Construction Equipment
  Managers (ACEM)
PO. Box 43859
Louisville, KY 40243
(502) 244-2574

Association of Diesel Specialists (ADS)
9140 Ward Parkway, Suite 200
Kansas City, MO 64114
(816) 444-3500

Association of Energy Engineers
4025 Pleasantdale Road, Suite 420
Atlanta, GA 30340
(404) 447-5083

Association of Home Appliance Manufacturers
  (AHAM)
20 North Wacker Drive
Chicago, IL 60606
(312) 984-5800

Association of Manufacturing Technology
  (AMT)
7901 Westpark Drive
McLean, VA 22101-4269
(703) 827-5520

Association of Physical Plant Administrators,
  University and Colleges
1446 Duke Street
Alexandria, VA 22314
(703) 684-1446

Association of School Business Officials,
  International
11401 North Shore Drive
Reston, VA 22090-4232
(703) 478-0405

BCR National Laboratory (BCR)
500 William Pitt Way
Pittsburgh, PA 15238
(412) 826-3030

Better Heating-Cooling Council
P.O. Box 218
35 Russo Court
Berkeley Heights, NJ 07922
(201) 464-8200

Brick Institute of America (BIA)
11490 Commerce Park Drive, Suite 300
Reston, VA 22091
(703) 620-0010

Builders Hardware Manufacturers Association
  (BHMA)
60 East 42nd Street, Room 511
New York, NY 10016
(212) 682-8142

Building Officials & Code Administrators
  International (BOCA)
4051 West Flossmoor Road
Country Club Hills, IL 60477-5795
(708) 799-2300

Building Owners and Managers Association
International
1201 New York Avenue NW, Suite 300
Washington, DC 20005
(202) 289-7000

Building Research Board
2101 Constitution Avenue NW
Washington, DC 20418
(202) 334-3376

Bureau of Land Management
Interior Building
18th and C Streets
Washington, DC 20245
(202) 343-1801

Cast Iron Soil Pipe Institute (CISPI)
5959 Shallowford Road, Suite 419
Chattanooga, TN 37412
(615) 892-0137

Casting Industry Suppliers Association (CISA)
6990 Rieber Street
Worthington, OH 43085
(614) 848-8199

Center for Disease Control (CDC)
1600 Clifton Road
Atlanta, GA
(404) 639-3311

Certified Ballast Manufacturers Association
Hanna Building
1422 Euclid Avenue, Suite 772
Cleveland, OH 44115
(216) 241-0711

Chemical Manufacturers Association (CMA)
2501 M Street NW
Washington, DC 20037
(202) 887-1100

Chemical Specialties Manufacturers
Association (CSMA)
1001 Connecticut Avenue NW
Washington, DC 20036
(202) 872-8110

Coastal Engineering Research Board (CERB)
P.O. Box 631
Vicksburg, MS 39180
(601) 634-2513

Cold Regions Research & Engineering
Laboratory (CRREL)
U.S. Department of Defense
72 Lyme Road
Hanover, NH 03755
(603) 646-4200

Color Association of the United States (CAUS)
343 Lexington Avenue
New York, NY 10016
(212) 683-9531

Combustion Institute
5001 Baum Blvd.
Pittsburgh, PA 15213
(412) 687-1366

Compressed Air and Gas Institute (CAGI)
1230 Kieth Building
Cleveland, OH 44115
(216) 241-7333

Compressed Gas Association (CGA)
1235 Jefferson Davis Highway
Arlington, VA 22202
(703) 979-0900

Concrete Reinforcing Steel Institute (CRSI)
933 Plum Grove Road
Schaumburg, IL 60195
(708) 517-1200

Construction Engineering Research Laboratory
U.S. Department of Defense
P.O. Box 4005
Champaign, IL 61820
(217) 373-7201

Construction Industry Manufacturers
Association (CIMA)
111 East Wisconsin Avenue, Suite 1700
Milwaukee, WI 53202
(414) 272-0943

Construction Specifications Institute (CSI)
601 Madison Street
Alexandria, VA 22314
(703) 884-0300

Consumer Product Safety Commission (CSPC)
5401 Westbard Avenue
Bethesda, MD 20814
(301) 492-6800

Consumer Products Safety Council (CPSC)
1111 18th Street NW
Washington, DC 20207
(202) 634-7700

Controlled Release Society (CRS)
16 Nottingham Drive
Lincolnshire, IL 60069
(708) 940-4277

Cooling Tower Institute (CTI)
P.O. Box 73383
Houston, TX 77273
(713) 583-4087

Copper Development Association (CDA)
Greenwich Office Park 2, Box 1840
Greenwich, CT 06836
(203) 625-8210

Corps of Engineers (COE)
U.S. Department of Defense
Washington, DC 20314
(202) 272-0660

Cosmetic, Toiletry & Fragrance Association
1110 Vermont Avenue NW, Suite 800
Washington, DC 20005 (202) 331-1770

Council of American Building Officials
5203 Leesburg Pike, Suite 708
Falls Church, VA 22041
(703) 931-4533

Cryogenic Society of America
c/o Huget Advertising
1033 South Blvd.
Oak Park, IL 60302
(312) 383-8848

Deep Foundations Institute (DFI)
P.O. Box 281
Sparta, NJ 07871
(201) 729-9679

Delaware River Basin Commission
1100 L Street NW
Washington, DC 20240
(202) 343-5761

Ductile Iron Research Association (DIRA)
245 Riverchase Parkway East, Suite D
Birmingham, AL 35244 (205) 988-9870

Economic Development Administration
15th & Constitution Avenues
Washington, DC 20203
(202) 377-5081

Edison Electric Institute (EEl)
1111 19th Street NW
Washington, DC 20207
(202) 778-6400

Electric Power Research Institute (EPRI)
P.O. Box 10412
Palo Alto, CA 94303
(415) 855-2000

Electrical Apparatus Service Association
(EASA)
1331 Baur Boulevard
St. Louis, MO 63132
(314) 993-2220

Electronic Industries Association (EIA)
1722 I Street NW
Washington, DC 20006
(202) 457-4900

Engine Manufacturers Association (EMA)
111 East Wacker Drive
Chicago, IL 60601
(312) 644-6610

Environmental Industry Council (EIC)
1825 K Street NW, Suite 210
Washington, DC 20006
(202) 331-7706

Environmental Management Association
(EMA)
1019 Highland Avenue
Largo, FL 34640
(813) 586-5710

Environmental Protection Agency (EPA)
Waterside Mall
401 M Street NW
Washington, DC 20460
(202) 382-2080

Environmental Resource Center (ERC)
3679 Rosehill Road
Fayetteville, NC 28311
(800) 537-2372

Equipment Maintenance Council (EMC)
113 Highland Lake Road
Lewisville, TX 70567
(214) 436-9257

Expansion Joint Manufacturers Association
25 North Broadway
Tarrytown, NY 10591
(914) 332-0040

Factory Mutual Engineering and Research
(FM)
1151 Boston-Providence Tpk.
Norwood, MA 02062
(617) 762-4300

Federal Aviation Administration (FAA)
800 Independence Avenue SW
Washington, DC 20591
(202) 267-3111

Federal Communications Commission (FCC)
1919 M Street NW
Washington, DC 20554
(202) 632-7000

Federal Highway Administration (FHA)
400 7th Street SW
Washington, DC 20590
(202) 366-0650

Federal Specifications (FS)
Superintendent of Documents
U.S. Government Printing Office
Washington, DC 20402
(202) 783-3238

Fire Apparatus Manufacturers Association
c/o Spartan Motors Inc.
P.O. Box 440
Charlotte, MI 48813
(517) 543-6400

Fire Equipment Manufacturers and Services
Association
1776 Massachusetts Avenue NW
Washington, DC 20036
(202) 659-0600

Fire Information Research and Education
Center (FIRE)
550 Cedar Street
St. Paul, MN 55101
(612) 296-6516

Fish and Wildlife Service
Interior Building
18th and C Streets
Washington, DC 20245
(202) 343-4717

Fluid Controls Institute (FCI)
P0. Box 9036
Morristown, NJ 07960
(201) 829-0990

Fluid Power Institute (FPI)
31 South Street, Suite 303
Morristown, NJ 07960
(201) 829-0990

Fluid Sealing Association (FSA)
2017 Walnut Street
Philadelphia, PA 19103
(215) 569-3650

Food and Drug Administration (FDA)
5600 Fishers Lane
Rockville, MD 20853
(301) 443-3170

Forest Service
14th and Independence Avenues SW
Washington, DC 20250
(202) 446-6661

Forging Industry Association (FIA)
1121 Illuminating Building
Cleveland, OH 44115
(216) 781-6260

Foundation for Cross Connection Control and
Hydraulic Research
University of Southern California
KAP-200 University Park, MC-2531
Los Angeles, CA 90089-2531
(213) 743-2032

Gas Appliance Manufacturers Association
(GAMA)
1901 North Moore Street, Suite 1100
Arlington, VA 22209
(703) 525-9565

Gas Processors Association (GPA)
6526 East 60th Street
Tulsa, OK 74145
(91 8), 493-3872

Gas Research Institute (GRI)
8500 West Bryn Mawr Avenue
Chicago, IL 60631
(312 399-8100

General Services Administration (GSA)
18th and F Streets, NW
Washington, DC 20405
(202) 566-0628

Geological Survey Department National Center
12201 Sunrise Valley Drive
Reston, VA 22091
(703) 860-7411

Geothermal Resources Council (GRC)
P.O. Box 1350
Davis, CA 95617
(916) 758-2360

Government Printing Office
North Capitol and G Streets NW
Washington, DC 20402
(202) 783-3238

Gypsum Association (GA)
101 South Wacker Drive
Chicago, IL 60606
(312) 606-4000

Hardwood Manufacturers Association (HMA)
805 Sterick Building
Memphis, TN 38103
(901) 525-8221

Hardwood Plywood Manufacturers Association
1825 Michael Faraday Drive, P.O. Box 2789
Reston, VA 22090
(703) 435-2900

Hazardous Waste Research and Information
  Center
c/o University of Illinois, Urbana-Champaign
1 East Hazelwood Drive
Champaign, IL 61820
(217) 333-8940

Health Industry Manufacturers Association
1030 15th Street NW
Washington, DC 20036
(202) 452-8240

Historical Construction Equipment Association
6604 Breeds Hill Road
Indianapolis, IN 46237
(317) 782-3612

Hydraulic Institute
14600 Detroit Avenue
Cleveland, OH 44107
(216) 226-7700

Hydraulic Tool Manufacturers Association
  (HTMA)
P.O. Box 1337
Milwaukee, WI 53201
(414) 639-6770

Hydronics Institute
35 Russo Place, Box 218
Berkeley Heights, NJ 07922
(201) 464-8200

Illuminating Engineering Society of North
  America (IESNA)
235 East 47th Street
New York, NY 10017
(212) 705-7900

Industrial Biotechnology Association (IBA)
1625 K Street NW
Washington, DC 20006
(202) 875-0244

Industrial Health Foundation (IHF)
34 Penn Circle West
Pittsburgh, PA 15206
(412) 363-6600

Industrial Heating Equipment Association
  (IHEA)
1901 North Moore Street
Arlington, VA 22209
(703) 525-2513

Industrial Risk Insurers (IRI)
85 Woodland Street
Hartford, CT 06012
(203) 520-7300

Industrial Safety Equipment Association
1901 North Moore Street, Suite 501
Arlington, VA 22209-1706
(703) 525-1695

Institute of Business Designers (IBD)
341 Merchandise Mart
Chicago, IL 60654
(708) 467-1950

Institute of Electrical and Electronic Engineers
345 East 47th Street
New York, NY 10017
(212) 705-7900

Institute of Environmental Sciences (IES)
940 East Northwest Highway
Mt. Prospect, IL 60056
(312) 255-1561

Institute of Gas Technology
3424 South State Street
Chicago, IL 60616
(708) 768-0500

Institute of Industrial Engineers
25 Technology Park
Atlanta, GA 30092
(404) 449-0460

Instrument Society of America (ISA)
P.O. Box 12277
67 Alexander Drive
Research Triangle Park, NC 27709

Insulated Cable Engineers Association (ICEA)
P0. Box P
South Yarmouth, MA 02664
(617) 394-4424

Insulating Glass Certification Council (IGCC)
Route 11, Industrial Park
Cortland, NY 13045
(607) 753-6711

Insurance Information Institute
10 Williams Street
New York, NY 10038
(212) 669-9200

Insurance Services Office (ISO)
160 Water Street
New York, NY 10038
(212) 487-5000

International Air Transport Association (IATA)
1001 Pennsylvania Avenue NW, Suite 285 N
Washington, DC 20004
(202) 624-2977

International Association of Fire Chiefs
(IABPFF)
1329 18th Street NW
Washington, DC 20036
(202) 833-3420

International Association of Heat and Frost
Insulators and Asbestos Workers
1300 Connecticut Avenue NW, Suite 505
Washington, DC 20036
(202) 785-2388

International Association of Mechanical and
Plumbing Officials (IAPMO)
20001 Walnut Drive, South
Walnut, CA 9 1789-2825
(714) 595-8449

International Compressor Remanufacturers
Association (ICRA)
P.O. Box 33092
Kansas City, MO 64114
(816) 822-8818

International Conference of Building Officials
(ICBO)
5360 South Workman Mill Road
Whittier, CA 90601
(213) 699-0541

International District Heating and Cooling
Association (IDHCA)
1101 Connecticut Avenue NW Suite 700
Washington, DC 20036
(202) 429-5111

International Electrical Testing Association
221 Red Rocks Vista P.O. Box 687
Morrison, CO 80465
(303) 467-0526

International Facility Management Association
1 East Greenway Plaza, 11th Floor
Houston, TX 77046-0194
(800) 359-4362
(713) 623-4362

International Institute of Ammonia
Refrigeration (IIAR)
111 East Wacker Drive, Suite 600
Chicago, IL 60601
(312) 644-6610

International Management Institute (IMI)
P.O. Box 266695
Houston, TX 77207
(713) 481-0869

International Mobile Air Conditioning
Association
3003 LBJ Freeway, Suite 219
Dallas, TX 75324
(214) 484-5750

International Municipal Signal Association
(IMSA)
P.O. Box 539
Newark, NY 14513
(315) 331-2182

International Ozone Association (IOA)
83 Oakwood Avenue
Norwalk, CT 06850
(203) 847-8169

International Sanitary Supply Association
(ISSA)
7373 North Lincolnwood Avenue
Lincolnwood, IL 60646-1799
(708) 982-0800

International Society of Fire Service Instructors
30 Main Street
Ashland, MA 01721
(508) 881-5801

International Society of Pharmaceutical
Engineers (ISPE)
3816 West Linebaugh Avenue, Suite 412
Tampa, FL 33624
(813) 960-2105

International Thermal Storage Advisory
Council (ITSAC)
3769 Eagle Street
San Diego, CA 92103
(619) 295-6267

Interstate Commerce Commission (ICC)
12th Street & Constitution Avenue NW
Washington, DC 20036 (202) 275-7119

Irrigation Association
1911 North Ft. Meyer Drive
Arlington, VA 22209
(703) 524-1200

Joint Commission on Accreditation of
Healthcare Organizations (JCAHO)
1 Renaissance Boulevard
Oakbrook Terrace, IL 60181
(708) 916-5600

Joint Council of Health Care Organizations
1 Renaissance Boulevard
Oakbrook Terrace, IL 60181
(708) 916-5600

Land Improvement Contractors of America
(LICA)
1300 Maybrook Drive, P.O. Box 9
Maywood, IL 60153
(708) 344-0700

Lead Industries Association (LIA)
292 Madison Avenue
New York, NY 10017
(212) 578-4750

Lightning Protection Institute (LPI)
P.O. Box 458
Harvard, IL 60033
(815) 943-7211

Manufacturers Standardization Society of the
Valve and Fittings Industry (MSS)
127 Park Street NE
Vienna, VA 22180
(703) 281-6613

Manufacturing Chemical Association (MCA)
1825 Connecticut Avenue NW
Washington, DC 20009
(202) 887-1100

Mechanical Contractors Association of America
(MCAA)
1385 Picard Drive
Rockville, MD 20832
(301) 869-5800

Metal Building Manufacturers Association
(MBMA)
1230 Keath Building
Cleveland, OH 44115
(216) 241-7333

Metal Lath/Steel Framing Association
(ML/SFA)
600 South Federal Street, Suite 400
Chicago, IL 60605
(312) 346-1600

Mineral Insulation Manufacturers Association
(MIMA)
1420 Wing Street
Alexandria, VA 22314
(703) 684-0084

Mississippi River Commission
P.O. Box 80
Vicksburg, MS 39180
(601) 634-5750

National Academy of Sciences (NAS)
2101 Constitution Avenue NW
Washington, DC 20418
(202) 334-2100

National Aeronautics and Space Administration
400 Maryland Avenue SW
Washington, DC 20546
(202) 453-1010

National Asbestos Council (NAB)
1777 Northeast Expressway, Suite 150
Atlanta, GA 30329
(404) 633-2622

National Asphalt Pavement Association
6811 Kenilworth Avenue, Suite 620
Riverdale, MD 20737
(301) 779-4880

National Association of Architectural Metal
  Manufacturers (NAAMM)
600 South Federal Street, Suite 400
Chicago, IL 60605
(312) 922-6222

National Association of Corrosion Engineers
P0. Box 218340
Houston, TX 77218
(713) 492-0535

National Association of County Engineers
326 Pike Road
Ottum, WA 52501
(515) 684-6928

National Association of Demolition
  Contractors (NADC)
4415 West Harrilson Street
Hillside, IL 60162
(708) 449-5959

National Association of Fire Equipment
  Distributors
111 East Wacker Drive, 1 Illinois Center
Chicago, IL 60601
(312) 644-6610

National Association of Home Builders
  Technical Services
15th and M Streets
Washington, DC 20005
(202) 822-0200

National Association of Oil Heating Service
  Managers
P.O. Box 380
Elmwood Park, NJ 07407
(201) 796-8121

National Association of Mutual Insurance
   Companies
7931 Castleway Drive
Indianapolis, IN 46250
(317) 875-5250

National Association of Plumbing, Heating &
   Cooling Contractors (NAPHCC)
P.O. Box 6068
180 South Washington Street
Falls Church, VA 22046
(703) 237-8100

National Association of Sewer Service
   Companies (NASSC)
101 Wymore Road, Suite 101
Altamonte, FL 32714
(305) 774-0304

National Association of Trade and Technical
   Schools (NATTS)
2251 Wisconsin Avenue NW, Suite 200
Washington, DC 20007 (202) 333-1021

National Association of Women in Construction
327 South Adams Street
Fort Worth, TX 76104
(817) 877-5551

National Board of Boiler and Pressure Vessel
   Inspectors
1055 Crupper Avenue
Columbus, OH 43229
(614) 888-8320

National Building Material Distributors
   Association (NBMA)
1417 Lake Cook Road, Suite 130
Deerfield, IL 60015
(708) 945-7201

National Bureau of Standards (NBS)
Gaithersburg, MD 20899
(301) 975-2000

National Cargo Bureau (NCB)
1 World Trade Center, Suite 2757
New York, NY 10048
(212) 571-5000

National Certified Pipe Welding Bureau
5410 Grosvenor Lane, Suite 120
Bethesda, MD 20814
(301) 897-0770

National Clay Pipe Institute (NCPI)
P.O. Box 759
Lake Geneva, WI 53147
(414) 248-9094

National Coal Association (NCA)
1130 17th Street NW
Washington, DC 20036
(202) 463-2625

National Computer Graphics Association
   (NCGA)
2722 Merrilee Drive, Suite 200
Fairfax, VA 22031
(703) 698-9600

National Concrete Masonry Association
   (NCMA)
P.O. Box 781
Herndon, VA 22070
(703) 435-4900

National Construction Software Association
   (NCSA)
7430 East Caley Avenue, Suite 350
Engelwood, CO 80111
(303) 740-8647

National Council of Acoustical Consultants
   (NCAC)
66 Morris Avenue, Box 359
Springfield, NJ 07081
(201) 379-1100

National Council on Radiation Protection and
   Measurement (NCRPM)
7910 Woodmont Avenue, Suite 1016
Bethesda, MD 20814
(301) 657-2652

National Corrugated Steel Pipe Association
2011 I Street NW
Washington, DC 20006
(202) 223-2217

National Electrical Contractors Association
7315 Wisconsin Avenue
Bethesda, MD 20814
(301) 657-3110

National Electrical Manufacturers Association
2101 L Street NW, Suite 300
Washington, DC 20037
(202) 457-8400

National Elevator Industry (NEIL)
630 Third Avenue
New York, NY 10016
(212) 986-1545

National Energy Information Center (NEIC)
1000 Independence Avenue SW
Washington, DC 20585
(202) 586-8800

National Energy Specialist Association (NESA)
518 NW Gordon
Topeka, KS 66608
(913) 232-1702

National Environmental Health Association
South Tower, Suite 970
720 South Colorado Blvd.
Denver, CO 80222
(303) 756-9090

National Fire Protection Association (NFPA)
P.O. Box 9101
One Batterymarch Park
Quincy, MA 02269-9101
(617) 770-3000

National Fire Sprinkler Association (NFSA)
P.O. Box 1000
Robin Hill Corporate Park, Route 22
Patterson, NY 12563
(914) 878-4200

National Fluid Power Association
3333 North Mayfair Road
Milwaukee, WI 53222
(414) 778-3344

National Food Processors Association
1401 New York Avenue NW
Washington, DC 20005
(202) 639-5900

National Forest Products Association (NFPA)
1250 Connecticut Avenue NW
Washington, DC 20036
(202) 463-2700

National Geothermal Association (NGA)
P.O. Box 1350
Davis, CA 95617
(916) 758-2360

National Hardwood Lumber Association
P.O. Box 34518
Memphis, TN 38184
(901) 377-1818

National Institute for Certification in
   Engineering Technologies (NICET)
1420 King Street
Alexandria, VA 22314-2715
(703) 684-2835

National Institute for Occupational Safety and
   Health (NIOSH)
NIOSH Building
Morgantown, WV 26505-2888
(304) 291-4126

National Institute of Building Sciences (NIBS)
1015 15th Street NW, Suite 700
Washington, DC 20005
(202) 347-5710

National Institute of Environmental Health
   Sciences
Research Triangle Park, NC 27709
(919) 541-3345

National Institute of Standards and Technology
   (NIST)
Gaithersburg, MD 20899
(301) 975-2000

National Institutes of Health (NIH)
9000 Rockville Pike
Bethesda, MD 20816
(301) 496-4000

National Insulation Contractors Association
   NICA)
1025 Vermont Avenue NW, Suite 410
Washington, DC 20005
(202) 783-6277

National Insulation and Abatement Contractors
   Association (NIAC)
99 Canal Center Plaza, Suite 222
Alexandria, VA 22314-1538
(703) 683-6422

National Liquified Petroleum Gas Association
   (NLPGA)
1600 Eisenhower Lane
Lisle, IL 60532
(708) 515-0600

National Oceanic and Atmospheric
   Administration (NOAA)
3300 Whitehaven Street NW
Washington, DC 20235
(202) 842-7460

National Paint and Coatings Association
1500 Rhode Island Avenue NW
Washington, DC 20005
(202) 462-6272

National Park Service
Interior Building
18th and C Streets
Washington, DC 20245
(202) 343-4621

National Petroleum Refiners Association
1899 L Street, NW
Washington, DC 20036
(202) 457-0480

National Pool and Spa Institute (NPSI)
2111 Eisenhower Avenue
Alexandria, VA 22314
(703) 838-0083

National Research Council (NRC)
2101 Constitution Avenue NW
Washington, DC 20418
(202) 334-2000

National Restaurant Association (NRA)
1200 17th Street NW
Washington, DC 20001
(202) 331-5900

National Roofing Contractors Association
6250 River Road
Rosemont, IL 60018
(708) 318-6722

National Safety Council (NSC)
444 North Michigan Avenue
Chicago, IL 60611
(312) 527-4800

National Sanitation Foundation (NSF)
P.O. Box 1408
3475 Plymouth Road
Ann Arbor, MI 48106
(313) 769-8010

National Science Foundation (NSF)
800 G Street NW
Washington, DC 20550
(202) 357-9489

National Society of Professional Engineers
1420 King Street
Alexandria, VA 22314
(703) 684-2810

National Solid Waste Management Association
1730 Rhode Island Avenue NW, Suite 1000
Washington, DC 20036 (202) 659-4813

National Swimming Pool Foundation (NSPF)
10803 Gulfdale, Suite 300
San Antonio, TX 78216
(512) 525-1227

National Technical Information Service (NTIS)
5285 Port Royal Road
Springfield, VA 22161
(703) 487-4650

National Transportation Safety Board (NTSB)
800 Independence Avenue SW
Washington, DC 20594
(202) 382-6500

National Truck Equipment Association (NTEA)
38705 Seven Mile Road, Suite 345
Livonia, MI 48152
(313) 462-2190

National Utility Contractors Association
1235 Jefferson Davis Highway, Suite 606
Arlington, VA 22202
(703) 486-5555

National Water Well Association (NWWA)
6375 Riverdale Drive
Dublin, OH 43017
(614) 761-1711

Naval Facilities Engineering Command
200 Stoval Street
Alexandria, VA 22332-2300
(703) 325-0589

Naval Publications and Forms Center
5801 Tabor Avenue
Philadelphia, PA 19120
(215) 697-2000

Nuclear Regulatory Commission (NRC)
1717 H Street NW
Washington, DC 20555
(301) 492-7000

OCCUPATIONAL SAFETY AND HEALTH
ADMINISTRATION (OSHA)
REGIONAL OFFICES:

OSHA Region 1 (MI, MA, NH, RI, VT)
U.S. Department of Labor, OSHA
133 Portland Street
Boston, MA 02114
(617) 565-7164

OSHA Region 2 (NY, NJ, PR)
U.S. Department of Labor, OSHA
201 Varick Street, Room 670
New York, NY 10014
(212) 337-2325

OSHA Region 3 (PA, DE, DC, MD, VA, WV)
U.S. Department of Labor, OSHA
Gateway Building, Suite 2100
3535 Market Street
Philadelphia, PA 19104
(215) 596-1201

OSHA Region 4 (AL, FL, GA, KY, MS, NC,
    SC, TN)
U.S. Department of Labor, OSHA
1375 Peachtree Street NE, Suite 587
Atlanta, GA 30367
(404) 347-3573

OSHA Region 5 (IN, IL, MN, MI, OH, WI)
U.S. Department of Labor, OSHA
230 South Dearborne Street,
32nd Floor, Suite 3244 Chicago, IL 60604
(312) 353-2220

OSHA Region 6 (AR, LA, NM, OK, TX)
U.S. Department of Labor, OSHA
525 Griffin Street, Room 602
Dallas, TX 75202
(214) 767-4731

OSHA Region 7 (KS, IA, MO, NE)
U.S. Department of Labor, OSHA
911 Walnut Street, Room 406
Kansas City, MO 64106
(816) 426-5861

OSHA Region 8 (CO, MO, ND, SD, UT, WY)
U.S. Department of Labor, OSHA
Federal Building, Room 1576
1961 Stout Street
Denver, CO 80204
(303) 844-3016

OSHA Region 9 (AZ, CA, NV)
U.S. Department of Labor, OSHA
71 Stevenson Street, 4th Floor
San Francisco, CA 94105
(415) 995-5896

OSHA Region 10 (AK, ID, OR, WA)
U.S. Department of Labor, OSHA
Federal Office Building, Room 6003
909 1st Avenue
Seattle, WA 89174
(206) 553-5930

Parenteral Drug Association (PDA)
1 Penn Plaza, Suite 640
Philadelphia, PA 19103
(215) 564-6466

Petroleum Equipment Institute (PEI)
P.O. Box 2380
Tulsa, OK 74101
(918) 494-9696

Petroleum Marketing Education Foundation
(PMEF)
5600 Rosewell Road, Prado North # 318
Atlanta, GA 30342
(404) 255-7600

Pharmaceutical Manufacturers Association
(PMA)
1100 15th Street NW
Washington, DC 20036
(202) 835-3400

Pipe Fabricators Institute (PFI)
P.O. Box 173
Springdale, PA 15144
(412) 274-4722

Pipe Line Contractors Association (PLCA)
4100 First City Center
Dallas, TX 75201-4618
(214) 969-2700

Plastic Pipe and Fittings Association (PPFA)
Building C, Suite 20
800 Roosevelt Road
Glen Ellyn, IL 60137
(708) 858-6540

Plastics Pipe Institute (PPI)
(202) 371-5306

Plumbing and Drainage Institute International
c/o Austin 0. Roche Jr.
5342 Boulevard Place
Indianapolis, IN 46208
(317) 251-5298

Plumbing, Heating and Cooling Contractors
P.O. Box 6808
Falls Church, VA 22046
(703) 237-8100

Plumbing Manufacturers Institute (PMI)
Building C, Suite 20
800 Roosevelt Road
Glen Ellyn, IL 60137
(708) 858-9172

Portland Cement Association (PCA)
5420 Old Orchard Road
Skokie, IL 60077
(708) 966-6200

Prestressed Concrete Institute (PCI)
201 North Wells Street
Chicago, IL 60606
(312) 346-4071

Project Managers Institute (PMI)
P.O. Box 43
Drexel Hill, PA 19026
(215) 622-1796

Property Loss Research Bureau (PLRB)
1501 East Woodfield Avenue
Schaumburg, IL 60194
(708) 330-8650

Public Health Service (PHS)
200 Independence Avenue SW
Washington, DC 20201
(202) 245-7000

Refrigerating Engineers and Technicians
Association (RETA)
111 East Wacker Drive, Suite 600
Chicago, IL 60601
(312) 644-6610

Refrigeration Research Foundation (RRF)
7315 Wisconsin Avenue, Suite 1200
North Bethesda, MD 20814 (301) 652-5674

Refrigeration Service Engineers Society
1666 Rand Road
Des Plaines, IL 60016
(708) 297-6464

Resilient Floor Covering Institute (RFCI)
966 Hungerford Drive, Suite 12-B
Rockville, MD 20805
(301) 340-8580

Rivers and Harbors Board of Engineers
Kingman Building
Fort Belvoir, VA 22060
(202) 355-2453

Robotic Industries Association
P.O. Box 3624
900 Vickers Way
Ann Arbor, MI 48106
(313) 994-6088

Rubber Manufacturers Association (RMA)
1400 K Street NW
Washington, DC 20005
(202) 682-4800

Safe Building Alliance (SBA)
655 15th Street NW, Suite 1200
Washington, DC 20005
(202) 879-5120

Safety Equipment Distributors Association
111 East Wacker Drive, Suite 600
Chicago, IL 60610
(312) 644-6610

Safety Glazing Certification Council (SGCC)
Route 11, Industrial Park
Cortland, NY 13045
(607) 753-6711

Scaffolding, Shoring and Forming Institute
1230 Kieth Building
Cleveland, OH 44152
(216) 241-7333

Scientific Apparatus Makers Association
(SAMA)
1101 16th Street NW
Washington, DC 20036
(202) 223-1360

Sealed Insulating Glass Manufacturers
Association (SIGMA)
111 East Wacker Drive
Chicago, IL 60601
(312) 644-6610

Sheet Metal and Air Conditioning Contractors
National Association (SMACNA)
P.O. Box 70
Merrifield, VA 22116
(703) 790-9890

Single Ply Roofing Institute (SPRI)
104 Wilmot Road, Suite 201
Deerfield, IL 60016
(708) 940-8800

Society of American Value Engineers (SAVE)
60 Revere Drive, Suite 500
Northbrook, IL 60062
(708) 480-1730

Society of Automotive Engineers (SAE)
400 Commonwealth Drive
Warrendale, PA 15096
(412) 776-4841

Society of Fire Protection Engineers (SFPE)
60 Batterymarch Street
Boston, MA 02110
(617) 482-0686

Society of Industrial Microbiology (SIM)
P.O. Box 12534
Arlington, VA 22209
(703) 941-5373

Society of Insurance Research (SIR)
P.O. Box 933
Appleton, WI 54912
(414) 730-8858

Society of Manufacturing Engineers (SME)
One SME Drive
Dearborn, MI 48121
(313) 271-1500

Society of Petroleum Engineers (SPE)
P.O Box 833836
Richardson, TX 75083
(214) 669-3377

Society of Plastics Engineers (SPE)
14 Fairfield Drive
Fairfield, CT 06804
(203) 775-0471

Society of Plastics Industries Composites
Institute
355 Lexington Avenue
New York, NY 10017
(212) 351-5410

Society of the Plastics Industry (SPI)
1275 K Street NW, Suite 400
Washington, DC 20005
(202) 371-5200

Society of Tribologists and Lubrication
  Engineers
840 Busse Highway
Park Ridge, IL 60068-2367
(708) 825-5536

Society of Women Engineers (SWE)
345 East 47th Street, Room 305
New York, NY 10017
(212) 705-7855

Soil Conservation Service (SCS)
14th and Independence Avenues SW
Washington, DC 20250 (202) 447-4525

Solar Energy Industries Association (SEIA)
1730 North Lynn Street
Arlington, VA 22209
(703) 524-6100

Southern Building Code Congress International
900 Montclair Road
Birmingham, AL 35213-1206
(205) 591-1853

Southwest Research Institute (SWRI)
6220 Culebra Road
San Antonio, TX 78284
(512) 684-5111

Standards Engineering Society (SES)
11 West Monument Drive, Suite 510
Dayton, OH 45410
(513) 223-2410

Steel Deck Institute (SDI)
P.O. Box 9506
Canton, OH 44711
(216) 493-7886

Steel Door Institute (SDI)
c/o A. P. Wherry & Associates
712 Lakewood Center North
Cleveland, OH 44107
(216) 226-7700

Steel Joist Institute (SJI)
1205 48th Street, North, Suite A
Myrtle Beach, SC 29577
(803) 449-0487

Steel Structures Painting Council (SSPC)
4400 Fifth Avenue
Pittsburgh, PA 15213
(412) 578-3327

Steel Tank Institute (STI)
728 Anthony Trail
Northbrook, IL 60062
(708) 498-1980

Sump and Sewage Pump Manufacturers
  Association
P.O. Box 298
Winnetka, IL 60093
(312) 446-4434

Superintendent of Documents
U.S. Government Printing Office
North Capitol and G Streets NW
Washington, DC 20235
(202) 512-1800

Technical Association of the Pulp and Paper
  Industry (TAPPI)
15 Technology Parkway South NW
Atlanta, GA 30092
(404) 446-1400

Tennessee Valley Authority (TVA)
400 West Summit Hill Drive
Knoxville, TN 37902
(615) 623-3554

Thermal Insulation Manufacturer's Association
7 Kerby Plaza
Mt. Kisco, NY 10549
(914) 241-2284

Tile Council of America (TCA)
P.O. Box 326
Princeton, NJ 08542
(609) 921-7050

Tissue Culture Association (TCA)
19110 Montgomery Village Avenue, Suite 300
Gaithersburg, MD 20879
(301) 869-2900

Topographic Laboratory
U.S. Department of Defense
Cude Building
Fort Belvoir, VA 22060
(202) 355-2600

Truss Plate Institute (TPI)
583 D'Onofrio Drive, Suite 200
Madison, WI 53719
(608) 833-5900

Tubular Exchange Manufacturers Association
  (TEMA)
25 North Broadway
Tarrytown, NY 10591
(914) 332-0040

Underground Contractors Association (UCA)
2720 River Road, Suite 222
Des Plaines, IL 60018
(708) 299-6930

Underwriters Laboratories (UL)
333 Pfingsten Road
Northbrook, IL 60062
(708) 272-8800

Uni-Bell PVC Pipe Association (UNI)
2655 Villa Creek Drive, Suite 155
Dallas, TX 75234
(214) 243-3902

U.S. Army Chief of Engineers (COE)
20 Massachusetts Avenue NW
Washington, DC 20314
(202) 272-0001

U.S. Bureau of Mines
2401 E Street NW
Washington, DC 20245
(202) 634-1004

U.S. Department of Agriculture
14th and Independence Avenue SW
Washington, DC 20250
(202) 447-3631

U.S. Department of Commerce (DOC)
Commerce Building
15th and Constitution Avenue NW
Washington, DC 20230 (202) 377-2112

U.S. Department of Defense (DOD)
The Pentagon
Washington, DC 20301
(202) 695-5261

U.S. Department of Energy (DOE)
1000 Independence Avenue SW
Washington, DC 20585
(202) 586-6120

U.S. Department of Health and Human
   Services (DHHS)
200 Independence Avenue SW
Washington, DC 20201
(202) 245-7000

U.S. Department of Housing and Urban
Development (HUD)
HUD Building
541 7th Street SW
Washington, DC 20410
(202) 755-6417

U.S. Department of the Interior (DOI)
Interior Building
18th and C Streets NW
Washington, DC 20245
(202) 343-7351

U.S. Department of Justice
10th and Constitution Avenue NW
Washington, DC 20530
(202) 633-2001

U.S. Department of State
2201 C Street NW
Washington, DC 20520
(202) 647-4910

U.S. Department of Transportation (DOT)
400 7th Street SW
Washington, DC 20590
(202) 366-4000

U.S. Fire Administration (USFA)
16825 South Seton Avenue
Emmitsburg, MD 21727
(301) 447-1080

U.S. Metric Association (USMA)
10245 Andesol Avenue
Northridge, CA 91325
(818) 363-2254

United States Pharmacopoeia
(301) 881-0666

Urban Mass Transportation Administration
400 7th Street SW
Washington, DC 20590
(202) 366-4040

Utility Location and Coordination Council
1313 East 60th Street
Chicago, IL 60637
(312) 667-2200

Valve Manufacturers Association of America
1050 17th Street NW, Suite 701
Washington, DC 20036
(202) 331-8105

Vibration Institute
6262 South Kingery Highway
Willowbrook, IL 60514
(708) 654-2254

Vinyl Institute
155 Route 46 West
Wayne, NJ 07470
(201) 890-9299

Wall Covering Manufacturers Association
66 Morris Avenue
Springfield, NJ 07081
(201) 379-1100

Water Pollution Control Federation (WPCF)
601 Wythe Street
Alexandria, VA 22314
(703) 684-2400

Water Quality Association (WQA)
4151 Naperville Road
Lisle, IL 60532
(708) 505-0160

Water Systems Council (WSC)
600 South Federal Street, Suite 400
Chicago, IL 60605
(312) 922-6222

Western Society of Engineers (WSE)
176 West Adams Street, Suite 1734
Chicago, IL 60603
(312) 372-3760

Wire Reinforcement Institute (WRI)
8361-A Greensboro Drive
McLean, VA 22102
(703) 790-9790

Wood Heating Alliance (WHA)
1101 Connecticut Avenue NW, Suite 700
Washington, DC 20036
(202) 857-1181

Woven Wire Products Association (WWPA)
2515 North Nordica Avenue
Chicago, IL 60635
(312) 637-1359

Zinc Institute
292 Madison Avenue
New York, NY 10017
(212) 578-475

APPENDIX B
VENT DIAGRAMS

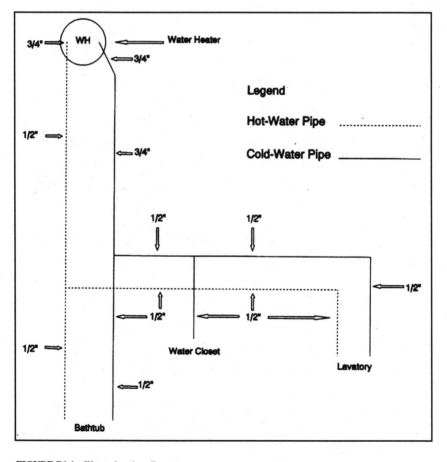

FIGURE B1.1   Water pipe riser diagram.

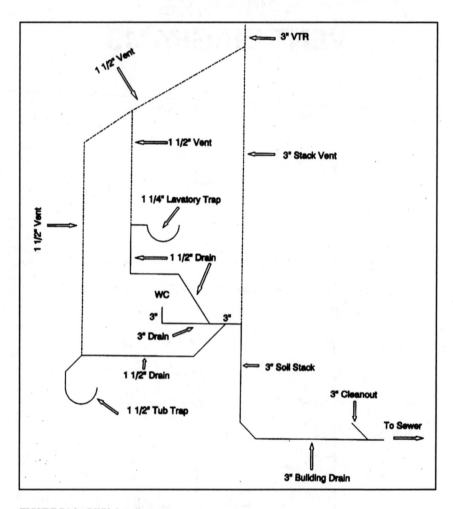

FIGURE B1.2 DWV riser diagram.

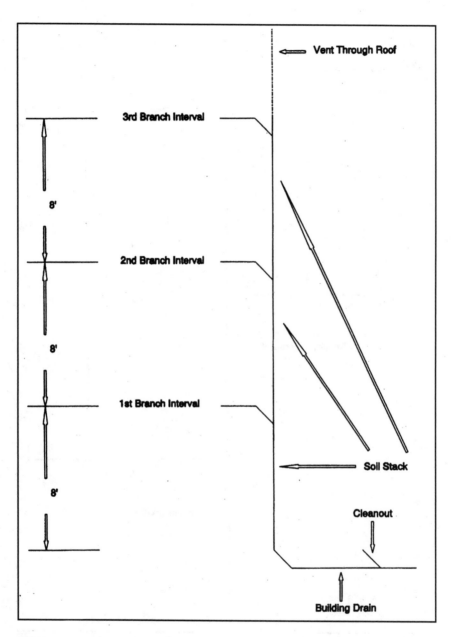

FIGURE B1.3 Branch-interval detail.

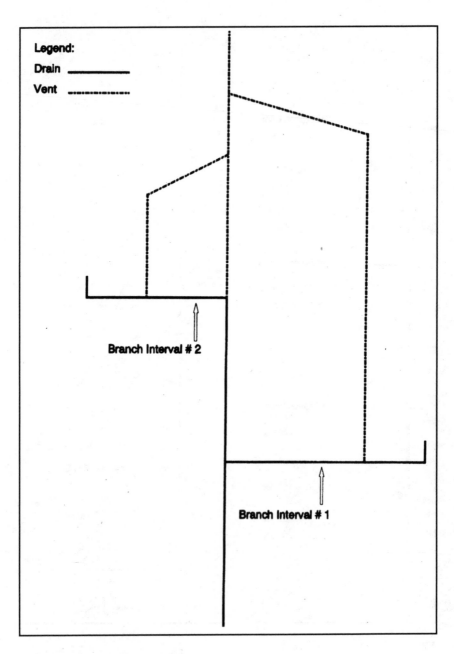

FIGURE B1.4   Stack with two branch intervals.

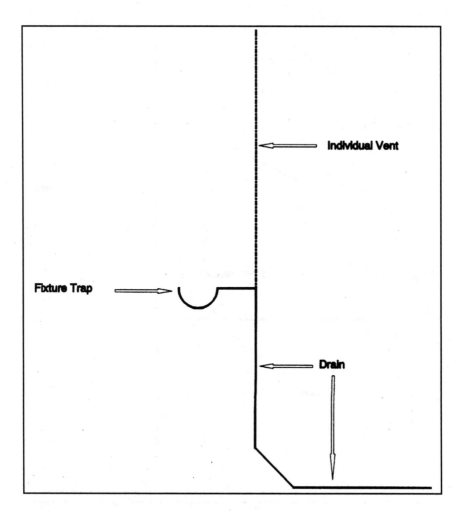

FIGURE B1.5 Individual vent.

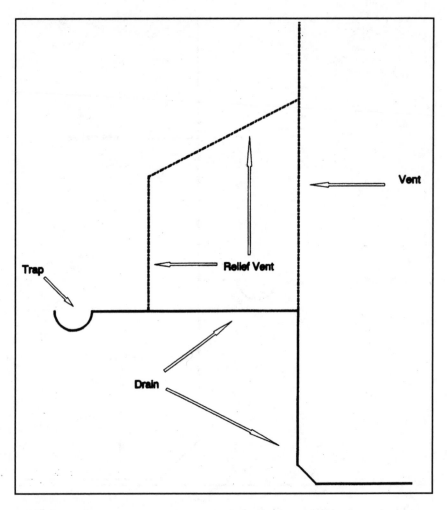

FIGURE B1.6 Relief vent.

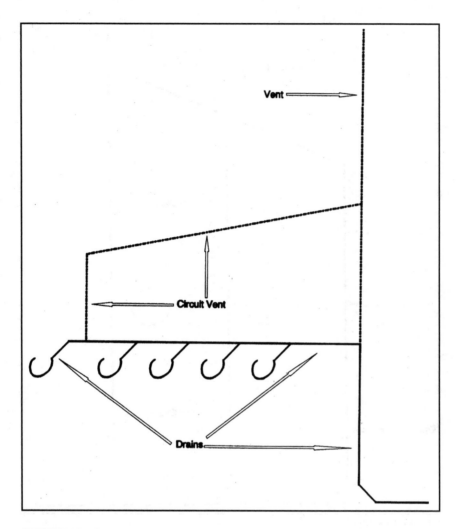

FIGURE B1.7 Circuit vent.

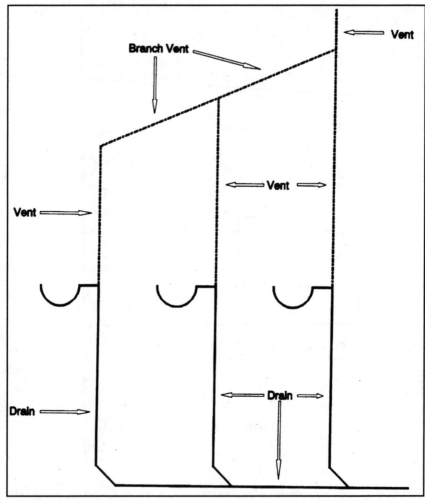

FIGURE B1.8 Branch vent.

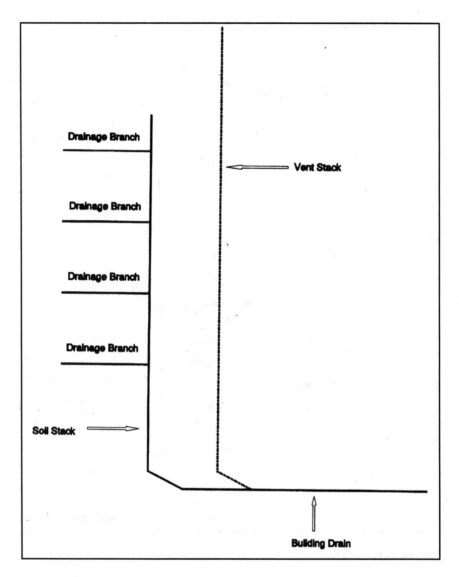

FIGURE B1.9   Vent stack.

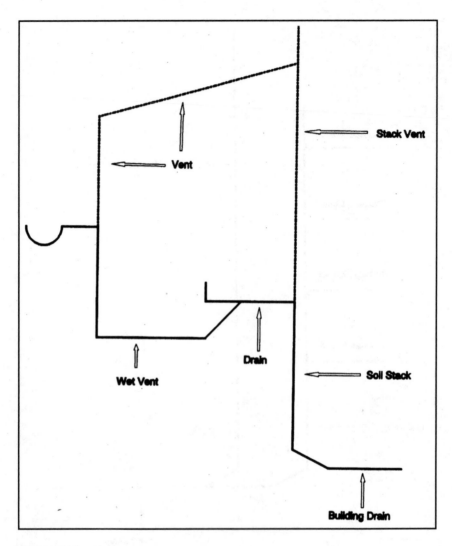

FIGURE B1.10 Stack vent.

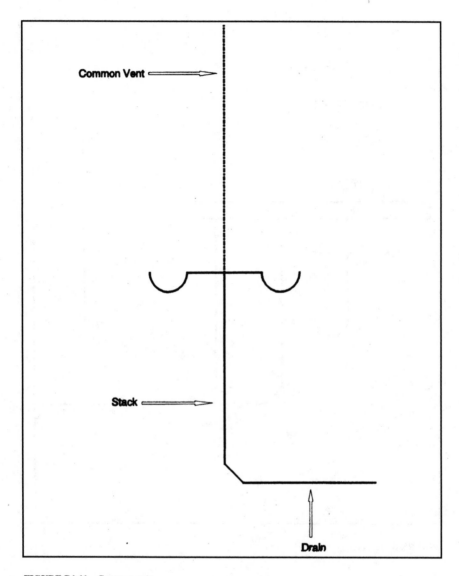

FIGURE B1.11 Common vent.

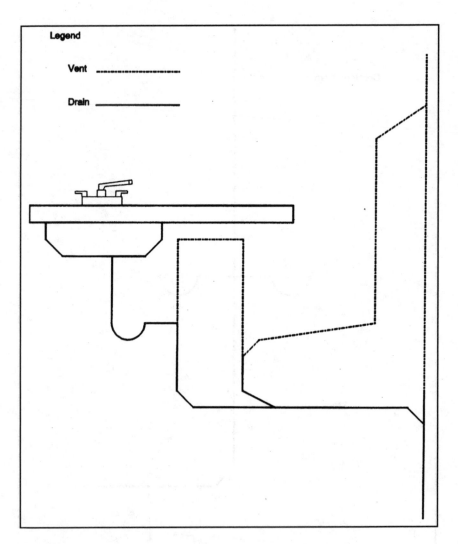

FIGURE B1.12 Island vent.

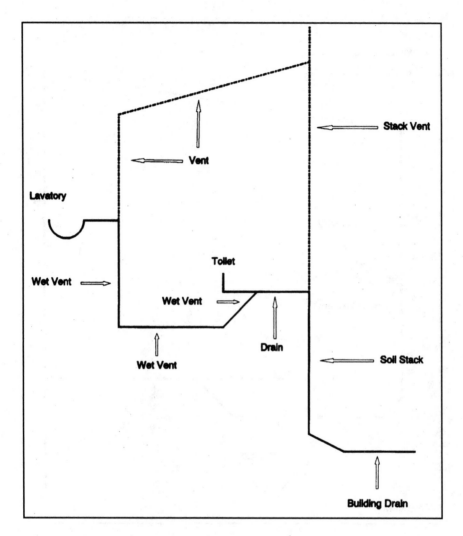

FIGURE B1.13 Wet-venting a toilet with a lavatory.

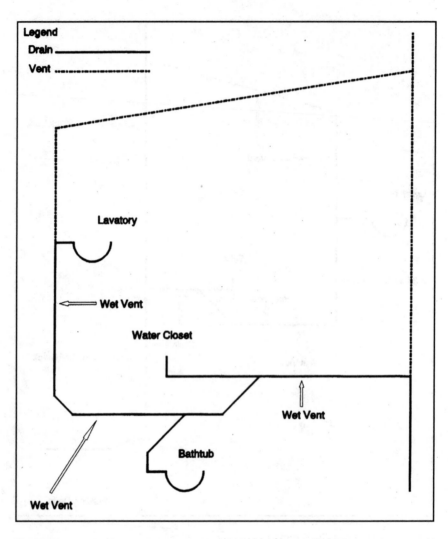

FIGURE B1.14 Wet-venting a bathroom group.

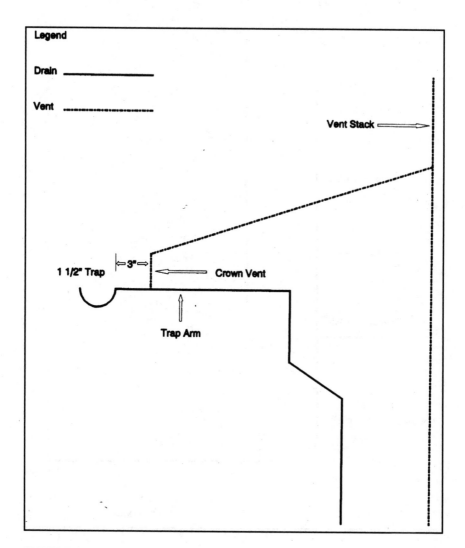

FIGURE B1.15 Crown venting.

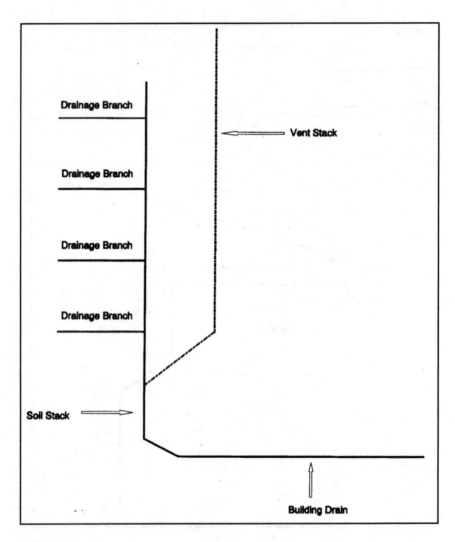

FIGURE B1.16 Vent stack.

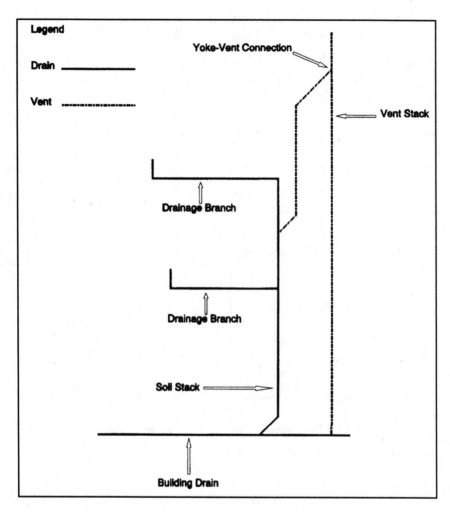

FIGURE B1.17 Yoke vent.

B1.17

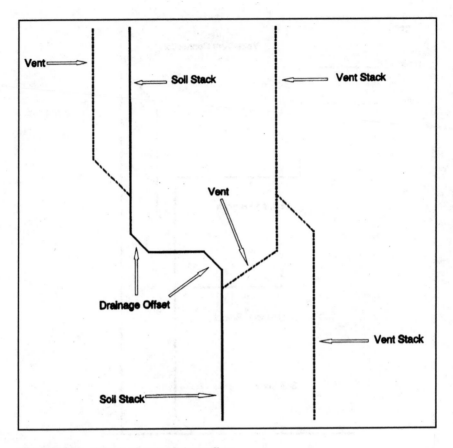

FIGURE B1.18 Example of venting drainage offsets.

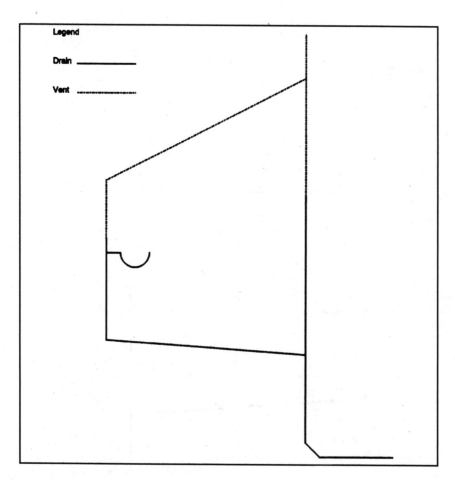

FIGURE B1.19 Graded-vent connection.

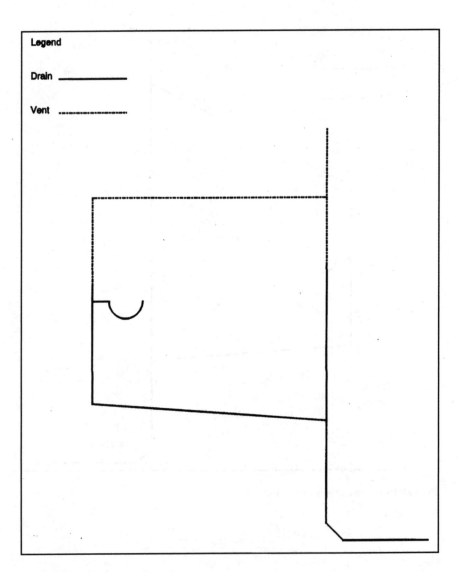

FIGURE B1.20 Zone 1's level-vent rule.

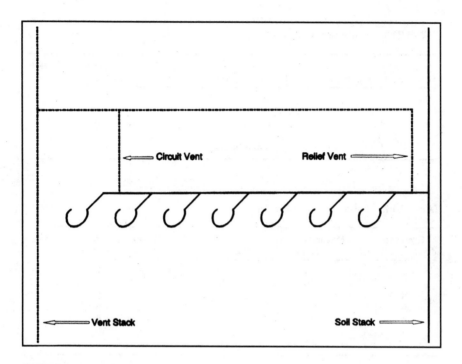

FIGURE B1.21   Circuit vent with a relief vent.

TABLE B1.1 Trap-to-vent distances in Zone 1

Grade on drain pipe (in)	Size of trap arm (in)	Maximum distance between trap and vent (ft)
¼	1¼	2½
¼	1½	3½
¼	2	5
¼	3	6
¼	4 and larger	10

TABLE B1.2 Trap-to-vent distances in Zone 2

Grade on drain pipe (in)	Fixture's drain size (in)	Trap size (in)	Maximum distance between trap and vent (ft)
¼	1¼	1¼	3½
¼	1½	1¼	5
¼	1½	1½	5
¼	2	1½	8
¼	2	2	6
⅛	3	3	10
⅛	4	4	12

TABLE B1.3 Trap-to-vent distances in Zone 3

Grade on drain pipe (in)	Fixture's drain size (in)	Trap size (in)	Maximum distance between trap and vent (ft)
¼	1¼	1¼	3½
¼	1½	1¼	5
¼	1½	1½	5
¼	2	1½	8
¼	2	2	6
⅛	3	3	10
⅛	4	4	12

Index

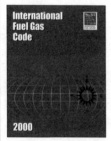